ASTRONOMY AND ASTROPHYSICS LIBRARY

Series Editors: G. Börner, Garching, Germany
A. Burkert, München, Germany
W. B. Burton, Charlottesville, VA, USA and
    Leiden, The Netherlands
A. Coustenis, Meudon, France
M. A. Dopita, Canberra, Australia
B. Leibundgut, Garching, Germany
A. Maeder, Sauverny, Switzerland
P. Schneider, Bonn, Germany
V. Trimble, College Park, MD, and Irvine, CA, USA

For further volumes:
http://www.springer.com/series/848

Michael Soffel • Ralf Langhans

Space-Time Reference Systems

Springer

Michael Soffel
Institute for Planetary Geodesy
Dresden Technical University
Lohrmann Observatory
Dresden
Germany

Ralf Langhans
Institute for Planetary Geodesy
Dresden Technical University
Lohrmann Observatory
Dresden
Germany

ISSN 0941-7834
ISBN 978-3-642-30225-1   ISBN 978-3-642-30226-8 (eBook)
DOI 10.1007/978-3-642-30226-8
Springer Heidelberg New York Dordrecht London

Library of Congress Control Number: 2012948010

© Springer-Verlag Berlin Heidelberg 2013
This work is subject to copyright. All rights are reserved by the Publisher, whether the whole or part of the material is concerned, specifically the rights of translation, reprinting, reuse of illustrations, recitation, broadcasting, reproduction on microfilms or in any other physical way, and transmission or information storage and retrieval, electronic adaptation, computer software, or by similar or dissimilar methodology now known or hereafter developed. Exempted from this legal reservation are brief excerpts in connection with reviews or scholarly analysis or material supplied specifically for the purpose of being entered and executed on a computer system, for exclusive use by the purchaser of the work. Duplication of this publication or parts thereof is permitted only under the provisions of the Copyright Law of the Publisher's location, in its current version, and permission for use must always be obtained from Springer. Permissions for use may be obtained through RightsLink at the Copyright Clearance Center. Violations are liable to prosecution under the respective Copyright Law.
The use of general descriptive names, registered names, trademarks, service marks, etc. in this publication does not imply, even in the absence of a specific statement, that such names are exempt from the relevant protective laws and regulations and therefore free for general use.
While the advice and information in this book are believed to be true and accurate at the date of publication, neither the authors nor the editors nor the publisher can accept any legal responsibility for any errors or omissions that may be made. The publisher makes no warranty, express or implied, with respect to the material contained herein.

Printed on acid-free paper

Springer is part of Springer Science+Business Media (www.springer.com)

Preface

A central core of this textbook has emerged from a series of lectures that were regularly given at Dresden Technical University since 1995. Numerous discussions with international experts in the frame of working groups or international meetings such as those from the series *Les Journées, systèmes de référence spatio-temporels*, that take place every year (with exceptions) in Paris or another European metropolis motivated us to work out the subject of astronomical–geodetical space–time reference systems in detail and to present it in the form of a textbook containing solved exercises and computer programs. It is not difficult to realize that with increasing accuracy in the realization of astronomical spatial–temporal reference systems the complexity of the subject has grown considerably and only a couple of experts worldwide possess some comprehensive view. In the past, it was fairly easy to master concrete problems of astronomical reference systems by means of certain astronomical yearbooks. Now, some of the more significant yearbooks are no longer in print and one expects that every better observatory or relevant national institution has its own software to tackle such practical problems. Many aspects of highly precise spatial–temporal astronomical reference systems exist that need further discussions and international commitments, e.g., by the International Astronomical Union (IAU). In the last decades the subject under discussion underwent drastic changes related with the improvements in measuring accuracy. One aspect is relativity. Today not only the aspect of timescales but also that of spatial coordinates have to be formulated relativistically, i.e., in the framework of Einstein's theory of gravity (GRT), or at least, in the post-Newtonian approximation of GRT.

A central subject of such reference systems is the transformation between some space-fixed, celestial (CRS) and a certain terrestrial reference system (TRS). To this end, traditional elements such as the translational motion of the Earth (Earth–Moon barycenter) in form of the ecliptic, as well as elements related with the rotational motion of the Earth in space (instantaneous rotation pole (IRP), celestial equator), have been introduced. As intermediate quantities in the CRS-TRS transformation, the IRP as well as the intersection between ecliptic and celestial equator (the equinox) historically played a central role. However, with the rapid advance of Very Long Baseline Interferometry (VLBI) for the first time in history a celestial reference system can be realized without reference to the translational and rotational motion of the Earth. This has caused a change of paradigms for the CRS-TRS transformation. The classical rotation pole was meanwhile replaced by some conventional celestial intermediate pole (CIP) and the classical equinox was replaced by some celestial intermediate origin (CIO). After these elements were officially accepted through IAU resolutions the necessity emerges to treat these new paradigms also in textbooks. From all of these reasons, we think that a new textbook devoted to the subject of astronomical spatial–temporal reference systems and frames is of importance, which serves as introduction to the two basic references to the subject: the *IERS Conventions* (IERS C03/C10) and the *Supplement to the Astronomical Almanac*.

For all subjects treated in this volume we offer a series of Maple^{TM} files for the reader to get some feeling for the various models and orders of magnitude. It is obvious that the symbol manipulating language Maple^{TM} is easy to use and has a very attractive graphic surface; however, it is definitively not designed to produce large data sets on a routine basis. The various Maple^{TM} files that the reader may want to download from http://astro.geo.tu-dresden.de/astroref/ merely serve didactical purposes. They were not developed for extensive numerical calculations.

At this place we would like to mention that a few textbooks have appeared recently that cover parts of the central subject. *The Measurement of Time: Time, Frequency and the Atomic Clock* by Audoin and Guinot (2001) provides a comprehensive introduction to the physics of time and time measurement. Many diverse aspects of this book are also treated in *Relativistic Celestial Mechanics of the Solar System* by Kopeikin et al. (2011) where the emphasis clearly lies on the aspect of relativity. Another very useful reference for many parts is Kaplan (2005). The book *Astrometry for Astrophysics: Methods, Models and Applications* by W. van Altena, Cambridge University Press, Cambridge should be published in 2012.

This book (Space-Time Reference Systems) presents an introduction to the problem of astronomical–geodetical space–time reference systems. For the beginner (e.g., students of geodesy or physically oriented sciences) the classical aspects, especially of the spatial coordinates, are introduced very exhaustively for didactical reasons (clearly for experts in the field that part looks very conservative and old-fashioned). We hope that the various (solved) exercises and Maple^{TM} files in the *AstroRef* package support the learning process and provide some insight into orders of magnitude. In this sense the book might serve as an introduction to the IERS Conventions (IERS C03/C10) and the professional astronomical software: SOFA or NOVAS, as discussed in Chap. 10 and the software provided by the IERS. For experts in that field coming from non-relativistic astronomy or space geodesy the book might help to better understand the relativistic aspects related with our central subject.

Finally, it is a pleasure for us to thank the large number of experts who have contributed to this book. Without the many suggestions for improvements, extensions, and actualizations the book would be still in a much poorer state. We especially would like to express our gratitude to Felicitas Arias (BIPM, Paris), Johannes Böhm (TU Vienna), Agnès Fienga (Besancon observatory), Erik Høg (U Copenhagen), Chopo Ma (GSFC), Hervé Manche (Paris observatory),

Jürgen Müller (TU Hannover), Elena Pitjeva (IPA, St.Petersburg), Cyril Ron (Astronomical Institute, Prague), Ulli Schreiber (Wettzell), Harald Schuh (TU Vienna), Irina Tupikova (TU Dresden), Lambert Wanninger (TU Dresden), Norbert Zacharias (USNO), and Sven Zschocke (TU Dresden). We are deeply indebted to the two scientists who likely have contributed most to the problem of space–time reference systems: Nicole Capitaine and Patrick Wallace. They have read practically all parts of the book and suggested a huge number of improvements that we have implemented finally. A large portion of their valuable time went into this book.

Dresden, Germany
Michael Soffel
Ralf Langhans

Contents

1	**Introduction**		1
	1.1	Astronomical Geodetic Reference Systems	1
		1.1.1 The Problem of Time	1
		1.1.2 Various Space–Time Reference Systems	5
		1.1.3 Classical Apparent, True, and Mean Places	7
	1.2	Astronomical Geodetic Applications	9
		1.2.1 Navigation	9
		1.2.2 Global Geodynamics and the System Earth	10
2	**Time**		11
	2.1	Stability of Oscillators	11
	2.2	Bias and Drift	15
	2.3	Quartz and Atomic Clocks	17
	2.4	Cesium Clocks	21
	2.5	Rubidium Clocks	21
	2.6	H-Maser	22
	2.7	Fountain Clocks	24
	2.8	Optical Clocks	24
	2.9	Application of Highly Precise Clocks	27
	2.10	Relativity and the Metric Tensor	28
	2.11	Lorentz Transformation	32
	2.12	Geocentric Timescales TCG, TT, TAI, and UTC	34
	2.13	Time Zones	41
	2.14	Julian Date	41
	2.15	Barycentric Timescales TCB, T_{eph}, and TDB	43
	2.16	Fairhead–Bretagnon Series	44
	2.17	Problems of Time Dissemination	45
	2.18	Time Dissemination by Means of Satellites	46

3	**Space–Time**		49
	3.1	Reference Systems and Frames	49
	3.2	Canonical Barycentric Metric	53
	3.3	Equations of Motion of Astronomical Bodies	55
4	**Barycentric Dynamical Reference System**		61
	4.1	Concepts	61
	4.2	Observational Methods	61
		4.2.1 Optical Ground-Based Astrometry	61
		4.2.2 Lunar Laser Ranging	62
		4.2.3 Radar Measurements to Planets	68
		4.2.4 Radar Tracking to Spacecraft	69
	4.3	Solar System Ephemerides	70
		4.3.1 Numerical Ephemerides	70
		4.3.2 Semianalytical Ephemerides	85
5	**Classical Astronomical Coordinates**		91
	5.1	Apparent Motion of Stars and Sun	91
	5.2	Spatial Coordinates	94
		5.2.1 Labels on the Celestial Sphere	97
		5.2.2 Cartesian and Spherical Coordinates	100
		5.2.3 Horizon Coordinates	101
		5.2.4 Equatorial Coordinates of the First Kind	101
		5.2.5 Equatorial Coordinates of the Second Kind	102
		5.2.6 Ecliptic Coordinates	103
	5.3	Relations Between Astronomical Coordinates	105
		5.3.1 Relations between (A, z) and (h, δ)	105
		5.3.2 Relations between (h, δ) and (α, δ)	107
		5.3.3 Rotational Matrices	108
		5.3.4 Relations Between (α, δ) and (λ, β)	111
	5.4	Sidereal Times	111
6	**Astrometry**		115
	6.1	Refraction	115
		6.1.1 The Model of Saastamoinen	117
		6.1.2 Refraction Corrections as Integrals	118
		6.1.3 Applying Refraction Corrections	122
	6.2	Parallax	123
		6.2.1 Annual Parallax	123
		6.2.2 Geocentric Parallax	127
	6.3	Aberration	128
		6.3.1 Annual Aberration	131
		6.3.2 Diurnal Aberration	132
	6.4	Space Motion of Stars	133
	6.5	Astronomical Precession	135
		6.5.1 Approximate Treatment of Precession	138

6.6	Astronomical Nutation		140
6.7	Apparent Places		144
6.8	High-Precision Astrometry		144
	6.8.1	Gravitational Light Deflection	144
	6.8.2	The Klioner Formalism	149

7 Celestial Reference System 155

7.1	Concepts		155
	7.1.1	Barycentric Celestial Reference System	155
	7.1.2	Geocentric Celestial Reference System	156
7.2	Observational Methods		160
	7.2.1	Very Long Baseline Interferometry	160
	7.2.2	Astrometric Space Missions	164
7.3	Classical Celestial Reference System		168
	7.3.1	Fundamental Catalogs FK3, FK4, and FK5	169
	7.3.2	Astrographic Catalog and Derived Catalogs	170
	7.3.3	Hipparcos and Tycho Catalogs	170
	7.3.4	2MASS	171
	7.3.5	Schmidt Plate Survey Catalogs	171
	7.3.6	UCAC	172
7.4	International Celestial Reference System		172
	7.4.1	Orientation of the ICRS	173
	7.4.2	Realization of the ICRF in the Optical	174

8 Terrestrial Reference System 175

8.1	Concepts		175
	8.1.1	Local Terrestrial Systems	175
	8.1.2	Global Terrestrial Systems	177
8.2	Observational Methods		177
	8.2.1	Classical Methods	177
	8.2.2	Satellite Laser Ranging	181
	8.2.3	Global Positioning System	183
	8.2.4	GLONASS	187
	8.2.5	DORIS	188
	8.2.6	GALILEO	188
	8.2.7	Gyroscopes	189
8.3	International Terrestrial Reference System		192

9 From the GCRS to the ITRS 197

9.1	Polar Motion		197
9.2	Universal Time, UT1, and Length-of-Day Variations		202
	9.2.1	Universal Time	203
	9.2.2	UT1 and Its Variations	206
9.3	Celestial Intermediate Pole		207
	9.3.1	Instantaneous Rotation Pole	208
	9.3.2	Definition of the CIP	211

		9.3.3	Motion of the CIP in the GCRS; CP Offsets	214

        9.3.3   Motion of the CIP in the GCRS; CP Offsets 214
        9.3.4   IAU 2000 Precession–Nutation Model 215
        9.3.5   Fundamental Nutation Angles 219
        9.3.6   Frame Bias Matrix 221
        9.3.7   The Series by Capitaine and Wallace 221
        9.3.8   Motion of the CIP in the ITRS 222
   9.4   CIO, TIO, ERA, and GAST 226
        9.4.1   Celestial Intermediate Origin 226
        9.4.2   Terrestrial Intermediate Origin 229
        9.4.3   Earth Rotation Angle θ and GAST 229
   9.5   Transformations from the GCRS to the ITRS 231
        9.5.1   Quasi-Classical Equinox-Based Transformation 231
        9.5.2   CIO-Based Transformation 232

10 Astronomical Software - Yearbooks 235
   10.1   Software Implementations 235
        10.1.1 SOFA 235
        10.1.2 NOVAS 236
   10.2   Astronomical Yearbooks 236
        10.2.1 APFS 237
        10.2.2 Astronomical Almanac 237

11 Astronomical Constants 239
   11.1   Natural Constants 239
   11.2   Defined and Measurable Natural Constants 240
   11.3   Problems Related with Natural Units 240
   11.4   Body Constants and Framework 241
   11.5   Initial Values and Model 242
   11.6   Current Best Estimates 242

References 245

List of Acronyms 257

A: Solutions to Exercises 261

B: Description of the AstroRef Package 277
   B.1   General Information 277
   B.2   Download 277
   B.3   Installing and Using the Package 278
   B.4   Function Reference 278
        B.4.1   Overview 278
        B.4.2   Auxiliary Functions 279
   B.5   Package Variables 308

Index 311

List of Symbols

List of symbols used in the text (acronyms can be found at the end of the book)

- t: Physical time coordinate; especially TCB
- T: Physical time coordinate; especially TCG
- f: Frequency
- ϕ: Phase of an oscillator
- $y(t)$: Relative frequency fluctuation
- $x(t)$: Phase time
- σ_y: Allan-Variance
- $S_y(f)$: Spectral power density
- F: Total spin quantum number
- I: Spin quantum number of an atomic nucleus
- m_F: Magnetic quantum number
- τ: Proper time (of a clock)
- c: Vacuum speed of light
- ds: Infinitesimal length element
- $g_{\mu\nu}$: Fundamental metric tensor
- x^μ: Coordinates with $\mu = 1, \ldots, N$ (N: dimension of space or space–time)
- **x**: Cartesian spatial coordinates; especially spatial BCRS coordinates
- **X**: Cartesian spatial coordinates; especially spatial GCRS coordinates
- U: Newtonian gravitational potential
- U_{geo}: Geopotential including the centrifugal potential
- $g(\psi)$: Gravitational acceleration
- k_E: Scale factor that relates TT with TCG
- L_C: Scale factor that relates the secular rates of TCB and TCG
- L_B: Scale factor that defines TDB in terms of TCB
- w, w^i: Metric potentials in the BCRS
- W, W^a: Metric potentials in the GCRS

- $\Gamma^\alpha_{\mu\nu}$: Christoffel symbols
- $\mathbf{z}_A$: Coordinate position of body A in the BCRS
- m_I: Inertial mass
- m_G: Gravitational mass
- η_N: Nordtvedt-parameter
- G: Newtonian gravitational constant
- z: Zenith distance
- a: Elevation
- A: Azimuth
- h: Hour angle
- δ: Declination
- λ: Ecliptic longitude
- β: Ecliptic latitude
- $\mathcal{R}_x$: Rotational matrix about the x-axis
- ϵ: Obliquity of the ecliptic
- $\Delta\epsilon$: Nutation in obliquity
- $\Delta\psi$: Nutation in longitude
- Λ: (Eastward) astronomical longitude of an observer
- Φ: Astronomical latitude
- $R_{12}(\tau)$: Cross-correlation function
- n: Index of refraction
- ζ: Refraction correction
- Π: Annual parallax
- Π_G: Geocentric parallax
- $\boldsymbol{\beta}$: $\boldsymbol{\beta} = \mathbf{v}/c$
- ζ_A, θ_A, z_A: Precession angles
- $\mathcal{P}$: Precession matrix
- $\mathcal{N}$: Nutation matrix
- x_p, y_p: Pole coordinates
- ψ_A, ω_A, χ_A: Additional precession angles
- l, l', F, D, Ω: Fundamental angles
- z: Redshift
- $\mathbf{L}$: Angular momentum
- $T(\sigma)$: Transfer function
- X, Y: GCRS coordinates of the CIP
- $\mathcal{B}$: Frame bias matrix
- $s(t)$: Angle defining the position of the CIO
- $s'(t)$: Angle defining the position of the TIO

Chapter 1
Introduction

1.1 Astronomical Geodetic Reference Systems

This textbook deals with astronomical (geodetic) reference systems. These present special kinds of spatial-temporal reference systems that play a fundamental role, especially in the vicinity of the Earth.

The phrase *reference system* implies more than a mere mathematical coordinate system in a space–time manifold: it points towards a labeling of space–time events in the near Earth space that has to be achieved with dedicated observing and analysis systems. Such events are supposed to have a small extent in space (here) and time (now). Examples are the phase centers in GPS or radio antennas, or the center of mass of a planet at a certain instant of time.

1.1.1 The Problem of Time

"What is time? If no one ask of me, I know; if I wish to explain to him who asks, I know not."

This sentence by St. Augustine starkly illuminates the difficulties related with an analysis of the notion of time (or any other fundamental scientific notion like space, matter, life, etc.). Here we would like to focus only on two aspects of the notion of time. The first is of a theoretical, the second more of a practical, nature. On the theoretical side, Newton confronted the "relative, apparent, and common time" of ordinary life with an "absolute, true, and mathematical time" that by itself and due to its own nature flows steadily without respect to some material object. This theoretical concept later was overthrown by Einstein's theory of gravity, due to conflicts with practical aspects.

The practical aspects are related simply to how time is measured. The famous Nobel Prize winner Richard Feynman once said: not what time is of interest in physics, but how it is measured. In this sense, the conceptual aspect of time is related

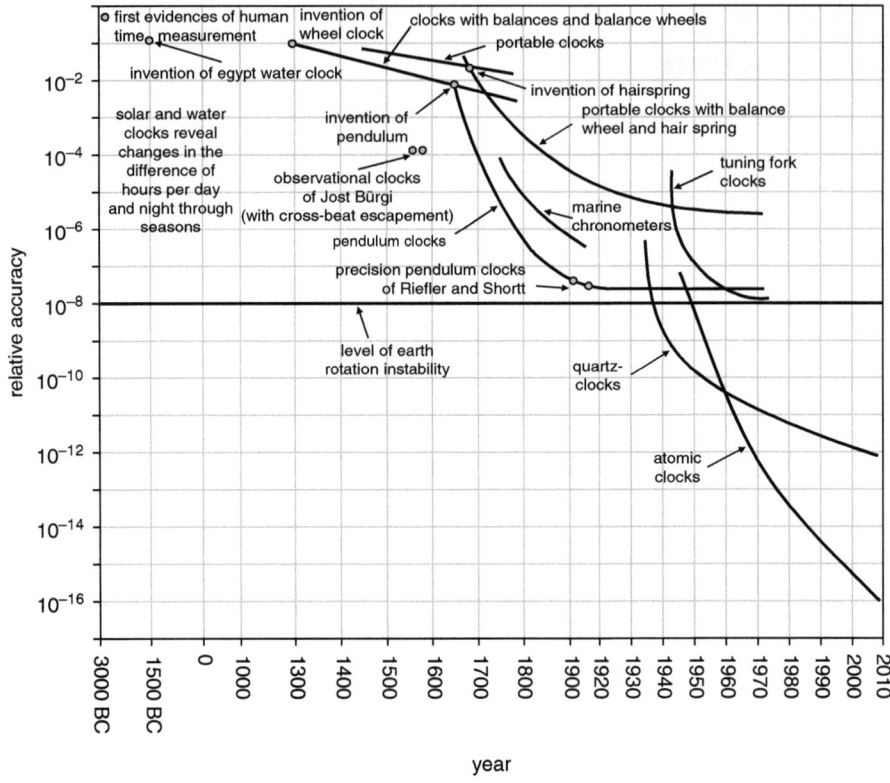

Fig. 1.1 The art of clockmaking over the course of time

with the art of clockmaking. Any clock uses a physical process whose laws are known and whose progression can be indicated. In practice, periodic processes are used exclusively, i.e., time measurement usually is a frequency determination.

Figure 1.1 provides an overview over the art of clock making from 3000 BC till now. One sees that the accuracy of clocks during that period of time increased by about fourteen orders of magnitude.

The sundial dates back to the Egyptian period, around 1500 BC. A stick or pillar was fixed to the ground, and from its shadow time was deduced. Water clocks were in use by the Egyptians, Romans, and Chinese. Significant progress in accuracy was achieved by the introduction of the pendulum as an oscillator by Galileo Galilei some time around 1600 and realized by Christian Huygens in 1656. The seafarer's *longitude problem* is well known: determining a ship's latitude by astronomical means was straightforward, but not the longitude. One way to determine geographic longitude at sea is by means of a marine chronometer that carries the time at a given port and can be compared with local time as determined by astronomical observations. However, for a loss or gain of only a minute per day, we end up with an error of half an hour for a sea passage of 1 month. As during 30 min the Earth

Fig. 1.2 John Harrison's marine chronomtere. Courtesy of the National Maritime Museum, Greenwich, London

revolves by an angle of 7.5°, this corresponds to a navigational error of several hundred miles.

At the beginning of the eighteenth century, a *Board of Longitude* appointed by the British parliament initiated a scientific competition to find a practical method of solving the longitude problem (Sobel and Andrewes 2003). Two possibilities were considered: the lunar distance method and a precise clock, i.e. a marine chronometer. The prize for the chronometer was awarded to the London clockmaker John Harrison. Harrison's 250-year-old masterpiece, his H4 (Fig. 1.2), stands comparison with a modern mechanical clock. The spring systems produced an almost constant drive even during winding, and inaccuracies caused by temperature changes during the sea trip were practically eliminated by a regulator controlled by a bimetallic strip. In the official test, the clock was three times as precise as the commission had requested: after a passage of 22 days from Portsmouth to Barbados, the referees compared the readings from the marine chronometer with the orbits of Jupiter's Galilean moons. The error was found to be less than 39 s! This implies that the relative accuracy of Harrison's H4 was about 1.8 s in a day or a relative accuracy of 2×10^{-5}.

At the beginning of the twentieth century, the accuracy of marine chronometers was about a tenth of a second per day; pendulum clocks were built with an accuracy of better than a hundredth of a second per day. A huge advance in the accuracy of clocks was achieved by employing oscillations of certain crystals instead of pendulum or hairspring. Even more accurate are modern atomic clocks, like cesium clocks, rubidium clocks, or hydrogen maser clocks. A new version of such classical atomic clocks in form of fountain clocks became popular in recent decades. Such classical atomic clocks are based upon atomic transition frequencies that lie in the microwave region; due to their stochastic character, one needs to average times over minutes and days to achieve the highest accuracies and stabilities. Meanwhile, new kinds of atomic clocks have been developed that use atomic transitions with much higher frequencies, i.e., optical clocks, for which averaging times are significantly shorter. Clock stabilities (accuracies) are now approaching 10^{-17}, a number that should be compared with the age of the universe of about 4×10^{17} s.

Now, according to relativity, the time indicated by a particular clock depends upon the clock's velocity with respect to the observer who tries to read the clock and also upon the gravitational potential that prevails at the clock's position. The readout of a certain clock for some co-moving observer (the clock's proper time) is not a good tool for other observers. For that reason, the introduction of standardized timescales is of great importance for practical applications. The readings of atomic clocks are the basis for all astronomical timescales now in use (note that some "times," notably universal time and sidereal time, are in fact angles that describe Earth rotation and are not timescales in the strict sense). A timescale is a carefully selected time coordinate that can be used for any event that occurs in a certain spatial region such as the vicinity of the Earth; it should have a simple and well-defined relation to the proper time of a real clock in that spatial region.

Due to relativity, one has in particular to distinguish geocentric from barycentric (the barycenter of the solar system) timescales. The basic geocentric timescale is geocentric coordinate time (TCG). TCG is the time coordinate of the geocentric celestial reference system (GCRS). Its practical realization is achieved through certain formulae that relate the proper times of atomic clocks and their spatial GCRS coordinates **X** to TCG. TCG is an important tool for the synchronization of clocks: two clocks showing proper time τ_1 and τ_2 are called synchronous if their corresponding TCG values agree, i.e., if $\text{TCG}(\tau_1, \mathbf{X}_1) = \text{TCG}(\tau_2, \mathbf{X}_2)$. Terrestrial time (TT) differs from TCG only by a constant rate that was chosen such that TT approximates proper time of a clock on the geoid. Finally, international atomic time (TAI) agrees with TT except for a constant offset that is there only for historical reasons. From a practical point of view, it is TAI that is directly derived from the readings of a large number of atomic clocks distributed around the globe, and from TAI the timescales TT and TCG are generated. In this sense, a practically useful timescale is generated by the readings of actual clocks (this is the essential difference from a pure mathematical time coordinate).

A timescale to be used as a reference is characterized by its reliability, frequency stability and frequency accuracy, and accessibility. Arias (2010) characterized these aspects as follows:

"The *reliability* of a timescale is closely linked to the reliability of the clocks involved in its construction. Reliability is also associated with redundancy; in the case of TAI, a large number of clocks are required.

The *frequency stability* of a timescale is its capacity to maintain a fixed ratio between its unitary scale interval and its theoretical counterpart (one measure of the frequency stability is the Allan variance that is discussed below).

The *frequency accuracy* of a timescale is the aptitude of its unitary scale interval to reproduce its theoretical counterpart. After the calculation of a timescale on the basis of an algorithm conferring the requested frequency stability, frequency accuracy can be improved by comparing the frequency of the timescale with that of primary frequency standards, and by applying, if necessary, frequency corrections.

The *accessibility* of a worldwide timescale is its aptitude to provide a means of dating events for everyone. This depends upon the precision which is required. In the case of TAI the ultimate precision requires a delay of a few tens of days in order to reach the long-term frequency stability required for a reference timescale."

Especially for interplanetary spacecraft navigation, barycentric timescales are needed. Barycentric coordinate time (TCB) plays a similar role as TCG but in the barycentric celestial reference system (BCRS). Meanwhile, barycentric dynamical time (TDB) used in certain planetary ephemerides has been defined to differ from TCB essentially only by a constant rate that was fixed by IAU 2006 Resolution B3. The relation between barycentric and geocentric timescales such as TCB and TCG involves complicated 4-dimensional space–time transformations (generalizations of the special relativistic Lorentz transformations).

1.1.2 Various Space–Time Reference Systems

Today, not only the problem of time but also that of spatial coordinates has to be formulated within Einstein's theory of gravity (GRT). For most applications in the solar system, the first post-Newtonian approximation to GRT will be sufficient. Space and time no longer are separate entities but are just aspects of a 4-dimensional space–time.

In Einstein's GRT, the gravitational field is described by some space–time metric tensor $g_{\mu\nu}$. This metric tensor is an extremely useful tool because it relates a coordinate picture of observations with observables. It provides a tool for the description of (1) (idealized) clock rates, (2) light propagation, and (3) the motion of astronomical bodies in a gravitational N-body problem.

A central conceptual starting point is the canonical barycentric space–time metric tensor. This canonical metric is the conceptual basis for the barycentric dynamical reference system (BDRS; officially still called "Conventional dynamical realization of the ICRS"), the BCRS, as well as the GCRS. All of these systems present some sort of quasi-inertial or "space-fixed" reference system.

Our discussion of concrete space–time reference systems will start with a BDRS on the basis of observations of massive solar system bodies (Sun, Earth, Moon, planets, asteroids, etc.). Thus, a BDRS is realized by means of an ephemeris compass. Today, a BDRS is realized by some modern numerical ephemeris for the solar system (DE, EPM, INPOP).

A celestial reference system (CRS) is based upon a stellar compass. In the early days, a fundamental catalog of stellar positions presented the realization of a CRS. In other words, a stellar catalog was the realization of the astronomical reference system; it presented the basis for the corresponding reference frame. At present, a catalog of radio sources, mostly quasars, forms the basic structure of the international celestial reference system (ICRS). If the forthcoming astrometric space mission Gaia is successful, an optical catalog with positional accuracy of order ten microarcseconds will be the realization of the astronomical quasi-inertial reference system. Such precise astrometry will also help to answer fundamental questions of our cosmological world view. The ultraprecise astrometric data may allow us to infer whether our universe shows a closed, flat, or open geometry and whether the cosmic expansion, possibly accelerated, will last forever.

In the framework of relativity, we have to distinguish a BCRS with coordinates $(t, \mathbf{x})$ from a GCRS with coordinates $(T, \mathbf{X})$. The space–time coordinates $(t, \mathbf{x})$ and $(T, \mathbf{X})$ are related by some complicated 4-dimensional space–time transformation, generalizing the simple *Newtonian relations*: $T = t$ and $\mathbf{X} = \mathbf{x} - \mathbf{z}_E$, where $\mathbf{z}_E$ denotes the barycentric position of the geocenter.

In practical applications, various different astronomical spatial coordinates play a role. For the problem of telescope pointing, azimuth, A (the angle in the horizon plane reckoned from north towards the east), and elevation, a, or zenith distance, z, are the usual choice. In the case of a position catalog, the usual equatorial coordinates right ascension, α, and declination, δ, are used, and the hour angle, h, plays an important intermediate role.

This classical astronomical reference system conceptually carries elements from the Earth's rotational motion (its z-axis is defined by the instantaneous rotation vector of the Earth) as well as its translational motion: the classical x-axis is given by the vernal equinox, the ascending node of the ecliptic on the celestial equator. The motion of such a classical astronomical reference frame in some quasi-inertial system is given by the precessional and nutational motion of its pole. The corresponding motion in a terrestrial system is the so-called polar motion that leads to variations in astronomical longitude, latitude, and azimuth. Together with the length-of-day variations of some milliseconds per day, the parameters for precession, nutation, and polar motion form the set of Earth orientation parameters (EOP).

After that, terrestrial reference systems (TRS) will be discussed. For their realization mainly the various satellite methods (satellite laser ranging, GPS, GLONASS, etc.) are employed. Of fundamental importance is the international terrestrial reference system (ITRS).

With the presently achievable accuracy of modern very long baseline interferometry (VLBI) measurements, for the first time in history, it became advantageous to define the basic astronomical reference system, *not* with respect to celestial equator (perpendicular to the instantaneous rotation pole (IRP) of the Earth) and the ecliptic but simply directly with respect to the quasar sky. Moreover, it can be shown that no geodetic space technique is sensitive to the IRP directly. For that reason, a new celestial intermediate pole (CIP) was introduced that replaces the IRP conceptually. It splits the transformation from the GCRS to the ITRS into two parts: precession/nutation as the space motion of the CIP on the one hand and polar motion as the terrestrial motion of the CIP on the other hand. Also the origin for right ascension, the classical equinox concept, has meanwhile been replaced by the so-called celestial intermediate origin (CIO), originally called the *nonrotating origin*. It is defined as a point that at no time has a velocity component along the CIP equator itself.

It is noteworthy that the precise astronomical geodetic methods for the determination of spatial coordinates are all based on precise time measurements. This is true for the geodetic space methods such as global navigation satellite systems (GNSS) like GPS, GLONASS, or GALILEO; laser distance measurements to satellites (satellite laser ranging, SLR) or to the Moon (lunar laser ranging, LLR); as well as VLBI.

Historically, positioning on Earth (the determination of longitude and latitude) with a classical astronomical method has been of great importance but is inevitably in decline in the GPS era. Nevertheless, some classical methods, such as the lines of position method or azimuth determination by means of solar observations, are still in use today.

The modern methods just mentioned have achieved remarkable accuracy: for VLBI at present (2012), it is below 7 mm for baselines of some 1,000 km. The distance to some retroreflectors on the lunar surface can be measured by LLR with a precision of a few centimeters.

1.1.3 Classical Apparent, True, and Mean Places

What realizes the fundamental astronomical reference system is a catalog containing the positions of astronomical objects such as quasars and stars. The corresponding astronomical observations were conducted from Earth or from some astrometric satellite and were subject to a number of geometrical and physical effects that affected the observed astronomical positions in a time-varying way. Consequently, a series of rotations and corrections are needed to transform the observed coordinates into the corresponding catalog positions.

If a star or quasar is observed from the Earth, the observed direction of the incoming radiation is affected by atmospheric refraction; correcting for refraction yields an *in vacuo* topocentric place. The observer's terrestrial location can then be used to deduce position and (from Earth rotation) velocity in space, putting the observation on the same footing as one made from Earth orbit. The next step is to make the direction geocentric by correcting for aberration and, for nearby objects, parallax. In the classical formulation, the direction at this stage is called the *geocentric apparent place*. The changing orientation of the Earth's axis due to precession–nutation can then be removed by applying successive coordinate rotations. The nutation correction gives the *mean place of date*; the precession correction gives the mean place with respect to the precessing equator and equinox for a chosen reference epoch, typically J2000.0. This is sometimes called the *proper place*.

Next, the position is referred to the solar system barycenter by applying corrections for annual aberration and parallax (and in very high-precision work, light deflection by the Sun). This produces the *astrometric place*.

Finally, if the object has a known space motion (the transverse component of which in the stellar case is called *proper motion*), the position is extrapolated to the chosen standard epoch (Fig. 1.3). Occasionally, the catalog equinox and epoch differ. For example, the Hipparcos catalog is equinox J2000.0 but epoch J1991.25.

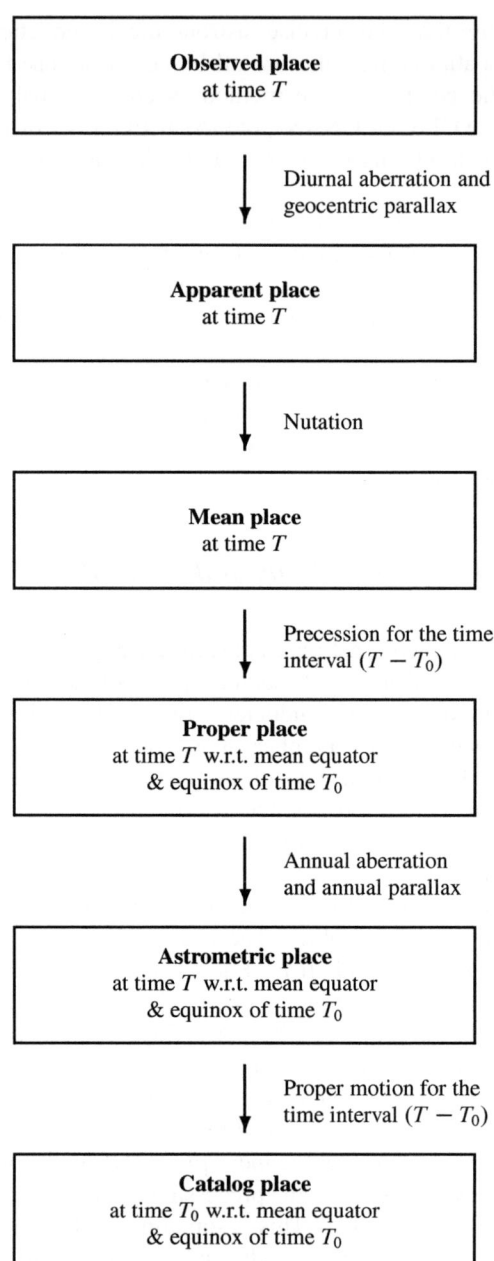

Fig. 1.3 Transformation path from observed place to catalog place

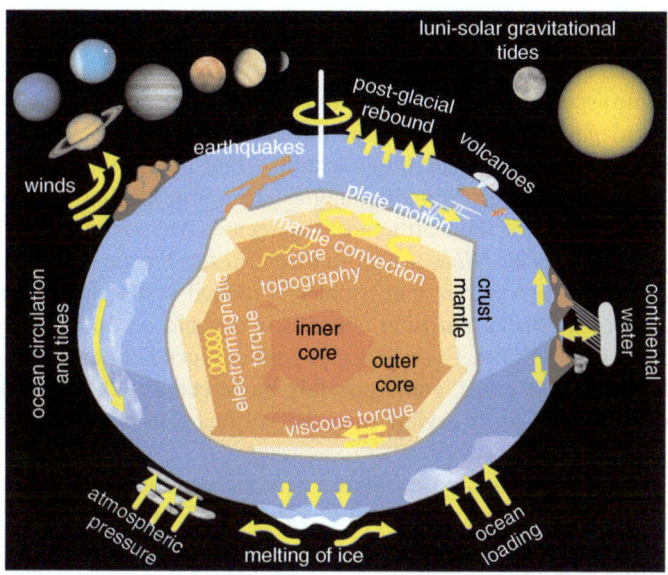

Fig. 1.4 Different processes that affect the Earth's rotation; from Global Geophysical Fluids Center, Goddard Space Flight Center, NASA

1.2 Astronomical Geodetic Applications

Due to the presently achievable accuracy of time and position measurements by means of astronomical geodetic reference systems, a variety of different applications can be realized.

1.2.1 Navigation

The classical application of astronomical geodetic reference systems clearly lies in the field of navigation. It cannot be concealed that the US satellite-based GPS was established primarily for military purposes and whose accuracy can be reduced for the civil user at any time. For that reason, the European community had decided to build up their own satellite-based navigation system GALILEO. For the successful accomplishment of interplanetary spacecraft navigation, highly precise astronomical reference systems are a compelling necessity.

1.2.2 Global Geodynamics and the System Earth

One remarkable aspect of modern astronomical geodetic measuring methods is their contributions to gain information about the system Earth. The EOPs nowadays can be measured with sheer incredible accuracy that lies in the region of several microarcseconds for the corresponding angles. This implies that tiny variations in the various subsystems of the Earth can be detected. An example is the recurrent phenomenon known as *El Niño*, a warm current in the Pacific Ocean towards the west coast of southern America that usually is related with a fish drop along the coastline of Peru (the fish's nutrition depends upon cold ocean currents) and that can be detected in the EOPs. Even anthropogenic-induced climate variations, e.g., related with greenhouse gases, could possibly be detected in the EOPs in the near future. Figure 1.4 gives an overview over the various causes for variations in the Earth's rotation.

The EOPs are integral quantities related with the system Earth, e.g., via its total angular momentum. However, methods like satellite laser ranging, SLR, are able to yield significant information about local mass transports that might be of relevance, e.g., for a future earthquake prediction.

Also for high-precision modeling of global geodynamics, an adequate space–time reference system clearly is a necessary prerequisite.

Chapter 2
Time

2.1 Stability of Oscillators

In the following it will be of relevance to have a good idea about the quality of an atomic clock (oscillator). One quality criterion is the stability of the oscillator, determined by the stochastic fluctuations in frequency space about the norm frequency. Usually the stability of an oscillator is characterized by the so-called *2-point Allan variance* (named after the American physicist David W. Allan; Allan et al. 1988) that will be discussed now.

Every precise frequency generator like a quartz oscillator or an atomic frequency standard shows frequency fluctuations both of random as well as systematic character. Systematic fluctuations are, e.g.,

- Frequency drifts, e.g., in quartz oscillators caused by aging (by the forming of dislocations in the lattice)
- Frequency modulations caused by the low-frequency part of the supply current or
- Quasiperiodic frequency modulations resulting from temperature and pressure variations

Sporadically in such oscillators, also unexplained frequency jumps might occur.

For the following considerations let t be the (Newtonian) time variable. Let us describe the output signal (e.g., alternating current) of an ideal oscillator by

$$U(t) = U_0 \sin 2\pi f_0 t,$$

where U_0 (f_0) is some norm amplitude (frequency). Every realistic, nonideal oscillator will show time-dependent amplitude and phase fluctuations, i.e.,

$$U(t) = [U_0 + \epsilon(t)] \times \sin[2\pi f_0 t + \varphi(t)].$$

For the problem of time determination, amplitude variations $\epsilon(t)$ basically play no role, so we write

$$U(t) = U_0 \sin \phi(t) \tag{2.1}$$

with

$$\phi(t) = 2\pi f_0 t + \varphi(t). \tag{2.2}$$

With the actual frequency

$$f(t) = f_0 + \frac{1}{2\pi} \frac{d\varphi(t)}{dt}$$

we can define the frequency fluctuation as

$$\Delta f(t) \equiv f(t) - f_0 = \frac{1}{2\pi} \frac{d\varphi(t)}{dt}.$$

With the relative frequency fluctuation

$$y(t) \equiv \frac{\Delta f(t)}{f_0} \tag{2.3}$$

we can define the phase time $x(t)$ as

$$x(t) \equiv \int_{t_0}^{t} y(t) dt. \tag{2.4}$$

The value of $x(t)$ corresponds to the error in time, a clock shows at time t since the start at t_0. The relative frequency fluctuation $y(t)$ cannot be measured directly. Only a mean value taken over a time interval τ is measurable:

$$y(t_k, \tau) = \frac{1}{\tau} \int_{t_k}^{t_k + \tau} y(t) \, dt = \frac{1}{\tau} [x(t_k + \tau) - x(t_k)]. \tag{2.5}$$

The phase time $x(t)$ of some test oscillator can be measured by means of a reference oscillator. Simple conditions prevail if the frequency fluctuations of the reference oscillator can be neglected (Fig. 2.1). For the determination of the phase time an impulse counter is employed. If, e.g., the reference oscillator and the test oscillator both produce a well-defined impulse every second, then the phase time between the second pulses can be determined if a second pulse of the reference oscillator starts the counter at time t and the following second pulse of the test oscillator stops the counter. Then the counter reading determines the phase time $x(t)$.

Often, we encounter the situation that $y(t)$ can be written as

$$y(t) = y_r(t) + at + y_0,$$

2.1 Stability of Oscillators

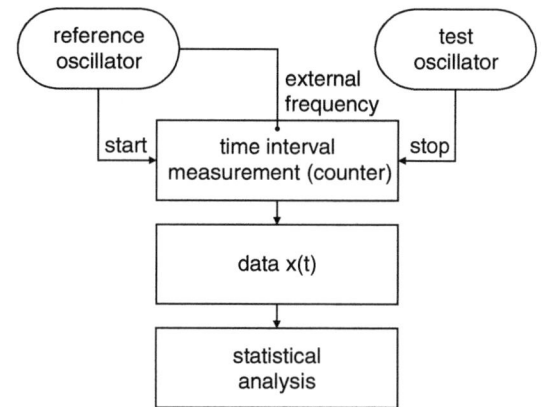

Fig. 2.1 Simple measuring assembly for the determination of the phase time $x(t)$

where $y_r(t)$ denotes a random fluctuation, a is the aging coefficient, and y_0 denotes a constant frequency shift. The resulting phase time $x(t)$ is then given by

$$x(t) = x_r(t) + \frac{1}{2}at^2 + y_0 t$$

with

$$x_r(t) = \int y_r(t)\,dt + x_0.$$

Under standard conditions,

$$\int_{t_0}^{t} y_r(t)\,dt = \int_{t_0}^{t} x_r(t)\,dt = 0.$$

We will now define a *measuring series* $\mathcal{M}(N, \Delta t, \tau)$ for $y(t, \tau)$. In such a series N, different y_k values are obtained by averaging over time intervals of length τ in regular intervals of time $\Delta t = t_{k+1} - t_k$ (Fig. 2.2). From such a measuring series we can derive a mean value

$$<y_i>_N \equiv \frac{1}{N}\sum_{k=1}^{N} y_k \qquad (2.6)$$

and the variance (Fig. 2.2)

$$\sigma_y^2(N, \Delta t, \tau) \equiv \frac{1}{N}\sum_{k=1}^{N}(y_k - <y_i>_N)^2. \qquad (2.7)$$

To define a statistically meaningful quantity we could consider a very large number of individual measurements, $N \to \infty$. Another possibility is to repeat the measuring series $\mathcal{M}(N, \Delta t, \tau)$ very often (at convenient times). We will now assume that the

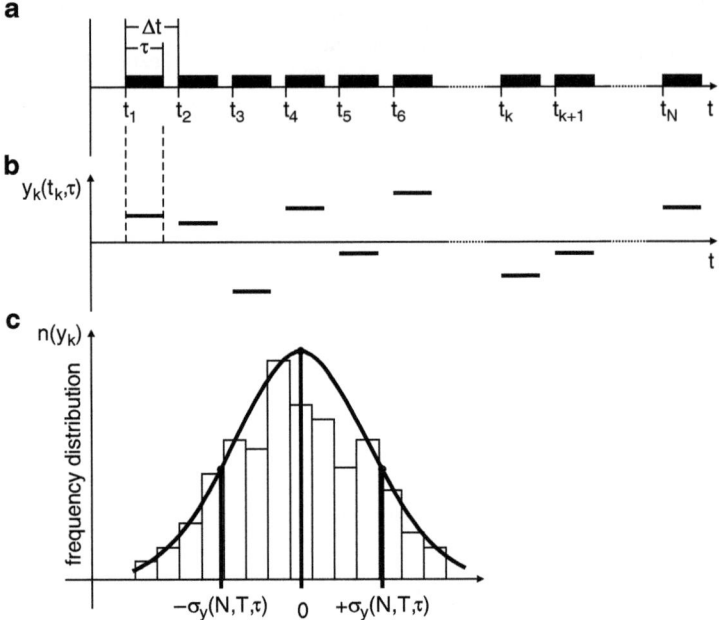

Fig. 2.2 (a) Measurement intervals for the determination of frequency fluctuations; (b) measurements averaged over the time interval τ; (c) frequency distribution of $y_k(t_k,\tau)$

measuring series is performed M times. As stability measure of the oscillator we can then define the Allan variance as

$$<\sigma_y^2(N, \Delta t, \tau)> \equiv \lim_{M\to\infty} \frac{1}{M} \sum_{i=1}^{M} \sigma_{y_i}^2(N, \Delta t, \tau). \quad (2.8)$$

For the special case of $N = 2$ we thereby get the 2-point Allan variance:

$$<\sigma_y^2(2, \Delta t, \tau)> \equiv \lim_{M\to\infty} \frac{1}{M} \sum_{i=1}^{M} \sigma_{y_i}^2(2, \Delta t, \tau). \quad (2.9)$$

In that case

$$\sigma_{y_i}^2 = \left[\left(y_{i,1} - \frac{y_{i,1} + y_{i,2}}{2}\right)^2 + \left(y_{i,2} - \frac{y_{i,1} + y_{i,2}}{2}\right)^2\right]$$

$$= \frac{1}{2}(y_{i,2} - y_{i,1})^2$$

and, therefore,

$$<\sigma_y^2(2, \Delta t, \tau)> = \lim_{M\to\infty} \frac{1}{M} \sum_{k=1}^{M} \frac{1}{2}(y_{k+1} - y_k)^2. \quad (2.10)$$

Let $F(f)$ be the Fourier transform of $y(t)$. Then, the spectral power density

$$S_y(f) = |F(f)|^2$$

is related with the autocovariance or autocorrelation function

$$C(\tau) = \int_{-\infty}^{+\infty} y(t) y(t+\tau) \, d\tau \tag{2.11}$$

by

$$C(\tau) = \int_{-\infty}^{+\infty} e^{2\pi i f \tau} |F(f)|^2 \, df \tag{2.12}$$

or

$$|F(f)|^2 = \int_{-\infty}^{+\infty} e^{-2\pi i f \tau} C(\tau) \, d\tau, \tag{2.13}$$

according to the Wiener–Khinchin theorem.

In most cases the spectral power density $S_y(f)$ of an oscillator can be represented in the form

$$S_y(f) = h_{-2} f^{-2} + h_{-1} f^{-1} + h_0 + h_1 f + h_2 f^2 \tag{2.14}$$

and the various f^s-terms can be related with different kinds of noises:

- h_{-2}: random walk in frequency (y).
- h_{-1}: flicker noise in frequency (y), caused, e.g., by fluctuations in the power supply unit.
- h_0: white frequency noise corresponding to random walk in phase.
- h_1: flicker phase noise caused by certain electronic devices.
- h_2: white phase noise caused by additive noise in the electronics.

Schematically this noise behavior is depicted in Fig. 2.3.

For every oscillator the Allan variance depends crucially upon the averaging time τ. Figure 2.4 shows the typical behavior of the Allan variance of an atomic clock as a function of the averaging time τ.

2.2 Bias and Drift

Besides stochastic fluctuations an actual frequency standard shows additional effects that can be modeled to some degree. If f_0 denotes the nominal frequency of the oscillator, then often the approach

$$f(t) - f_0 = \Delta f + \dot{f}(t - t_0) + f_r(t) \tag{2.15}$$

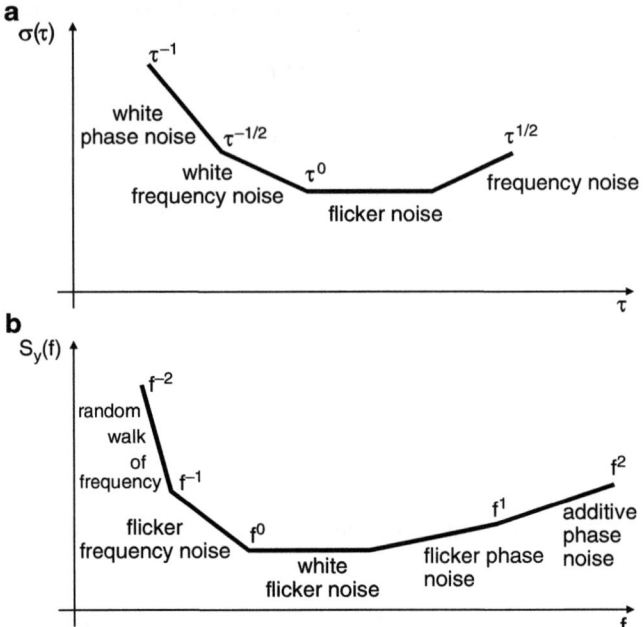

Fig. 2.3 Different kinds of noise in the time (panel **a**) and frequency (panel **b**) domain

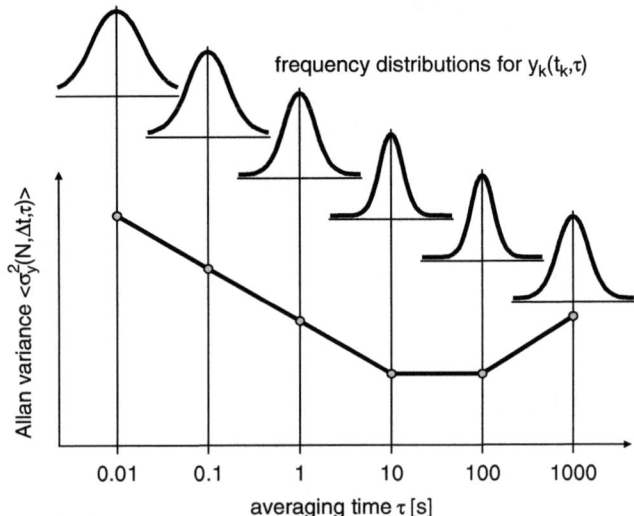

Fig. 2.4 Dependency of the Allan variance $<\sigma_y^2(N,\Delta t,\tau)>$ upon the averaging time τ for an atomic clock

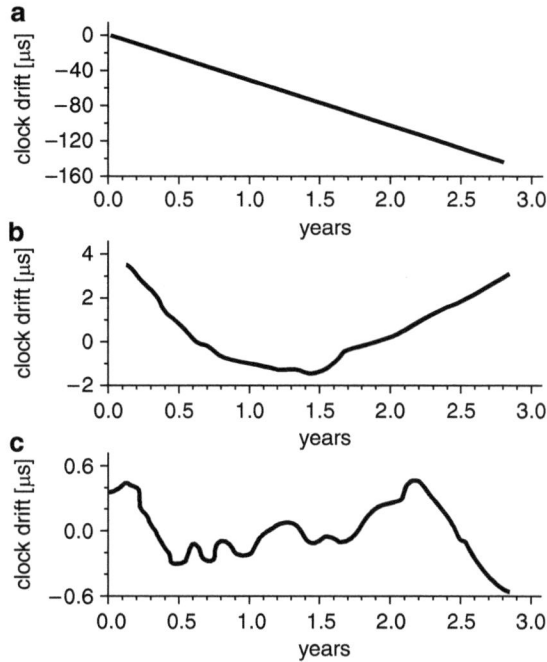

Fig. 2.5 Comparison of the behavior of a commercial Cs clock with those of an ensemble of atomic clocks. (**a**) Raw difference, (**b**) a linear trend was removed, (**c**) also a quadratic part was removed; from Jones and Tryon (1987). Copyright ©1987 Society for Industrial and Applied Mathematics. Reprinted with permission. All rights reserved

is sufficient. Here, Δf is called the bias, and the $\dot{f}$-term describes the drift of the oscillator; f_r describes the stochastic fluctuations.

Figure 2.5 shows a comparison of the behavior of a commercial Cs clock with those of the whole ensemble of atomic clocks from the US Naval Observatory (USNO; Washington). In the middle part a linear trend was removed; in the lower panel, also a quadratic part (Jones and Tryon 1987).

2.3 Quartz and Atomic Clocks

In a quartz oscillator one employs the mechanical vibrations of a quartz crystal showing the piezoelectrical effect: if the crystal is deformed, a voltage is induced. If on the other hand the crystal is placed between two electrodes and a voltage is applied, it will be deformed. If crystal vibrations are stimulated, they lead to a corresponding alternating current that can be used for the construction of a quartz clock. The clock frequency depends crucially upon shape and size of the crystal, and usually the electrodes are mounted directly on the quartz body.

Figure 2.6 schematically shows a typical block diagram of a quartz oscillator and Fig. 2.7 a typical variance behavior of commercial quartz oscillators.

In atomic oscillators certain atomic hyperfine levels in selected atoms, like hydrogen (^{1}H), rubidium (^{87}Rb), or cesium (^{133}Cs), are employed to significantly improve the stability of quartz oscillators. These hyperfine transitions result from

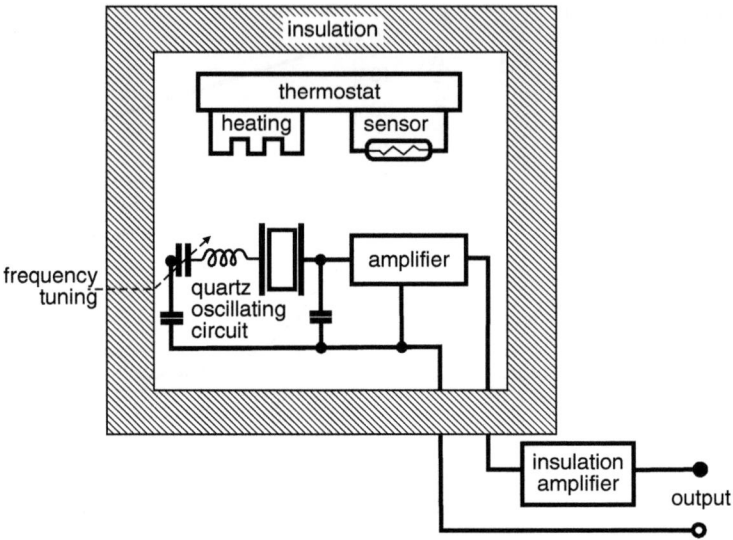

Fig. 2.6 Block diagram of a quartz oscillator

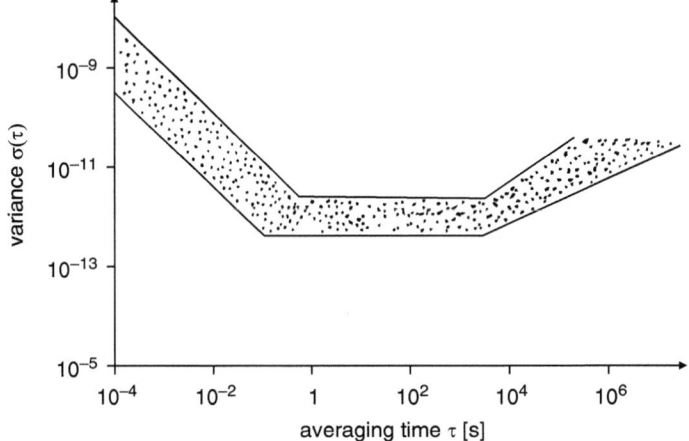

Fig. 2.7 Typical variance behavior of commercial quartz oscillators

the magnetic interaction between the spin of the atomic nucleus and the valence electron in the shell. The hyperfine transition frequencies lie in the microwave region where corresponding analysis techniques are available.

The hyperfine levels are characterized by a quantum number F describing the total spin of the atom. It results from the spin quantum number of the nucleus I and the valence electron with a value of $1/2$ (in units of the reduced Planck's constant $\hbar$):

$$F = I \pm \frac{1}{2}.$$

2.3 Quartz and Atomic Clocks

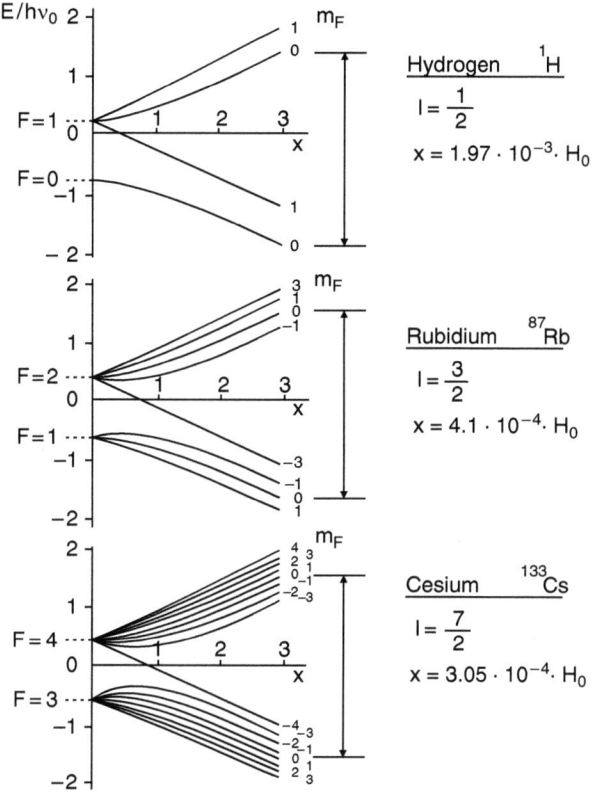

Fig. 2.8 Atomic hyperfine transitions are depicted that are employed in the construction of classical atomic clocks. The quantity x is a measure of the magnetic field strength

The two signs indicate that the direction of the electron's spin could be parallel or antiparallel to the spin of the nucleus.

From the values $I = 1/2$ (H), $3/2$ (Rb), and $7/2$ (Cs) correspondingly, one gets the following possibilities: ^{1}H: $F = 0$ and $F = 1$, ^{87}Rb: $F = 1$ and $F = 2$, ^{133}Cs: $F = 3$ and $F = 4$.

If a static magnetic field is applied, these levels are split into magnetic sublevels because of the Zeeman effect. These sublevels are further characterized by a magnetic quantum number m_F describing the component of the total spin in the direction of the applied magnetic field; it can take the values:

$$m_F = 0, \pm 1, \pm 2, \ldots, \pm F.$$

Thus, each F-level splits into $2F+1$ magnetic sublevels. Quantum mechanical rules allow only transitions with $\Delta F = 0, \pm 1$, and $\Delta m_F = 0, \pm 1$. For the realization of atomic frequency standards, transitions with $\Delta F = \pm 1$, $\Delta m_F = 0$ are best suited.

Atomic transitions in hydrogen, rubidium, and cesium standards are depicted in Fig. 2.8. Details of the fine and hyperfine splitting of the relevant energy levels in the cesium atom are shown in Fig. 2.9.

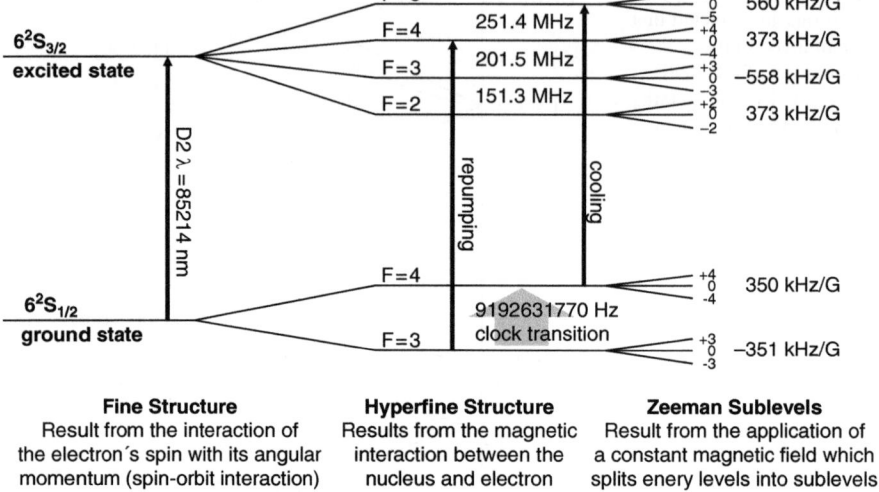

Fig. 2.9 The splitting of energy levels in the cesium atom. The fine structure splitting of spectral lines results from the spin–orbit coupling of the valence electron. The hyperfine splitting results from the interaction of the electron's spin with that of the atomic nucleus. A weak external magnetic field produces the Zeeman splitting

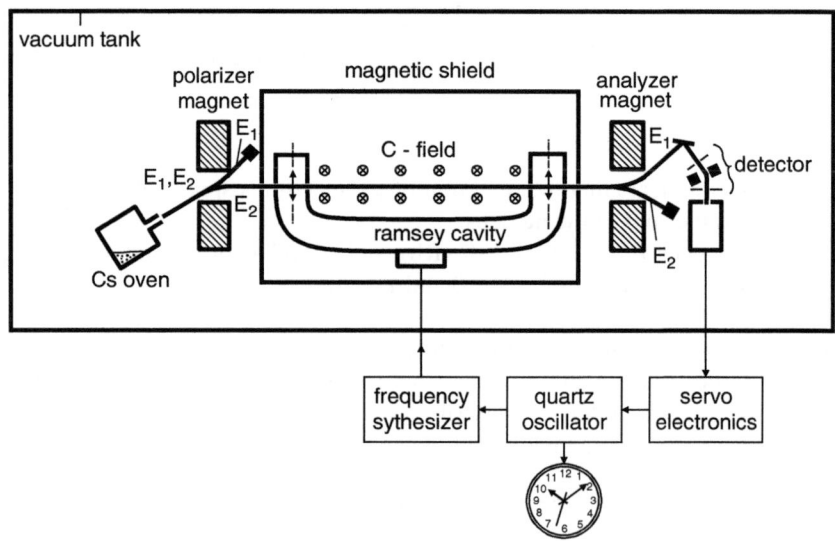

Fig. 2.10 Schematic diagram of a Cs frequency oscillator; from Riehle (2001). Courtesy of F. Riehle

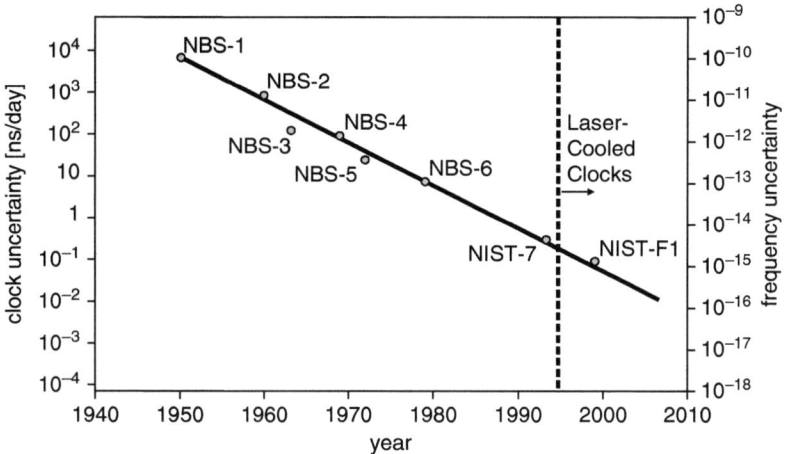

Fig. 2.11 Frequency stability of primary frequency standards at NIST, USA

2.4 Cesium Clocks

An atomic cesium clock is based upon the principle of atomic beam magnetic resonance as developed by I. Rabi before 1940. The essential parts of a cesium oscillator are shown in Fig. 2.10. A cesium oven at 100–200°C creates a cesium vapor that escapes through a small orifice. It enters an inhomogeneous magnetic field that splits the two energy levels of interest ($F = 3, F = 4$) due to their different magnetic dipole moments. Typically only the atoms in the upper ($F = 4$) level are allowed to enter the microwave resonator which is magnetically shielded, and the hyperfine splitting is achieved with a constant magnetic field. Inside the resonator, perpendicular to the beam axis (to reduce Doppler effects) a microwave signal derived from a quartz oscillator in the vicinity of the transition frequency of 9.2 GHz is applied. In case of resonance between this signal and the hyperfine levels in the cesium atom induced transitions will occur and the corresponding atoms having left the resonator will be deflected by a second magnet onto a particle detector. With a feedback loop the frequency of the quartz clock is steered until a maximal particle current is seen by the detector. In this way the stability of the quartz clock is improved significantly (Figs. 2.11 and 2.12).

2.5 Rubidium Clocks

Like the Cs generator the rubidium frequency oscillator is a passive system (Fig. 2.13). The clock transition of the ^{87}Rb isotope between the ($F = 2$) and ($F = 1$) state leads to a radiation at 6.835 GHz. Light from a Rb lamp, filtered

Fig. 2.12 The cesium standard NIST-7 (*left*) and the fountain clock NIST-F1 (*right*) with improved accuracy (NIST, USA)

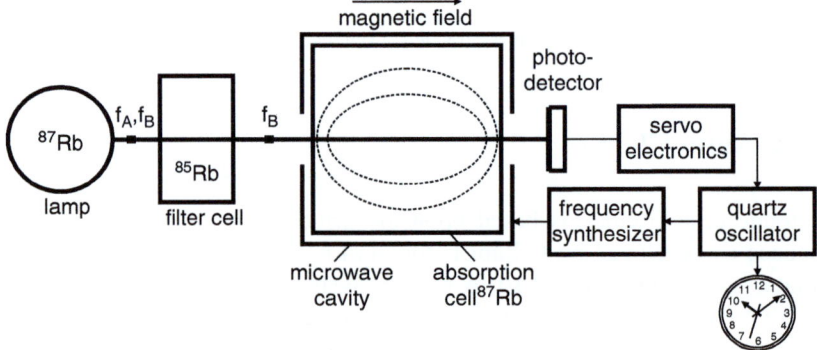

Fig. 2.13 Functional principle of a rubidium standard; from Riehle (2001). Courtesy of F. Riehle

by a ^{85}Rb gas cell by a mechanism called optical pumping, induces a different occupation of the two relevant energy levels in the rubidium gas inside a microwave resonator. The almost empty $F = 1$ level is again populated by the action of a high-frequency field derived from a quartz clock inducing $F = 2$ to $F = 1$ transitions. This leads again to an absorption of the pump signal. The corresponding change in transparency of the gas cell is monitored with a photo diode, and by means of a feedback loop, the stability of the quartz clock is improved. Frequency stabilities of commercial rubidium frequency standards are shown in Fig. 2.14.

2.6 H-Maser

Figure 2.15 shows schematically the principle of a hydrogen maser clock. In an H maser molecular hydrogen is dissociated, and the atoms in the upper clock level are forwarded into a storage bulb. This bulb, usually made from quartz glass, is

2.6 H-Maser

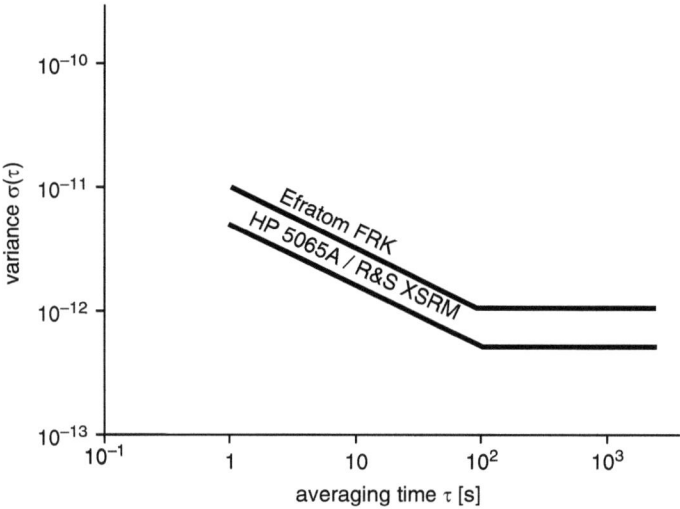

Fig. 2.14 Frequency stability of commercial Rb standards

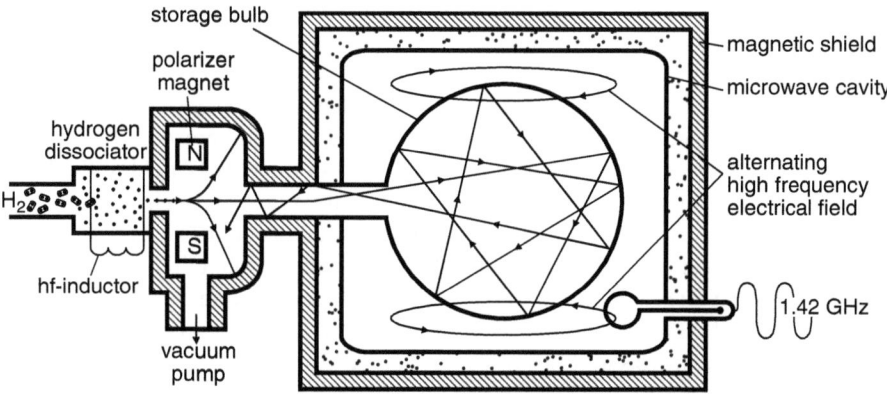

Fig. 2.15 Schematical diagram of a hydrogen maser

coated, e.g., with Teflon to increase the lifetime of atoms in the upper level after being reflected by the wall of the bulb. The vessel is placed inside of a magnetically shielded microwave resonator where a high-frequency noise field containing the clock transition is created to excite induced emission. This process leads to maser activity at the resonance frequency that can be gripped from the resonator to stabilize a quartz clock. Figure 2.16 shows the typical stability behavior of an H maser.

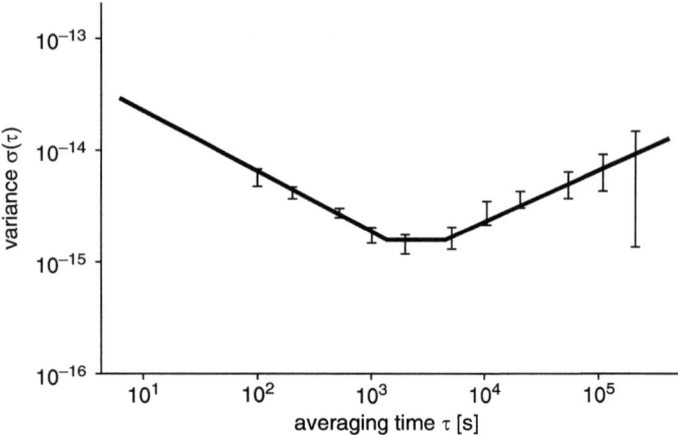

Fig. 2.16 Typical stability behavior of an H maser

2.7 Fountain Clocks

The precision and stability of atomic clocks can be improved by cooling the relevant reference atoms to sub-Kelvin temperatures and thereby reducing their thermal motion significantly. Neutral atoms can be kept by interaction with laser light in optical traps where they can be cooled and further manipulated. In that way it is possible to create fountain clocks where, e.g., Cs atoms are cooled to a few microkelvins and catapulted about a meter upwards in the gravity field of the Earth where in course of their parabolic orbit, they move twice through a microwave resonator. The advantage lies in the long duration time inside the resonator leading to a resonance width of less than 1 Hz. The first cesium fountain clock has been built by LNE-SYRTE at Paris Observatory in 1995 (see, e.g., http://tycho.usno.navy.mil/cesium.html). The clock at Paris Observatory is depicted in Fig. 2.17. Current fountain clocks achieve stabilities of better than 10^{-15}.

2.8 Optical Clocks

Though cesium clocks especially in the form of fountains achieve a remarkable level of stability, their limitations can clearly be seen. A stability of about 10^{-15} can be achieved only after averaging over a day which presents a problem for real-time applications. Now the achievable stability of an atomic clock is related with the clock frequency that for cesium clocks lies in the microwave region at 9.2 GHz. Meanwhile optical clocks have been manufactured with clock frequencies of about 10^{15} Hz, where comparable stabilities can be achieved already after a second. Likely in the near future optical clocks will be superior to conventional atomic clocks. The energy levels of an optical clock are depicted in Fig. 2.18 (left).

2.8 Optical Clocks

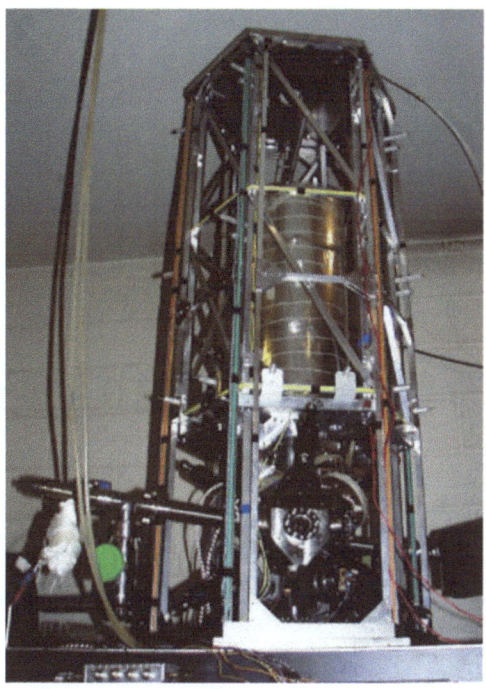

Fig. 2.17 Cs fountain clock at Paris Observatory

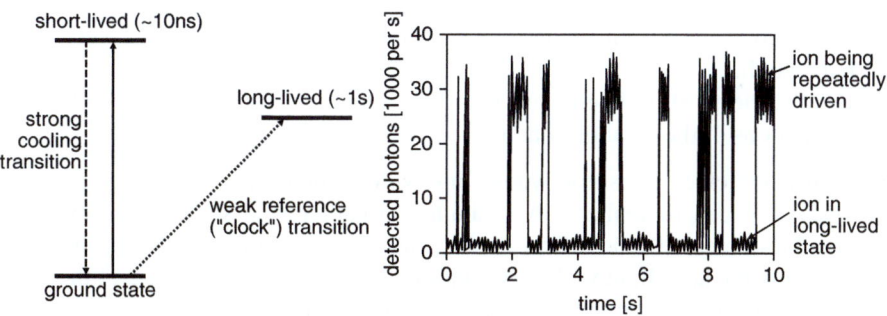

Fig. 2.18 Energy levels of an optical clock (*left*). The cooling transition is accompanied by the emission of fluorescence light (*right*); from http://physicsworld.com (4 May 2005)

An optical clock consists of three main elements: (1) an optical clock transition, (2) a local oscillator, and (3) some counter. In addition we face further elements like traps for atoms or ions, laser for cooling, etc.

1. The first basic element is a precise reference frequency delivered by some very narrow absorption line. Meanwhile a natural line width of less than 1 Hz has

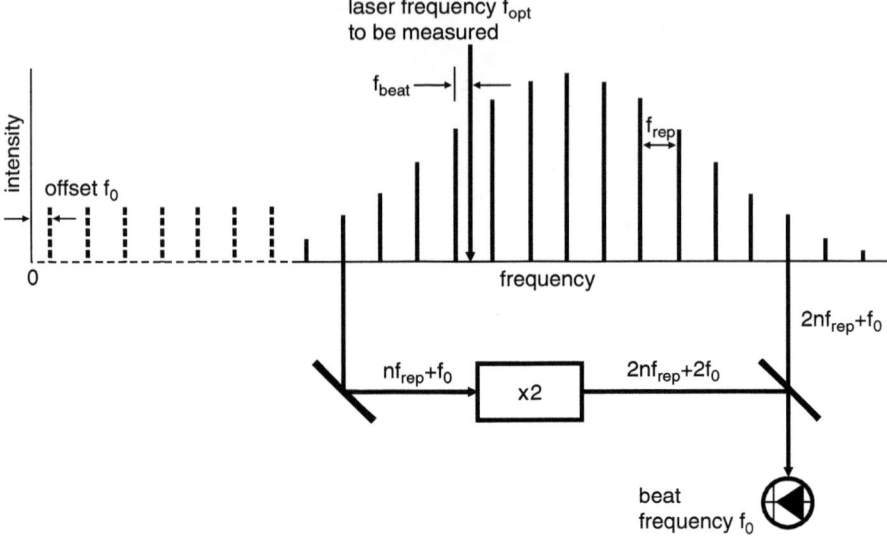

Fig. 2.19 The femtosecond comb in frequency range; from http://physicsworld.com, (4 May 2005)

been achieved for this clock transition, where ions such as ^{199}Hg$^+$ or ^{171}Yb$^+$ were used.
2. The second main element is a test laser (local oscillator) with a very small bandwidth.
3. Finally, the third basic element is a device that can count the extremely rapid oscillations of the local oscillator.

Critical is the clock transition whose line width should be as narrow as possible. Nowadays, it can be realized by a single ion captured in some electromagnetic trap and laser cooled to less than 1 mK. This clock transition is then probed with monochromatic laser light with a line width of less than 1 Hz. Because of the small line width of the clock transition, the absorption of test photons is very weak; for that reason, a technique called electron shelving comes into play. Laser cooling of the ion is realized by means of some short-living energy level, and during phases of cooling, the electron state oscillates rapidly between this state and the ground state which is accompanied with the emission of fluorescence photons (Fig. 2.18, right). If the test laser, however, hits the clock frequency, the electron enters a long-living state, and the fluorescence light discontinues which is easy to detect.

Since the local oscillator oscillates with a period in the region of a femtosecond, the construction of a counter really presents a challenge. Only in 1999, Ted Hänsch and colleagues succeeded to construct such a counter. It consists of a femtosecond laser that emits a whole train of pulses with a repetition frequency f_{rep} of some

2.9 Application of Highly Precise Clocks

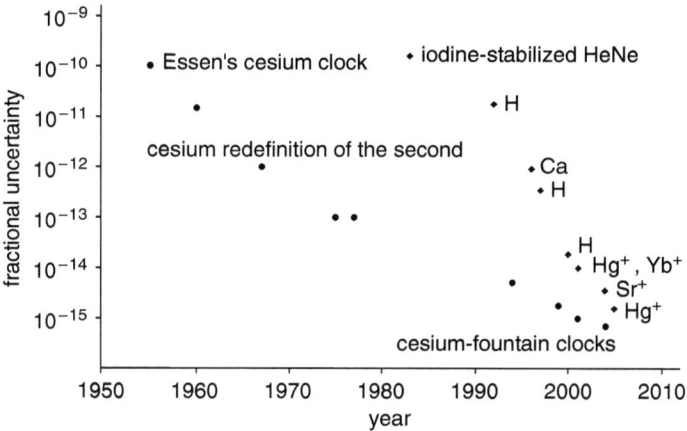

Fig. 2.20 Comparison of achievable stabilities of cesium clocks (*dots*) and optical clocks (*diamonds*)

100 MHz. In that frequency domain, this sequence of pulses appears as a series of equidistant frequencies with separation f_{rep} (Fig. 2.19). For that reason, such a counter is called femtosecond comb. The frequency of a line of such a comb is given by $n f_{rep}$ plus a displacement f_0. Both frequencies f_{rep} and f_0 can be measured by means of beats. Now the comb can be coupled with a cesium clock, and the beat frequency between f_{opt} and the neighboring comb mode can be determined. Meanwhile, it has been shown that optical frequencies can be compared with a femtosecond comb at the level of 10^{-19}. Figure 2.20 shows a comparison of achievable stabilities of cesium and optical clocks.

2.9 Application of Highly Precise Clocks

Figure 2.21 shows the requirements on frequency stability for different modern geodetical methods (EDM: electromagnetic distance measurement; SLR: satellite laser ranging; GPS: global positioning system; LLR: lunar laser ranging; VLBI: very long baseline interferometry). Due to their unrivaled long-term stability (3 h to months), industrial cesiums are used for the generation of national times. Moreover, they are used in mobile SLR systems and the GPS. Rb standards are often used as secondary frequency standards. Because of their good short-term stability, they are also used in SLR, LLR, Doppler, and GPS measurements. Due to their distinguished short-term stability, H masers are employed, e.g., for VLBI measurements.

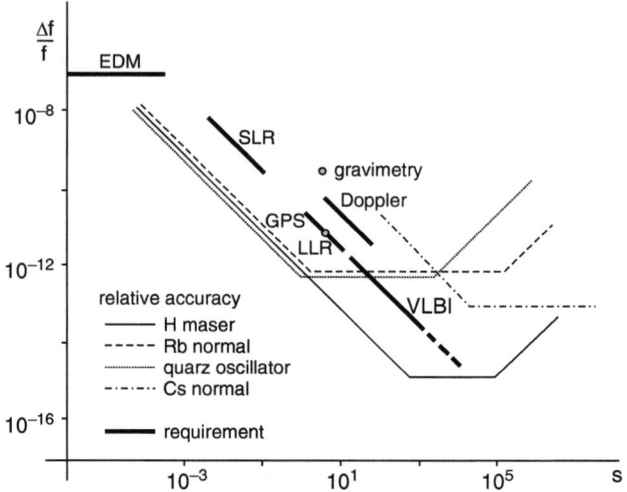

Fig. 2.21 Requirements on frequency stability of several geodetic measuring techniques

2.10 Relativity and the Metric Tensor

Based on the presently achievable stabilities and accuracies of atomic clocks, relativistic effects have to be considered inevitably. The theory of relativity is based upon certain principles that have been verified experimentally with highest precision. The principle of the constancy of the velocity of light in vacuum plays a central role. This principle comprises several independent aspects. First of all the velocity of light in vacuum is independent to the source's state of motion, similar to the sound speed that is also independent upon the velocity of the acoustic source. Then the velocity of light is independent of frequency and polarization. The vacuum is not dispersive. It is common to designate the speed of light in vacuum by c. The numerical value of $c = 299\,792\,458$ m/s is meanwhile used for a definition of the meter.

The meter is the base unit of length in the International System of Units (SI). Originally intended to be one ten-millionth of the distance from the Earth's equator to the North Pole at sea level, its definition has been periodically redefined (Wikipedia). In 1889 at the first General Conference on Weights and Measures (CGPM: Conférence Générale des Poids et Mesures), it was decided to define the SI meter via a prototype (Fig. 2.22) made from a platinum-iridium alloy which was kept at BIPM in Sèvres near Paris. This prototype defined the meter till 1960, when the meter definition was based upon a wavelength of krypton-96 radiation.

After 1983, according to the 17th General Conference on Weights and Measures (1983), the meter is defined as that distance that light travels in vacuum during a time interval of $1/299\,792\,458$ s.

2.10 Relativity and the Metric Tensor

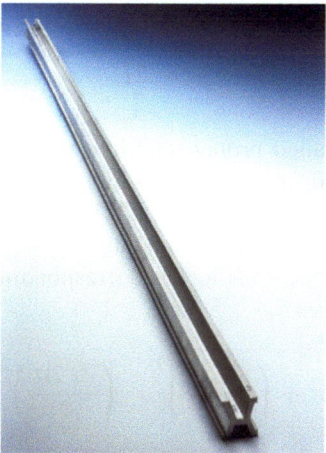

Fig. 2.22 The old international prototype of the meter: the famous platinum-iridium stick that defined the meter until 1960. Today the meter is defined by the speed of light in vacuum

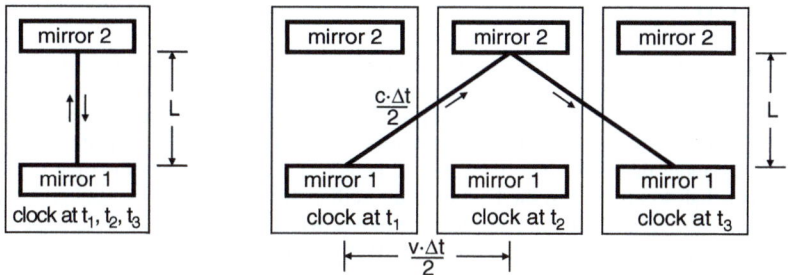

Fig. 2.23 A simple light clock where a light pulse travels between two unaccelerated mirrors back and forth. In the *left part* the observer is at rest with respect to the clock; in the *right part*, the clock moves with velocity v with respect to the observer

Now, the velocity of light in vacuum is also independent to the observer's state of motion (in complete contrast to the sanity and reason); otherwise the present definition of the meter would not make much sense. This aspect of light propagation, which has been experimentally confirmed by many Michelson–Morley-type experiments, has a profound consequence for the problem of clock reading that becomes dependent upon the relative state of motion between clock and observer. To demonstrate this we consider a simple light clock that consists only of two ideal mirrors that move unaccelerated without external forces with constant distance L and a light pulse that moves between the two mirrors back and forth. We first consider an observer at rest with respect to this simple clock (Fig. 2.23, left).

Let us denote the time interval between reflection of a pulse at one mirror and its next return by $\Delta\tau$:

$$\Delta\tau = \frac{2L}{c}.$$

This time interval $\Delta\tau$ is called *proper time interval*. In Fig. 2.23 (right), the same clock is depicted as seen from an observer that moves with the velocity v to the left in the direction of the mirror extensions. It is crucial that also for this observer, light moves with velocity c in vacuum just as for the co-moving observer. Using the Pythagorean theorem, one finds for the corresponding time interval Δt that the moving observer experiences

$$L^2 + \left(\frac{v \cdot \Delta t}{2}\right)^2 = \left(\frac{c \cdot \Delta t}{2}\right)^2,$$

leading to

$$\Delta t = \frac{\Delta\tau}{\sqrt{1 - v^2/c^2}}. \tag{2.16}$$

The observed indication of a clock depends upon the clock's velocity with respect to the observer.

Exercise 2.1: Bill, a 6-year-old boy, is a tennis wizard. Being bored in a railway car, he plays with his tennis ball. He hits the ball hard perpendicular to the floor so that it will be reflected not only by the floor but also from the ceiling in vertical direction and it (almost) comes back to the same point on the ground. His father Ernesto, a physics professor from Stanford University, meditates about the tennis ball: let L be the height of the wagon, then the time it takes for the tennis ball to move from the bottom to the ceiling and back is $\Delta\tau = 2L/v_T$, if v_T is the velocity of the tennis ball. Suppose the train moves with constant speed v with respect to the train station in Hayward. Show that some observer standing on the platform of the train station measures the same time interval as professor Ernesto, i.e., $\Delta t = \Delta\tau$.

The metric tensor has turned out to be a very efficient tool for the description of light propagation and clock rates. For its introduction we first describe a certain part of space, together with time, by means of 4-dimensional space–time coordinates $x^\mu = (ct, x^i)$. The variable t designates the time coordinate that is related with the space–time coordinate $x^0 = ct$.

We want to declare that a Greek index like μ indicates a space–time index that runs over the values 0, 1, 2, 3. The index 0 refers to the time coordinate ($x^0 = ct$); the indices 1, 2, 3 describe the three spatial coordinates (e.g., $x^1 = x, x^2 = y$ and $x^3 = z$). Such coordinates label certain events in space–time. An event is some instance that has only a small extension in space and time. Now we consider two infinitesimally close events e_1 and e_2 with coordinates $e_1 : x^\mu$ and $e_2 : x^\mu + dx^\mu$. The space–time distance ds between these two neighboring events is then given by

$$ds^2 = g_{\mu\nu}\, dx^\mu dx^\nu, \tag{2.17}$$

2.10 Relativity and the Metric Tensor

where $g_{\mu\nu}$ is the metric tensor. In (2.17), we have used the usual Einstein summation convention saying that over two identical (dummy) indices, a summation is implied automatically; in this case, a summation over μ and ν, $\mu, \nu = 0, 1, 2, 3$ is assumed automatically. The metric tensor $g_{\mu\nu}$ can be viewed as a 4×4 matrix with components $g_{00}, g_{01}, \ldots, g_{10}, g_{11}, \ldots$, etc. In the 3-dimensional Euclidean space with Cartesian coordinates $x^i = x, y, z$, the metric tensor is simply given by $g_{ij} = \text{diag}(1, 1, 1)$, i.e., the length element takes the form

$$ds^2 = dx^2 + dy^2 + dz^2,$$

which is nothing but the infinitesimal version of the Pythagorean theorem.

Without gravitational fields we can introduce inertial Cartesian coordinates $x^\mu = (ct, \mathbf{x})$ such that the metric tensor takes the form

$$g_{\mu\nu} = \text{diag}(-1, 1, 1, 1) \tag{2.18}$$

and the corresponding length element reads

$$ds^2 = -c^2 dt^2 + d\mathbf{x}^2. \tag{2.19}$$

We would like to stress that the length element is negative in the direction of time and positive for spatial separations. In this way there is always a clear distinction between time and space. With the line element we can describe the propagation of light rays and clock rates in a very simple manner. Light rays are simply curves with

$$ds^2 = 0. \quad \text{(light propagation)} \tag{2.20}$$

This ensures the complete principle of the constancy of the velocity of light in vacuum as discussed above. We now consider a clock with coordinates $\mathbf{x}_c(\tau)$ where τ is the proper time, i.e., the time an observer co-moving with the clock reads. The curve $(c\tau, \mathbf{x}_c(\tau))$ is called the world line of our (idealized) clock. Let ds be an infinitesimal element of this world line; then the proper time of the clock, $d\tau$, can be obtained by

$$d\tau^2 = -\frac{1}{c^2} ds^2 = -\frac{1}{c^2} g_{\mu\nu} dx_c^\mu dx_c^\nu. \tag{2.21}$$

For a clock moving with velocity $\mathbf{v}$, we get from (2.19)

$$d\tau^2 = dt_c^2 - \frac{1}{c^2} d\mathbf{x}_c^2 = dt_c^2 \left(1 - \frac{\mathbf{v}^2}{c^2}\right) \tag{2.22}$$

in accordance with relation (2.16).

Exercise 2.2: The coordinate time interval of a clock Δt_c is 1 s, and its velocity is 30 km/s. What is the difference between the proper time and coordinate time interval?

Exercise 2.3: Let us consider the 2-dimensional Euclidean plane with Cartesian coordinates (x, y) and basis vectors $\mathbf{e}_x$ and $\mathbf{e}_y$. Define $\mathbf{e}_{x'} = \mathbf{e}_x$ and $\mathbf{e}_{y'}$, a unit vector inclined by an angle α with respect to $\mathbf{e}_x$. Construct coordinates (x', y') such that

$$\mathbf{v} = x\mathbf{e}_x + y\mathbf{e}_y = x'\mathbf{e}_{x'} + y'\mathbf{e}_{y'}.$$

Compute the components of the metric tensor in the primed coordinates from $ds^2 = dx^2 + dy^2$.

2.11 Lorentz Transformation

We now consider some inertial system with coordinates $(ct, \mathbf{x})$ and a second one with coordinates $(cT, \mathbf{X})$ whose origin moves with velocity $\mathbf{v}$ with respect to the first system. According to the above, the line element in the first system is given by

$$ds^2 = -c^2 dt^2 + d\mathbf{x}^2$$

and in the second (moving) inertial system by

$$ds^2 = -c^2 dT^2 + d\mathbf{X}^2.$$

Now, ds^2 is a coordinate-independent geometrical length element, so the right hand sides of the last two equations have to agree. This form invariance of the metric tensor (i.e., identical components of the metric tensor) if we go from one inertial system to another leads to the form of the coordinate transformation from $x^\mu = (ct, \mathbf{x})$ to $X^\alpha = (cT, \mathbf{X})$:

$$X^\alpha = \Lambda^\alpha_\beta x^\beta \tag{2.23}$$

with

$$\begin{aligned} \Lambda^0_0 &= \gamma \\ \Lambda^0_i &= \Lambda^i_0 = -\gamma \beta^i \\ \Lambda^i_j &= \delta_{ij} + \frac{v^i v^j}{v^2}(\gamma - 1) \end{aligned} \tag{2.24}$$

and

$$\gamma \equiv (1 - \beta^2)^{-1/2}, \qquad \beta^i = v^i/c. \tag{2.25}$$

The transformation (2.23) and (2.24) is called *Lorentz transformation*. Explicitly we have

$$cT = \Lambda^0_\beta x^\beta = \Lambda^0_0 ct + \Lambda^0_i x^i$$

2.11 Lorentz Transformation

or

$$T = \gamma\left(t - \frac{\mathbf{v}\cdot\mathbf{x}}{c^2}\right). \tag{2.26}$$

The spatial coordinates transform according to

$$X^i = \Lambda^i_\beta x^\beta = \Lambda^i_0 ct + \Lambda^i_j x^j$$

or

$$\mathbf{X} = \mathbf{x} - \gamma\mathbf{v}t + (\gamma - 1)\frac{\mathbf{v}(\mathbf{v}\cdot\mathbf{x})}{v^2}. \tag{2.27}$$

For the special case that the velocity $\mathbf{v}$ points into x-direction, we have

$$T = \gamma\left(t - \frac{vx}{c^2}\right)$$
$$X = \gamma(x - vt) \tag{2.28}$$

or

$$t = \gamma\left(T + \frac{vX}{c^2}\right)$$
$$x = \gamma(X + vT) \tag{2.29}$$

and $Y = y, Z = z$. The following exercise shows that the Lorentz transformation leads indeed to the invariance of the line element.

Exercise 2.4: In some inertial system (the rest system) with coordinates $x^\mu \equiv (ct, \mathbf{x})$, the metric tensor is given by $ds^2 = -c^2 dt^2 + d\mathbf{x}^2$. A second inertial system with coordinates $X^\alpha \equiv (cT, \mathbf{X})$ and the metric tensor $dS^2 = -c^2 dT^2 + d\mathbf{X}^2$ moves with constant velocity along the common x-axes with velocity v. Let the moving coordinates be related with those of the rest system by a Lorentz transformation (2.28). Show that under these conditions $dS^2 = ds^2$.

Of interest is an expansion of the Lorentz transformation in powers of $1/c$. With

$$\gamma = 1 + \frac{1}{2}\beta^2 + \frac{3}{8}\beta^4 + \mathcal{O}(\beta^6)$$

we obtain

$$T = t\left(1 + \frac{1}{2}\beta^2 + \frac{3}{8}\beta^4\right) - \left(1 + \frac{1}{2}\beta^2\right)\frac{\mathbf{v}\cdot\mathbf{x}}{c^2} + \mathcal{O}(c^{-5}),$$
$$\mathbf{X} = \mathbf{x} - \left(1 + \frac{1}{2}\beta^2\right)\mathbf{v}t + \frac{1}{2}\frac{\mathbf{v}(\mathbf{v}\cdot\mathbf{x})}{c^2} + \mathcal{O}(c^{-4}). \tag{2.30}$$

2.12 Geocentric Timescales TCG, TT, TAI, and UTC

The frequency standards described above generate ultrastable oscillations that are counted; converted into hours, minutes, and seconds; and are finally displayed. The quality of a clock depends essentially upon two quantities: (1) its stability and (2) its accuracy. The stability usually described by means of the two-point Allan variance gives a measure of frequency fluctuations as function of averaging time. On the other hand, accuracy describes the ability to realize the SI second.

The SI second is defined as the duration of 9 192 631 779 periods of the radiation that the ^{133}Cs atom at rest emits during the corresponding hyperfine transition.

An individual clock primarily delivers a free timescale. The readings of several clocks can be related with a suitable synchronization procedure. In practice the geocentric coordinate time is used to synchronize clocks in the vicinity of the Earth. Let (T, X^a) be geocentric nonrotating quasi-inertial coordinates (the acceleration of the geocenter will not play an important role for the following). The geocentric coordinate time T is also called

$$T = \text{TCG}.$$

Actual clocks in the vicinity of the Earth do not show the time $T = \text{TCG}$ but their proper time that is related with TCG by

$$c^2 d\tau^2 = \left(1 - \frac{2U}{c^2}\right) c^2 dT^2 - d\mathbf{X}^2.$$

In this equation U denotes the Newtonian gravitational potential. By means of relation (2.21), we might interpret this relation by writing the metric tensor in geocentric quasi-inertial Cartesian coordinates in the form

$$ds^2 = -\left(1 - \frac{2U}{c^2}\right) c^2 dT^2 + d\mathbf{X}^2. \tag{2.31}$$

Neglecting higher-order terms in $(1/c)$, we can solve this equation for $d\tau/dT$:

$$c^2 d\tau^2 = c^2 dT^2 \left[1 - \frac{2U}{c^2} - \frac{1}{c^2}\left(\frac{d\mathbf{X}}{dT}\right)^2\right]$$

$$= c^2 dT^2 \left[1 - \frac{2U}{c^2} - \frac{\mathbf{V}^2}{c^2}\right]$$

or

$$\frac{d\tau}{dT} = \left[1 - \frac{2U}{c^2} - \frac{\mathbf{V}^2}{c^2}\right]^{1/2}. \tag{2.32}$$

2.12 Geocentric Timescales TCG, TT, TAI, and UTC

Here, $\mathbf{V}$ is the geocentric coordinate (GCRS) velocity of the clock. A Taylor expansion $(1-x)^{1/2} = 1 - x/2 + \cdots$ finally yields

$$\frac{d\tau}{dT} \simeq 1 - \frac{U}{c^2} - \frac{1}{2}\frac{\mathbf{V}^2}{c^2}. \tag{2.33}$$

The U-term in this relation results from the *gravitational redshift* (the wavelength of a photon increases if it moves upwards in a gravitational field) implying that an atomic clock is slowed down in a gravitational field. The $\mathbf{V}^2$-term results from the quadratic Doppler effect implying that a fast-moving clock apparently is slowed down. This last effect, however, is relative; for an observer that is co-moving with the clock, a clock at rest would appear to be slowed down.

We will now switch from our quasi-inertial coordinates to coordinates $\overline{\mathbf{X}}$ that are co-rotating with the Earth. With

$$\mathbf{V} = \mathbf{\Omega} \times \overline{\mathbf{X}} + \overline{\mathbf{V}},$$

where $\overline{\mathbf{V}}$ is the clock's velocity in the rotating system, we have

$$c^2 d\tau^2 = c^2 dT^2 \left[1 - \frac{2U_{\text{geo}}}{c^2} - 2(\mathbf{\Omega} \times \overline{\mathbf{X}}) \cdot \frac{\overline{\mathbf{V}}}{c^2} - \frac{\overline{\mathbf{V}}^2}{c^2} \right], \tag{2.34}$$

where

$$U_{\text{geo}} = U + \frac{1}{2}(\mathbf{\Omega} \times \overline{\mathbf{X}})^2$$

is the geopotential that consists of the gravitational potential of the Earth and the centrifugal potential. $\mathbf{\Omega}$ is the angular velocity of the Earth. For earthbound clocks with $d\overline{\mathbf{X}} = 0 = \overline{\mathbf{V}}$, we get

$$\frac{d\tau}{dT} = 1 - \frac{U_{\text{geo}}}{c^2} = 1 - \frac{U_0}{c^2} + \frac{g(\psi) \cdot h}{c^2} + \cdots. \tag{2.35}$$

Here $U_0 = U_{\text{geo}}(\text{geoid}) = \text{const.}$ denotes the geopotential on the geoid, ψ is geographic latitude, and $g(\psi)$ is the gravity acceleration on the geoid:

$$g(\psi) = (9.78027 + 0.05192 \sin^2 \psi) \times 10^2 \, \text{cm/s}^2 \quad \text{and}$$

h denotes the clock's height above the geoid. For two earthbound clocks, we have

$$\frac{(d\tau)_1}{(d\tau)_2} = \frac{f_2}{f_1} = \frac{(1 - U_{\text{geo}}/c^2)_1}{(1 - U_{\text{geo}}/c^2)_2} = \frac{(1 - U_0/c^2 + g(\psi) \cdot h/c^2)_1}{(1 - U_0/c^2 + g(\psi) \cdot h/c^2)_2}$$

$$\simeq \frac{(1 + g(\psi) \cdot h/c^2)_1}{(1 + g(\psi) \cdot h/c^2)_2}.$$

As an example we consider a first clock at the PTB (Physikalisch Technische Bundesanstalt) in Brunswick, Germany, with $h = 73$ m and $gh/c^2 = 8 \cdot 10^{-15}$ and a second one at NIST (US National Institute of Standards and Technology, Boulder, Colorado, USA) with $h = 1,634$ m and $gh/c^2 = 2 \cdot 10^{-13}$. The difference in clock rates is given by

$$f_{\text{PTB}} \simeq (1 + 2 \cdot 10^{-13}) f_{\text{NBS}}.$$

This leads to a difference of 5.4 µs that has accumulated during the time span of 1 year.

Exercise 2.5: If somebody put an atomic clock on top of Mount Everest (geographic latitude: 27°59'17"N, height: 8,844 m) and compares the clock readings with that of an atomic clock of PTB in Brunswick (geographic latitude: 52°17'40"N, height: 73 m), what would be the time offset after 1 year?

The definition of the SI second makes it clear that it is defined first of all for proper times. However, if we have a well-defined relation between some coordinate time and proper time, the SI second will be the induced time unit also for some coordinate time. This especially applies for terrestrial time (TT) and international atomic time (TAI).

Let TT be a timescale differing from $T = $ TCG only by a constant rate. This rate should agree at some level of approximation with that of a clock located on the geoid. For earthbound clocks, we had

$$\frac{d\tau}{dT} \simeq 1 - \frac{U_0}{c^2} + \frac{g(\psi) \cdot h}{c^2}.$$

On the geoid with $h = 0$, we have

$$d\tau = \left(1 - \frac{U_0}{c^2}\right) dT = k_E \, dT \equiv d(\text{TT})$$

with

$$k_E = 1 - \frac{U_0}{c^2}.$$

Thus we have the relation

$$\text{TT} = k_E \, T = k_E \, \text{TCG}. \tag{2.36}$$

For presently achievable clock accuracies the precise definition and realization of the geoid, however, presents a problem. For that reason, the timescales TT and TAI are related with TCG by a defining constant k_E:

$$k_E = 1 - 6.969290134 \times 10^{-10}.$$

2.12 Geocentric Timescales TCG, TT, TAI, and UTC

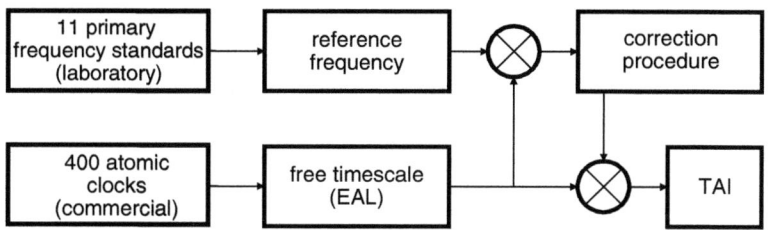

Fig. 2.24 Scheme how TAI is realized

Due to their stochastic properties actual clocks do not exactly show idealized proper time τ. International atomic time (Temps Atomique International) is therefore derived from a mean value over the readings of many individual atomic clocks. We write

$$\frac{d\tau}{dT} \simeq 1 - \frac{1}{c^2}(U_0 + U_R)$$

with

$$U_R = -g(\psi) \cdot h + \cdots.$$

Then TAI is defined as the mean value

$$\text{TAI} = \frac{1}{N} \sum_{i=1}^{N} (\tau_i + S_i/c^2)$$

with

$$S_i = \int_{T_0}^{T} U_R \, dT'.$$

Here, the index i labels the various atomic clocks used for the realization of TAI. The second term in the TAI definition simply indicates that the readings of some clock have to be transferred to the geoid before the averaging over the various clocks is taken.

Presently the readings of about 400 clocks in 68 laboratories are employed for the realization of TAI (Fig. 2.24). These clocks are:

- Standard Cs clocks with stabilities of 5×10^{-14} for an averaging time of about 5 days and accuracies of about 10^{-12},
- Highly precise Cs standards with stabilities of 1×10^{-14} for an averaging time of about 5 days and accuracies of 5×10^{-13}, as well as
- H maser with stabilities of about 7×10^{-16} for an averaging time of about 1 day.

These atomic clocks first define a free timescale called Echelle Atomique Libre (EAL). From the EAL, the frequency of TAI is finally tuned by a few number of primary frequency standards (PFS). The TAI zero point was fixed to TAI = UT1 for 1 January 1958, 0^h. Presently the PFS are located at LNE-SYRTE, Laboratoire National de Métrologie et d'Essais, Paris, France; INRIM, Istituto Nazionale di Ricerca Metrologia, Torino, Italy; NIST, National Institute of Standards and Technology, Boulder (Colorado), USA; NPL, National Physical Laboratory, Middlesex, GB; and PTB, Physikalisch Technische Bundesanstalt, Brunswick, Germany:

- LNE-SYRTE-F01 (Cs fountain)
- LNE-SYRTE-FO2 (Cs/Rb fountain)
- LNE-SYRTE-FOM (transportable Cs fountain)
- LNE-SYRTE-JPO (optically pumped Cs beam)
- IT-CSF1, INRIM (Cs fountain)
- NIST-F1 (Cs fountain)
- NPL-CSF2 (Cs fountain)
- PTB-CS1 (Cs beam, continuous operation)
- PTB-CS2 (Cs beam, continuous operation)
- PTB-CSF1 (Cs fountain)
- PTB-CSF2 (Cs fountain)

A single clock produces a free timescale. A relation between two such timescales can be established easily if the two clocks are placed in immediate vicinity so that the two clocks can be read off simultaneously. The notion of *simultaneity*, however, rests on a convention for spatially separated clocks. Einstein has discussed a special synchronization procedure for well-separated clocks by means of electromagnetic signals; e.g., by means of an observer that is placed in the middle between the two clocks and simultaneously emits two light signals towards the two clocks that by definition arrive there simultaneously. However, it can be shown that (due to the Sagnac effect in time that is discussed below) a clock synchronization according to Einstein is not possible on the rotating Earth. This point will now be discussed more extensively.

One defines that two clocks showing proper times τ_1 and τ_2 are called synchronous if the corresponding T = TCG agree, i.e., if $T_1 = T_2$. Since this kind of clock synchronization employs the coordinate time T = TCG, it is also called coordinate time synchronization. According to the above we have

$$d\tau \simeq dT \left[1 - \frac{U_{\text{geo}}}{c^2} - \frac{1}{2}\frac{\overline{\mathbf{V}}^2}{c^2} - \frac{(\mathbf{\Omega} \times \overline{\mathbf{X}}) \cdot \overline{\mathbf{V}}}{c^2} \right]$$

or with $(1+x)^{-1} = 1 - x + \cdots$

$$\Delta T \simeq \int d\tau \left[1 + \frac{U_{\text{geo}}}{c^2} + \frac{1}{2}\frac{\overline{\mathbf{V}}^2}{c^2} + \frac{(\mathbf{\Omega} \times \overline{\mathbf{X}}) \cdot \mathbf{V}}{c^2} \right]. \tag{2.37}$$

The synchronization of earthbound clocks can be achieved by means of a clock transport. For a slow transport ($\overline{\mathbf{V}}^2 \sim 0$) at low altitudes ($h \sim 0$), we have

2.12 Geocentric Timescales TCG, TT, TAI, and UTC

$$\Delta(\text{TT}) \simeq \left(1 - \frac{U_0}{c^2}\right) \Delta T \simeq \int d\tau \left[1 + (\mathbf{\Omega} \times \overline{\mathbf{X}}) \frac{\overline{\mathbf{V}}}{c^2}\right] \quad (2.38)$$
$$\simeq \Delta\tau + \frac{\Omega R_E}{c^2} \int d\tau \, v_{\text{east}} \cos \psi,$$

where v_{east} is the eastward component of $\overline{\mathbf{V}}$. For a transport along a meridian with $v_{\text{east}} = 0$, we simply get $\Delta(TT) = \Delta\tau$. For a transport along a parallel, we get

$$\Delta(\text{TT}) \simeq \Delta\tau + \frac{\Omega R_E}{c^2} L_{\text{east}} \cos \psi,$$

if L_{east} denotes the distance covered in eastward direction. We now consider two events of clock comparison: one before and one after the transport of a clock. For both events the TCG or TT value will be the same for the two clocks. However, for the stationary clock $\Delta\tau = \Delta(\text{TT})$, whereas the elapsed proper time of the transported clock will be shorter by 207 ns by the Sagnac effect in time if it was moved once around the Earth along the equator in eastward direction.

For practical reasons also, TT is derived from atomic time TAI. The IAU has decided that

$$\text{TT} = \text{TAI} + 32.184 \, \text{s}.$$

The constant 32.184 s is simply a convention that was chosen for historical reasons.

We now come to the important definition of coordinated universal time UTC. It is primarily an atomic time. Later we will introduce so-called Earth's timescales that strictly speaking are not physical timescales but angles related with the Earth's orientation in space. UT1 is such a timescale that will be introduced later. Now, UTC is defined to differ from TAI by a certain integral number of (leap) seconds, i.e.,

$$\text{TAI} = \text{UTC} + N \, \text{s},$$

if N denotes a positive integer. Since 1 January 1972 is defined by the introduction of leap seconds so that the difference $|\text{UTC} - \text{UT1}|$ is always smaller than 0.9 s,

$$|\text{UTC} - \text{UT1}| < 0.9 \, \text{s}.$$

This has the advantage that over a very long time span, UTC that is related with the Sun via the Earth's orientation in space will remain useful for ordinary life. Since 1972, a leap second is introduced with preference at 30 June or 31 December of a year, always at $23^h 59^m 59^s$ UTC. This way a new offset between UTC and UT1 is valid at $0^h 00^m 00^s$ of 1 July or 1 January respectively. Table 2.1 shows the instants of time when new leap second values became valid. The *AstroRef* function `GetLeapSeconds` returns the number of leap seconds valid for a given JD_{UTC}. The offset between UT1 and UTC is denoted

$$\Delta \text{UT1} = \text{UT1} - \text{UTC}$$

Table 2.1 Moments of time where new offsets N (TAI-UTC in seconds) between TAI and UTC became valid by the introduction of leap seconds

Julian date	Date	N	Julian date	Date	N
JD 2441317.5	1 Jan 1972	+10	JD 2446247.5	1 Jul 1985	+23
JD 2441499.5	1 Jul 1972	+11	JD 2447161.5	1 Jan 1988	+24
JD 2441683.5	1 Jan 1973	+12	JD 2447892.5	1 Jan 1990	+25
JD 2442048.5	1 Jan 1974	+13	JD 2448257.5	1 Jan 1991	+26
JD 2442413.5	1 Jan 1975	+14	JD 2448804.5	1 Jul 1992	+27
JD 2442778.5	1 Jan 1976	+15	JD 2449169.5	1 Jul 1993	+28
JD 2443144.5	1 Jan 1977	+16	JD 2449534.5	1 Jul 1994	+29
JD 2443509.5	1 Jan 1978	+17	JD 2450083.5	1 Jan 1996	+30
JD 2443874.5	1 Jan 1979	+18	JD 2450630.5	1 Jul 1997	+31
JD 2444239.5	1 Jan 1980	+19	JD 2451179.5	1 Jan 1999	+32
JD 2444786.5	1 Jul 1981	+20	JD 2453736.5	1 Jan 2006	+33
JD 2445151.5	1 Jul 1982	+21	JD 2454832.5	1 Jan 2009	+34
JD 2445516.5	1 Jul 1983	+22	JD 2456109.5	1 Jul 2012	+35

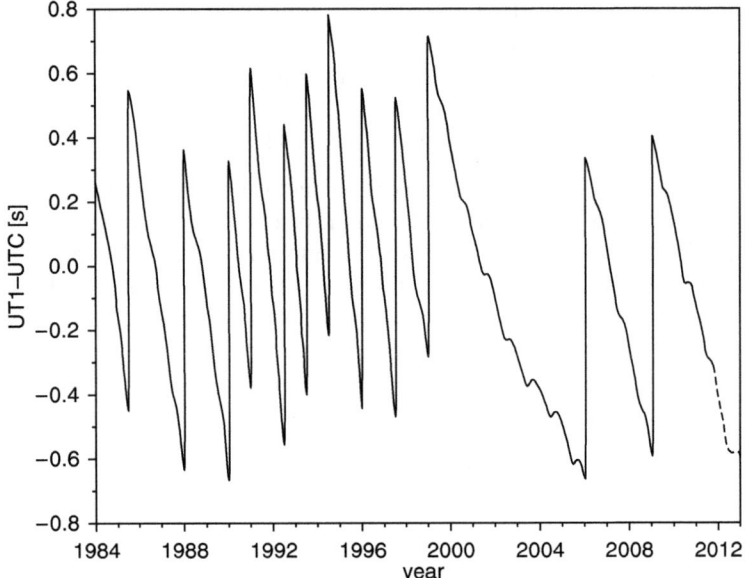

Fig. 2.25 UT1–UTC with extrapolated values (based on data from USNO)

and is continuously determined by the International Earth Rotation and Reference Systems Service (IERS). The *AstroRef* function `GetDeltaUT1` implements access to this data. Using the function `ConvertTime`, it is possible to convert one timescale into another (see appendix B). Finally, Fig. 2.25 shows the difference UT1–UTC for the years 1984 to mid-2012 (extrapolated).

Presently these leap seconds have to be considered for highly precise applications. However, from a practical point of view the introduction of a leap second is related with a considerable expenditure, not only for financial reasons. For that

2.14 Julian Date

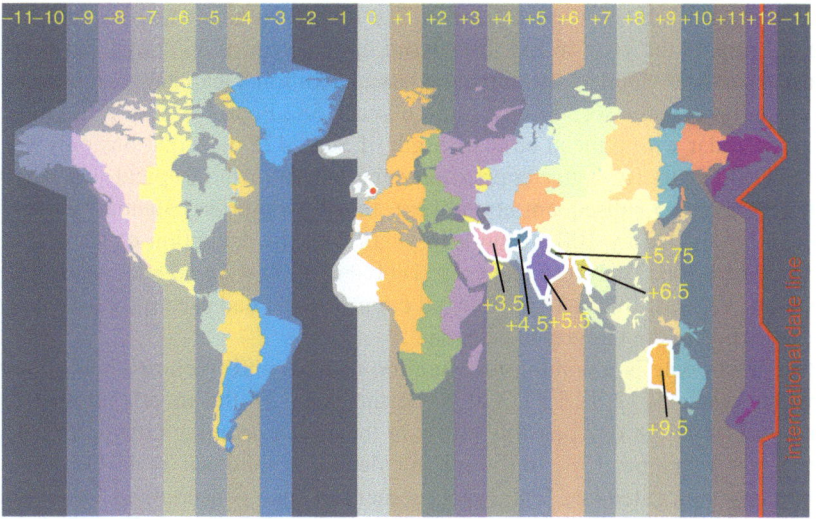

Fig. 2.26 The division of the Earth into time zones

reason the discussions about leap seconds will continue, and it can very well be that in the future, one might introduce leap hours at larger intervals of time for the difference TAI–UTC.

2.13 Time Zones

The national timescales as well as daylight saving times are based upon UTC. Central European time CET, e.g., is given by

$$\text{CET} = \text{UTC} + 1^{\text{h}} \tag{2.39}$$

and the corresponding daylight saving time (central European summer time) by

$$\text{CEST} = \text{UTC} + 2^{\text{h}}. \tag{2.40}$$

Different national times usually differ from UTC by an integer number of hours according to a division into time zones that are depicted in Fig. 2.26.

2.14 Julian Date

In astronomy it has become a usual practice to characterize a certain moment of time by the Julian date (JD). The definition is general enough so that it can be applied to any continuous timescale such as TCG or TT; at highest precision one has to distinguish JD_{TT} from JD_{TCG} etc. The Julian calendar was introduced by the Romans around the year 46 BC. It is based upon a year of 365 days, but every 4th year that can be divided by the number 4 has an additional leap day. A Julian century therefore has exactly 36,525 days. In the year 1582 Pope Gregory XIII suggested a new calendar with a mean annual length of 365.2425 days that has only 97 leap years in 400 years (to better approximate the length of the tropical year, the time span between two subsequent equinoxes). In the Gregorian calendar, those years that can be divided by 100 are no leap years with the exception of those that can be divided by 400. Therefore, the years 1800 and 1900 were no leap years, the year 2000 however was. In course of time, all countries that had used the Julian calendar have switched to the Gregorian calendar.

The Julian date is a continuous counting of days and fractions thereof starting from the year $-4712, 12^h$ TT. Let Y be the year, M the month number (1 for January), and D the day possibly either in the Julian or in the Gregorian calendar (notice that 4 October 1582 in the Julian calendar is followed by 15 October 1582 in the Gregorian calendar). The exact moment of time is accounted for by adding fractions of a day in the D value (as digits after the decimal point). The JD algorithm then reads:

1. For $M > 2$, Y and M are left unchanged. For $M = 1$ or 2, Y is replaced by $Y - 1$ and M by $M + 12$.
2. In the Gregorian calendar, one computes

$$A = \text{INT}(Y/100); \qquad B = 2 - A + \text{INT}(A/4),$$

where INT denotes the integer part of the number (e.g., INT(3.92) = 3). In the Julian calendar, $B = 0$.

3. JD is then given by

$$JD = \text{INT}(365.25 \cdot (Y + 4716)) + \text{INT}(30.6001 \cdot (M + 1)) + D + B - 1524.5.$$

Of great importance is the date J2000.0 for which

$$J2000.0 = 1 \text{ January } 2000, 12^h = JD\, 2\,451\,545.0.$$

In the literature one often finds also values for the *modified Julian date* (MJD) that is defined as

$$MJD = JD - 2\,400\,000.5.$$

The value $MJD = 0$ corresponds to 17 November 1858, 0^h.

The Maple™ function JDDate of the supplied *AstroRef* package computes the Julian date from a date in the Julian or Gregorian calendar. The function CalendarDate does the inverse transformation (see Appendix B).

Exercise 2.6: Calculate the Julian date JD and modified Julian date MJD for the following instants of time: 30 July 1853, 21 April 1937, 1 February 1977, 1 January 2000, 20 May 2006, 1 June 2011, and for 12^h TT.

Exercise 2.7: Calculate the day of the week when Edmund Halley observed a solar eclipse on 3 May 1715!

Exercise 2.8: Use the function `CalendarDate` to compute the calendar date and time for the following JD values: 2451545.00, 2450313.25, and 2450026.75.

Exercise 2.9: Calculate the Julian date for the ancient solar eclipse 28 May 585 BC!

2.15 Barycentric Timescales TCB, T_{eph}, and TDB

The timescales introduced in the last section are all related with the geocenter; they are geocentric timescales to be used in the vicinity of the Earth. For certain applications, however, such as solar system ephemerides or interplanetary spacecraft navigation, barycentric timescales have to be used where the barycenter refers to the solar system's center of mass. The basic barycentric timescale is barycentric coordinate time TCB. For the definition of TCB, one might image some fictitious clock at the geocenter neglecting the gravity field of the Earth. For this case, the first post-Newtonian relation between TCB and TCG reads

$$\frac{d(TCG)}{d(TCB)} \simeq 1 - \frac{U_{ext}(\mathbf{z}_E)}{c^2} - \frac{1}{2}\frac{v_E^2}{c^2}$$

or

$$\Delta(TCG) \simeq \Delta(TCB) - \frac{1}{c^2}\left[\int \left(U_{ext}(\mathbf{z}_E) + \frac{1}{2}v_E^2\right) dt\right].$$

Here, $t = $ TCB, U_{ext} is the gravitational potential of Sun and planets excluding the Earth and $\mathbf{z}_E$, $\mathbf{v}_E$ the barycentric coordinate position and velocity of the geocenter. For some point $\mathbf{x}$ close to the geocenter, this relation is given by

$$TCB - TCG = c^{-2}\left[\int_{t_0}^{t} \left(U_{ext}(\mathbf{z}_E) + \frac{1}{2}v_E^2\right) dt' + \mathbf{v}_E \cdot (\mathbf{x} - \mathbf{z}_E)\right]. \quad (2.41)$$

The second term on the right hand side results from the fact that two events that are simultaneous with respect to TCB generally are not simultaneous in a geocentric system (i.e., with respect to TCG). The moment of time t_0 where TCB = TCG is given by

$$t_0 = JD\,2\,443\,144.5\,.$$

TCB and TCG differ not only by periodic terms but also by a secular drift given by

$$(TCB - TCG)_{sec} = L_C(JD - t_0) \times 86,400\,s$$

with

$$L_C = 1.480827 \times 10^{-8}.$$

Some of the best solar system ephemerides, the DE ephemerides of the Jet Propulsion Laboratory (JPL), do not employ TCB as basic time variable. The original idea was to use a timescale that differs from TT practically only by periodic terms. Strictly speaking, because of arbitrarily long periods contained in the motion of the solar system, such a timescale cannot be realized with ultimate precision. Consequently, the timescales T_{eph} used in barycentric ephemerides show a relation with TCB or TCG depending slightly upon the ephemeris itself. In the past the DE ephemerides used the Fairhead–Bretagnon series (Fairhead and Bretagnon 1990) to derive T_{eph} from TT. Meanwhile another barycentric timescale called TDB has been defined by

$$\text{TDB} = \text{TCB} - L_B \times (\text{JD}_{\text{TCB}} - T_0) \times 86,400\,\text{s} + \text{TDB}_0, \qquad (2.42)$$

with $T_0 = \text{JD}\,2443144.5003725$,

$$L_B = 1.550519768 \cdot 10^{-8}, \qquad \text{TDB}_0 = -6.55 \times 10^{-5}\,\text{s}.$$

The value for L_B was chosen so to minimize the linear drift between TDB and TT for the ephemeris DE405. On the surface of the Earth, this difference is less than 2 ms for the forthcoming centuries.

In the planetary ephemerides developed at Paris Observatory, as described in Fienga et al. (2009), TDB is used for the description of planetary motion, and the differences between TT and TDB are estimated by solving (2.41) at the geocenter at each step of the integration. By using the INPOP values for TT–TDB in the data analysis and the fit of the planetary ephemerides, there is a complete consistency between the positions and velocities of the planets and TT–TDB, all provided to users since INPOP08.

2.16 Fairhead–Bretagnon Series

Analytical approximations for $\Delta T \equiv T_{\text{eph}} - \text{TT}$ are based upon (semi)analytical ephemerides of the solar system. The Fairhead–Bretagnon approximation is based upon the planetary theory VSOP82 (Bretagnon 1982) and the lunar theory ELP2000 (Chapront-Touzé and Chapront 1983). This series is written in the form

$$\begin{aligned}
\Delta T = &\; C_0 \text{TT}^2 \\
&+ \sum_i A_i \sin(\omega_{ai}\text{TT} + \phi_{ai}) + \text{TT} \sum_i B_i \sin(\omega_{bi}\text{TT} + \phi_{bi}) \\
&+ \text{TT}^2 \sum_i C_i \sin(\omega_{ci}\text{TT} + \phi_{ci}) + \text{TT}^3 \sum_i D_i \sin(\omega_{di}\text{TT} + \phi_{di}).
\end{aligned} \qquad (2.43)$$

Here, TT is counted in millennia since J2000.0 and $\Delta T \equiv T_{\text{eph}} - \text{TT}$ is in microseconds. The first ten coefficients are given in Table 2.2. The full set of 127 coefficients can be found in Fairhead and Bretagnon (1990).

Table 2.2 The ten largest terms in the Fairhead–Bretagnon series for ΔT

i	A_i (μs)	ω_{ai} (rad/1,000 year)	ϕ_{ai} (rad)	Period (year)
1	1656.674564	6283.075943033	6.240054195	1.0000
2	22.417471	5753.384970095	4.296977442	1.0921
3	13.839792	12566.151886066	6.196904410	0.5000
4	4.770086	529.690965095	0.444401603	11.8620
5	4.676740	6069.776754553	4.021195093	1.0352
6	2.256707	213.299095438	5.543113262	29.4572
7	1.694205	−3.523118349	5.025132748	1783.4159
8	1.554905	77713.772618729	5.198467090	0.0809
9	1.276839	7860.419392439	5.988822341	0.7993
10	1.193379	5223.693919802	3.649823730	1.2028

The *AstroRef* package supplies the function `TephOffset` which may be used to calculate ΔT.

2.17 Problems of Time Dissemination

In earlier days sometimes clocks were synchronized by means of clock transport. This, however, is a laborious method that has been ousted by the transmission of electromagnetic signals.

A simple method of time dissemination is the continuous emission of time signals by certain broadcasting stations. In Europe we face four of such stations emitting time signals in the range from 50 to 77.5 kHz. These are the stations: DCF 77 (Mainflingen, Germany), HBG 75 (Prangins, Switzerland), MSF 60 (Rugby, United Kingdom), and OMA 50 (Podebrady, Prague, Czech Republic). From the German *Physikalisch Technische Bundesanstalt* (PTB), Brunswick, via the broadcasting station DCF 77, time signals related with CET are emitted that can be received up to a distance of about 2,000 km. PTB is responsible for the generation of the time signals, whereas *Deutsche Telekom* is responsible for the handling of the antenna arrangements. As emitting antennas vertical 50-kW omnidirectional antennas with a height of 150 or 200 m are in use. The DCF 77 signals at the location of emission (Mainflingen near Frankfurt/M.) are derived from atomic clocks at PTB and are controlled by PTB. The carrier frequency of 77.5 kHz presents a highly stable normal frequency. Not only the second pulses can be used for time dissemination but also the phase leading to a synchronization accuracy of some microseconds.

At this place we would like to mention the navigation system LORAN-C (Long Range Navigation) that mainly served for the position determination of ships and to some extent also of airplanes. It presents an advancement of LORAN-A that was established by MIT during the Second World War. LORAN-C was reconstructed by

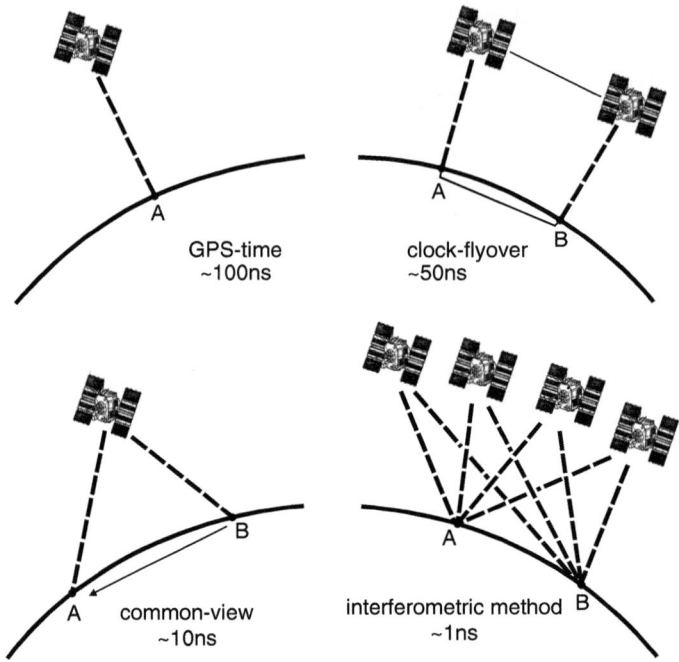

Fig. 2.27 Different methods of time dissemination with GPS

the US Coast Guard almost worldwide and served military as well as civil purposes. For a long time, it was the standard procedure of worldwide time comparisons with an accuracy of about a microsecond. Meanwhile LORAN-C was replaced by GPS as far as possible.

Finally, it should be noted here that in some cases, primary frequency standards are compared by the help of transportable cesium fountains like LNE-SYRTE-FOM.

2.18 Time Dissemination by Means of Satellites

With the adoption of artificial satellites new possibilities opened for the purposes of time dissemination with higher accuracies, larger operating distances, and faster availability. Among the satellite techniques we distinguish one-way and two-way methods. In the active one-way method one uses the time signals of the highly stable onboard clock. If the satellite orbit is known exactly, the signal propagation time to the ground station can be calculated, and the satellite time can be compared with that of the station clock. Uncertainties result from the lack of very precise orbital data, the stability of the satellite clock, and problems with atmospheric refraction. In the passive one-way method, time signals from some master clock on the ground

2.18 Time Dissemination by Means of Satellites

are emitted to the satellite that merely distributes the information. With the two-way methods time comparisons are accomplished between two station clocks by means of satellites. In this case the satellite has to be equipped with a corresponding transponder that instantaneously reflects the signal.

Of ever increasing prominence is the time dissemination with GPS. With GPS time can be disseminated practically worldwide with an accuracy of about a nanosecond. One distinguishes different methods of time comparisons with GPS (Fig. 2.27):

– Measuring a GPS satellite during a flyover to the ground station and comparison of the ground clock with GPS system time (± 100 ns).
– Flyover a GPS satellite over two ground stations (A and B) and exchange of the measuring results for the determination of the time difference between the clocks at A and B (± 50 ns).
– Simultaneous measurement of a GPS satellite from two ground stations A and B ("common view") and exchange of the measuring results (± 10 ns) and
– simultaneous measurements of several GPS satellites from two ground stations and correlation of signals ("interferometric method"; ± 1 ns).

Chapter 3
Space–Time

3.1 Reference Systems and Frames

A reference system in astronomy or geodesy is a necessary prerequisite for the description of motion of some astronomical body like the Earth and some artificial satellite or the motion of the solar system. Such a reference system first of all attributes three spatial coordinates $x^i = x_1, x_2, x_3 = x^1, x^2, x^3$ to some well-defined point of our astronomical body. In the framework of relativity where the readings of a clock depend upon the clock's velocity with respect to the observer and upon the clock's location in the gravitational field, one usually adds a time coordinate to the three spatial ones and speaks about coordinates of an event: something that happens at a certain instance of time at a certain point in space. Thus, in astronomy and geodesy, a *reference system* is a purely mathematical construction (a chart) giving "names" to space–time events characterized by four numbers $x^\mu = x_0, x_1, x_2, x_3$, where x_0 is related with a time coordinate t by $x_0 = ct$, where c is the vacuum speed of light. The coordinates of a reference system should cover the relevant parts of space–time; as an idealization, it can be global and cover the entire model universe or local, e.g., covering only a certain neighborhood of some astronomical body.

In principle, there are several ways to define some astronomical reference system depending upon the way the system will be materialized, i.e., how the corresponding frame will be realized. The system itself consists of a set of rules how the space–time coordinates of certain events are to be constructed by means of observations and the corresponding frame is materialized, e.g., in form of a catalog or ephemeris containing positions of some celestial objects relative to the reference system under consideration. Since each of such reference systems has its specific time coordinate, clear rules are needed to relate this coordinate with the readings of a certain (atomic) clock.

The three basic principles for the construction of an astronomical reference system are:

- An ephemeris compass
- A stellar compass
- An inertial compass

In addition to these three compasses, selected material marks (targets) can be employed for the realization of a local reference system, e.g., on the surface of the Earth. Also some gravitational compass (e.g., the direction of the plumb line) can locally be used on the surface of the Earth.

An ephemeris compass employs as basis the translational dynamics of a set of interacting astronomical bodies. Corresponding dynamical equations of motion are formulated in the basic reference system, and observations of the positions of these bodies as a function of time are used for the construction of the corresponding frame. A system based upon the ephemeris compass will be called *dynamical reference system* (DRS). Note that such a definition refers only to the way how the system will be materialized and implies nothing with respect to the appearance of inertial forces like Coriolis or centrifugal forces in the system. Depending on how such inertial forces are treated in the dynamical equations of motion, a DRS can be dynamically or kinematically nonrotating with respect to another reference system.

The international terrestrial reference system (ITRS) is mainly based upon an ephemeris compass, i.e., upon observations of artificial satellites. The ITRS is the basic reference system for geodesy and geophysics. A barycentric dynamical reference system (BDRS) is realized by means of some barycentric ephemeris for the motion of the astronomical solar system bodies, an indispensable tool for interplanetary spacecraft navigation.

The stellar compass uses light information from some remote astronomical bodies like quasars for the construction of some quasi-inertial reference system. Such a *celestial reference system* (CRS) will be considered to be *kinematically nonrotating*. Below we will introduce the BCRS (Barycentric CRS) and the GCRS (Geocentric CRS) and some related systems. Presently and in the near future, the BCRS is of fundamental importance since the astrometric observations of remote astronomical sources will be reduced to the barycenter in a first fundamental step. All other CRS discussed here will be derived from the BCRS. The fundamental system (here the BCRS) is considered to be dynamically and kinematically nonrotating. The GCRS is defined to be kinematically nonrotating with respect to the BCRS which implies that it is not inertial with respect to rotations and Coriolis forces appear in corresponding equations of motion.

Finally, instead of the translational motion of astronomical objects, some inertial system like a laser gyroscope can be used to define a reference system a direct way to construct a dynamically nonrotating system.

As we have seen above, the problem of time has to be formulated within the framework of Einstein's theory of gravity (general relativity theory, GRT). The necessity to use a relativistic framework, however, is not restricted to the problem of time but to the problem of spatial coordinates as well. For example, lunar laser ranging (LLR) measures the distance to the Moon with a precision of a few centimeters, thereby operating at the 10^{-10} level. At this level, several relativity effects are significant and observable. Relativity effects related with the motion of the Earth–Moon system about the Sun are of the order $(v_{\text{orbital}}/c)^2 \simeq 10^{-8}$. The Lorentz contraction of the lunar orbit about the Earth that appears in barycentric coordinates has an amplitude of about 100 cm, whereas in some suitably chosen (local) coordinate system that moves with the Earth–Moon barycenter, the dominant relativistic range oscillation reduces to a few centimeters only (Mashhoon 1985; Soffel et al. 1986).

The situation is even more critical in the field of astrometry. It is well known that the gravitational light deflection at the limb of the Sun amounts to 1″.75 and decreases only as $1/r$ with increasing impact parameter r of a light ray to the solar center. Thus, for light rays incident at about 90° from the Sun, the angle of light deflection still amounts to 4 mas. To describe the accuracy of astrometric measurements, it is useful to make use of the space-curvature parameter γ of the parametrized post-Newtonian (PPN) formalism. In the PPN formalism, the angle of light deflection is proportional to $(\gamma + 1)/2$ so that astrometric measurements might be used for a precise determination of the parameter γ. Meanwhile, VLBI has achieved accuracies of better than 0.1 mas, and regular geodetic VLBI measurements have frequently been used to determine the space-curvature parameter. A detailed analysis of VLBI data from the projects MERIT and POLARIS/IRIS gave $\gamma = 1.000 \pm 0.003$ (Robertson and Carter 1984; Carter et al. 1985), where a formal standard error is given. Recently, an advanced processing of VLBI data provided the best current estimates $\gamma = 0.9996 \pm 0.0017$ (Lebach et al. 1995), $\gamma = 0.99994 \pm 0.00031$ (Eubanks et al. 1997), and $\gamma = 0.99992 \pm 0.00012$ (Lambert and Le Poncin-Lafitte 2011). Current accuracy of modern optical astrometry as represented by the Hipparcos catalog is about 1 mas, which gave a determination of γ at the level of 0.997 ± 0.003 (Froeschlé et al. 1997). With the planetary ephemerides, it is also possible to estimate the sensitivity of spacecraft navigation data to any general relativity modification through the PPN parameters β and γ. INPOP08 (Fienga et al. 2009) and INPOP10a (Fienga et al. 2011) have given upper limits for such β and γ deviations from the Einstein value of one.

Future astrometric missions such as Gaia will push the accuracy to the level of a few microarcseconds (μas), and the expected accuracy of determinations of γ will be $10^{-6} - 10^{-7}$. The accuracy of 1 μas should be compared with the maximal possible light deflection due to various parts of the gravitation field: the post-Newtonian effect of 1″.75 due to the mass of the Sun, 240 μas caused by the oblateness of Jupiter (J_2), 10 μas due to Jupiter's J_4, the post-post-Newtonian effect of 11 μas due to the Sun, etc. This illustrates how complicated the relativistic modeling of future astrometric observations will be. It is clear that for such high accuracy, the corresponding model must be formulated in a self-consistent relativistic framework.

These examples show clearly that high-precision modern astronomical observations can no longer be described by Newtonian theory but require Einstein's theory of gravity. A reference system in reality is different from an idealized reference system in the classical Newtonian framework. In the Newtonian framework, there exist a bundle of preferred inertial coordinates which have a direct physical meaning. In relativity, such preferred coordinates no longer exist, and there is a complicated

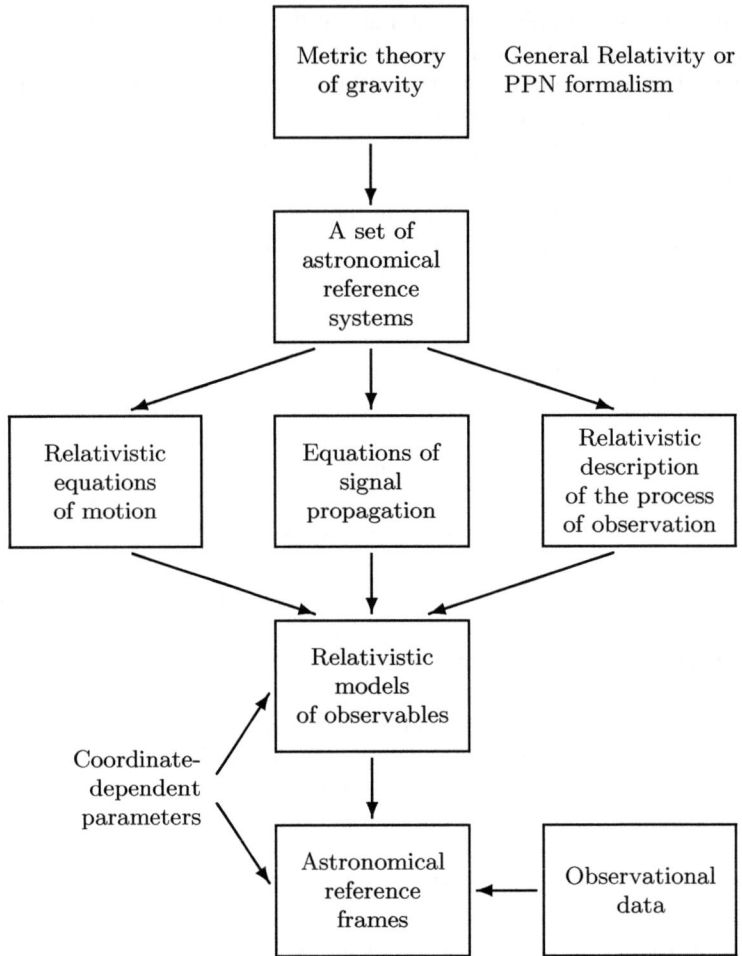

Fig. 3.1 General principles of relativistic modeling of astronomical observations (Klioner 2003a,b)

chain of relations between the systems, the corresponding frames, and observations. For the stellar and the ephemeris compass, this is illustrated in Fig. 3.1.

Starting from Einstein's theory of gravity (GRT), or any other metric theory of gravity, or the parametrized post-Newtonian formalism, one should define at least one relativistic 4-dimensional reference system covering the region of space–time where all the processes constituting the particular kind of astronomical observations are located. A typical astrometrical observation as depicted in Fig. 3.2 consists of four constituents: a moving observer with its space–time trajectory (world line), a moving observed object, the light propagation from the observed object to the observer, and the process of observation. Each of these four constituents has to be modeled in the relativistic framework. The equations of motion of both the

3.2 Canonical Barycentric Metric

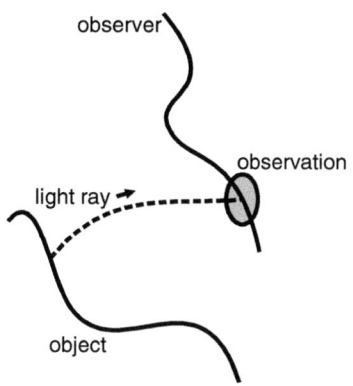

Fig. 3.2 Four constituents of an astronomical observation: (1) motion of the observed object, (2) motion of the observer, (3) propagation of an electromagnetic signal from the observed object to the observer, and (4) the process of observation

observed object and the observer in the chosen reference system have to be derived, and a method to solve these equations has to be found. Typically, the equations of motion are second-order ordinary differential equations, and numerical integration with suitable initial or boundary conditions can be used to solve them. Astrometric information of the object can be read off the electromagnetic signals propagating from the object to the observer. Therefore, the corresponding equations of light propagation in the chosen reference system have to be derived and solved. The equations of motion of the object and the observer and the equations of light propagation enable one to compute positions and velocities of the object, observer, and light rays in the particular reference system at a given moment of its coordinate time, provided that the positions and velocities at some initial epoch are known. These positions and velocities obviously depend on the employed reference system. On the other hand, the results of observations cannot depend on the reference system used to theoretically model the observations. Therefore, it is clear that one more step of the modeling is needed: a relativistic description of the process of observation. This part of the model allows one to compute a coordinate-independent theoretical prediction of the observables starting from the coordinate-dependent position and velocity of the observer and, in some cases, the coordinate velocity of the electromagnetic signal at the point of observation.

3.2 Canonical Barycentric Metric

As we have seen already in the context of presently achievable accuracies, relativity plays an important role not only for timescales but for spatial coordinates as well. Here, the concept of a 4-dimensional space–time metric tensor, $g_{\mu\nu}$, that was already introduced in chapter 2 is especially useful.

In the absence of gravitational fields in any system of inertial Cartesian coordinates, the metric tensor is given by

$$ds^2 = -c^2 dt^2 + d\mathbf{x}^2. \tag{3.1}$$

Because of the constancy of the speed of light, $ds^2 = 0$ for light rays in any inertial system. If gravitational fields are taken into account, the corresponding gravitational potentials are contained in the metric tensor. In Einstein's theory of gravity, the gravitational field is completely determined by the metric tensor, i.e., by the geometry of space–time. The metric tensor describes:

- The propagation of light rays as null geodesics.
- The motion of test bodies as geodesics of the (external) metric tensor.
- The proper time τ of an ideal clock by

$$d\tau^2 = -\frac{1}{c^2} ds^2,$$

where ds has to be taken along the clock's world line.

In Einstein's theory of gravity, the metric tensor obeys Einstein's field equations. In the following, we will discuss a canonical form of a barycentric space–time metric tensor. It will be constructed in the so-called post-Newtonian limit of Einstein's field equations. Formally, this canonical barycentric metric tensor agrees with that of the barycentric celestial reference system (BCRS) that will be discussed below. According to IAU 2000 Resolution B1.3 (e.g., Soffel et al. 2003), the form of the BCRS metric tensor reads:

$$g_{00} = -1 + \frac{2w}{c^2} - \frac{2w^2}{c^4} + \mathcal{O}(c^{-5}),$$

$$g_{0i} = g_{i0} = -\frac{4}{c^3} w^i + \mathcal{O}(c^{-5}) \qquad (3.2)$$

$$g_{ij} = \delta_{ij}\left(1 + \frac{2}{c^2} w\right) + \mathcal{O}(c^{-4}).$$

This canonical barycentric metric is completely determined by two gravitational potentials: a scalar potential w and a vector potential w^i. Without these potentials, the line element ds reduces to that of (3.1). The most important gravitational term is the w/c^2-term in g_{00} that leads to the gravitational redshift. In the Newtonian limit, the scalar potential w reduces to the Newtonian potential U. The corresponding post-Newtonian limit of Einstein's field equation reduces to

$$\left(-\frac{1}{c^2}\frac{\partial^2}{\partial t^2} + \Delta\right) w = -4\pi G \sigma, \qquad (3.3)$$

where σ denotes the gravitational mass density. The scalar potential w is also called the *gravitoelectric* potential. The *gravitomagnetic* potential w^i describes the gravitational effects of moving (e.g., rotating) matter. The corresponding post-Newtonian field equation reads

$$\Delta w^i = -4\pi G \sigma^i, \qquad (3.4)$$

where σ^i denotes the mass-current density.

The canonical barycentric metric is defined by considering only the bodies of our solar system. All other stars, galaxies, dark matter or dark energy, etc., are ignored. For that reason, far from the solar system one requires

$$\lim_{r \to \infty} g_{\mu\nu} = \text{diag}(-1, +1, +1, +1). \tag{3.5}$$

A solution satisfying this boundary condition reads

$$w(t, \mathbf{x}) = G \int d^3x' \frac{\sigma(t, \mathbf{x}')}{|\mathbf{x} - \mathbf{x}'|} + \frac{1}{2c^2} G \frac{\partial^2}{\partial t^2} \int d^3x' \sigma(t, \mathbf{x}')|\mathbf{x} - \mathbf{x}'|,$$

$$w^i(t, \mathbf{x}) = G \int d^3x' \frac{\sigma^i(t, \mathbf{x}')}{|\mathbf{x} - \mathbf{x}'|}. \tag{3.6}$$

For a system of mass monopoles (i.e., point masses), the barycentric metric potentials therefore read

$$w = \sum_A w_A; \qquad w^i = \sum_A w_A^i$$

with ($r_{BA} \equiv |\mathbf{x}_B - \mathbf{x}_A|$; $\mathbf{a}_A \equiv d\mathbf{v}_A/dt$)

$$w_A(t, \mathbf{x}) = \frac{GM_A}{r_A} \left[1 + 2\frac{v_A^2}{c^2} - \frac{1}{c^2} \sum_{B \neq A} \frac{GM_B}{r_{BA}} - \frac{1}{2c^2} \left(\frac{(\mathbf{r}_A \cdot \mathbf{v}_A)^2}{r_A^2} + \mathbf{r}_A \cdot \mathbf{a}_A \right) \right] \tag{3.7}$$

and

$$w_A^i(t, \mathbf{x}) = \frac{GM_A}{r_A} v_A^i. \tag{3.8}$$

For the derivation of these expressions for the barycentric metric potentials, post-Newtonian definitions for the masses are needed. Details can be found, e.g., in Damour et al. 1991 or Soffel 1989.

3.3 Equations of Motion of Astronomical Bodies

The form of the metric potentials (3.7) and (3.8) determines the equations of motion for a set of mass monopoles. As already mentioned above, mass monopoles move along geodesics in the external part of the metric tensor. It is not difficult to derive the form of geodesics if the metric tensor is given, e.g., with a variational method. Let $g_{\mu\nu}$ be the metric tensor and $g^{\alpha\sigma}$ the inverse metric tensor with

$$g^{\alpha\sigma} g_{\sigma\beta} = \delta_{\alpha\beta}. \tag{3.9}$$

We then consider two fixed points P_1 and P_2 and all kinds of curves $x^\mu(\lambda)$, where λ is some curve parameter, that connects these two points. We will now look for those curves (geodesics) that minimize (or maximize) the length of such a curve with the two endpoints fixed; it will obey a relation of the form

$$\delta \int_{P_1}^{P_2} \left(g_{\mu\nu} \frac{dx^\mu}{d\lambda} \frac{dx^\nu}{d\lambda} \right) d\lambda = 0. \tag{3.10}$$

This relation can be understood in the following way: let $x_\gamma^\mu(\lambda)$ be the desired geodesic. One then considers small variations of that curve, i.e., curves of the form $x_\gamma^\mu(\lambda) + \delta x^\mu(\lambda)$ with fixed endpoints:

$$\delta x^\mu(P_1) = \delta x^\mu(P_2) = 0.$$

With the rule

$$\delta A(x^\mu) = \frac{\partial A}{\partial x^\mu} \delta x^\mu \equiv A_{,\mu} \delta x^\mu \tag{3.11}$$

(i.e., the comma denotes a partial derivative) for any differentiable function $A(x^\mu)$ and the notation

$$\dot{x}^\mu \equiv \frac{dx^\mu}{d\lambda}$$

we get

$$0 = \int \left[(\delta g_{\mu\nu}) \dot{x}^\mu \dot{x}^\nu + 2 g_{\mu\nu} \dot{x}^\mu \delta \dot{x}^\nu \right] d\lambda$$

$$= \int \left[g_{\mu\nu,\rho} \delta x^\rho \dot{x}^\mu \dot{x}^\nu + 2 g_{\mu\rho} \dot{x}^\mu (\delta x^\rho)^{\cdot} \right] d\lambda$$

$$= \int \left[g_{\mu\nu,\rho} \dot{x}^\mu \dot{x}^\nu - (2 g_{\mu\rho} \dot{x}^\mu)^{\cdot} \right] \delta x^\rho d\lambda$$

$$= -2 \int \left[g_{\mu\rho} \ddot{x}^\mu + \dot{g}_{\mu\rho} \dot{x}^\mu - \frac{1}{2} g_{\mu\nu,\rho} \dot{x}^\mu \dot{x}^\nu \right] \delta x^\rho d\lambda$$

$$= -2 \int \left[g_{\mu\rho} \ddot{x}^\mu + g_{\mu\rho,\nu} \dot{x}^\nu \dot{x}^\mu - \frac{1}{2} g_{\mu\nu,\rho} \dot{x}^\mu \dot{x}^\nu \right] \delta x^\rho d\lambda$$

$$= -2 \int \left[g_{\mu\rho} \ddot{x}^\mu + \frac{1}{2} (g_{\mu\rho,\nu} + g_{\nu\rho,\mu} - g_{\mu\nu,\rho}) \dot{x}^\mu \dot{x}^\nu \right] \delta x^\rho d\lambda$$

$$= -2 \int g_{\alpha\rho} \left[\ddot{x}^\alpha + \frac{1}{2} g^{\alpha\sigma} (g_{\mu\sigma,\nu} + g_{\nu\sigma,\mu} - g_{\mu\nu,\sigma}) \dot{x}^\mu \dot{x}^\nu \right] \delta x^\rho d\lambda.$$

In the first line only the chain rule was used, in the second the variation of $g_{\mu\nu}$ was performed according to the rule (3.11), the 3rd line involved an integration by parts, in the 4th line a factor of 2 was taken in front of the integral and a dot derivative was written out, the dot derivative of $g_{\mu\nu}$ was written out in the 5th line, the middle term in the 5th line was written symmetrically with respect to the indices μ and ν in the 6th line, and finally a factor $g_{\alpha\rho}$ was taken out of the bracket in the last line.

3.3 Equations of Motion of Astronomical Bodies

The equation for a geodesic can therefore be written as

$$\ddot{x}^\alpha + \Gamma^\alpha_{\mu\nu} \dot{x}^\mu \dot{x}^\nu = 0 \tag{3.12}$$

with

$$\Gamma^\alpha_{\mu\nu} \equiv \frac{1}{2} g^{\alpha\sigma} (g_{\mu\sigma,\nu} + g_{\nu\sigma,\mu} - g_{\mu\nu,\sigma}). \tag{3.13}$$

The quantities $\Gamma^\alpha_{\mu\nu}$ are called *Christoffel symbols*.

3.3.0.1 Geodesics on the Unit Sphere

As a preliminary example, we will look at a very simple application: the geodesics on a unit sphere. To this end, we need an expression for the metric tensor on the unit sphere in suitable coordinates. We will start with 3-dimensional Euclidean flat space. In Cartesian coordinates $x^1 = x, x^2 = y, x^3 = z$ the metric tensor is given by

$$ds^2 = dx^2 + dy^2 + dz^2$$

(3-dimensional Pythagorean theorem). A transformation to spherical coordinates according to

$$x = r \sin\theta \cos\phi, \ y = r \sin\theta \sin\phi, \ z = r \cos\theta$$

and taking the differentials

$$dx = \frac{\partial x}{\partial r} dr + \frac{\partial x}{\partial \theta} d\theta + \frac{\partial x}{\partial \phi} d\phi = (\sin\theta \cos\phi) dr + (r \cos\theta \cos\phi) d\theta - r(\sin\theta \sin\phi) d\phi$$

etc., yields

$$ds^2 = dr^2 + r^2 (d\theta^2 + \sin^2\theta \, d\phi^2).$$

Hence, with respect to the coordinates $x^1 = r, x^2 = \theta, x^3 = \phi$, the nonvanishing components of the metric tensor read

$$g_{11} = g_{rr} = 1, \ g_{22} = g_{\theta\theta} = r^2, \ g_{33} = g_{\phi\phi} = r^2 \sin^2\theta.$$

For $r = 1; dr = 0$ we can write the desired metric tensor for the unit sphere in spherical coordinates $x^1 = \theta, \quad x^2 = \phi$:

$$g_{11} = g_{\theta\theta} = 1, \ g_{12} = g_{\theta\phi} = g_{21} = 0, \ g_{22} = g_{\phi\phi} = \sin^2\theta.$$

The inverse metric tensor then reads

$$g^{11} = 1, \ g^{12} = g^{21} = 0, \ g^{22} = \sin^{-2}\theta$$

and the nonvanishing Christoffel symbols are given by

$$\Gamma^1_{22} = -\sin\theta\cos\theta, \quad \Gamma^2_{12} = \Gamma^2_{21} = \cot\theta.$$

The geodesic equations on the unit sphere therefore take the form

$$\ddot\theta - \sin\theta\cos\theta\,\dot\phi^2 = 0,$$

$$\ddot\phi + 2\cot\theta\,\dot\phi\,\dot\theta = 0.$$

The geodesics through the two poles with $\theta_+ = 0, \theta_- = \pi$ therefore are the meridians

$$\phi = \text{const.}, \quad \theta = \lambda \quad \lambda \in [0, \pi].$$

3.3.0.2 Derivation of the EIH Equations of Motion

Next we will turn to the equations of motion of (spherical) astronomical bodies in the canonical metric with coordinates $x^\mu = (ct, \mathbf{x})$. The metric tensor then is given by (3.2). The following expansions for the inverse metric tensor and the Christoffel symbols are found ($\mathcal{O}_n \equiv \mathcal{O}(c^{-n})$):

$$g^{00} = -1 - \frac{2w}{c^2} - \frac{2w^2}{c^4} + \mathcal{O}_6$$

$$g^{0i} = g^{i0} = -\frac{4}{c^3}w_i + \mathcal{O}_5 \tag{3.14}$$

$$g^{ij} = \delta_{ij}\left(1 - \frac{2}{c^2}w\right) + \mathcal{O}_4$$

and

$$\Gamma^0_{00} = -\frac{w_{,0}}{c^2} + \mathcal{O}_5$$

$$\Gamma^0_{0i} = -\frac{w_{,i}}{c^2} + \mathcal{O}_6$$

$$\Gamma^0_{ij} = \delta_{ij}\frac{w_{,0}}{c^2} + \frac{4}{c^3}w_{(i,j)} + \mathcal{O}_5$$

$$\Gamma^i_{00} = -\frac{w_{,i}}{c^2} + 4\frac{w\,w_{,i}}{c^4} - \frac{4}{c^3}w_{i,0} + \mathcal{O}_6 \tag{3.15}$$

$$\Gamma^i_{0j} = -\frac{4}{c^3}w_{[i,j]} + \frac{w_{,0}}{c^2}\delta_{ij} + \mathcal{O}_5$$

$$\Gamma^i_{jk} = \delta_{ij}\frac{w_{,k}}{c^2} + \delta_{ik}\frac{w_{,j}}{c^2} - \delta_{jk}\frac{w_{,i}}{c^2} + \mathcal{O}_4$$

3.3 Equations of Motion of Astronomical Bodies

with

$$w_{(i,j)} \equiv \frac{1}{2}(w_{i,j} + w_{j,i})$$

$$w_{[i,j]} \equiv \frac{1}{2}(w_{i,j} - w_{j,i}).$$

Exercise 3.1: By explicit calculation, show that the expansions (3.14) and (3.15) for the inverse metric tensor and the Christoffel symbols for a metric of the form (3.2) are valid.

From the geodesic equation (3.12)

$$\frac{d^2 x^\mu}{d\lambda^2} + \Gamma^\mu_{\nu\sigma} \frac{dx^\nu}{d\lambda} \frac{dx^\sigma}{d\lambda} = 0,$$

we are in a position to derive the coordinate acceleration of the center of mass z_A of a spherical astronomical body A in the gravitational field of a system of other spherical bodies:

$$\frac{d^2 z^i_A}{dt^2} = \left(\frac{dt}{d\lambda}\right)^{-1} \frac{d}{d\lambda}\left[\left(\frac{dt}{d\lambda}\right)^{-1} \frac{dz^i_A}{d\lambda}\right]$$

$$= \left(\frac{dt}{d\lambda}\right)^{-2} \frac{d^2 z^i_A}{d\lambda^2} - \left(\frac{dt}{d\lambda}\right)^{-3} \frac{d^2 t}{d\lambda^2} \frac{dz^i_A}{d\lambda}$$

$$= -\Gamma^i_{\nu\sigma} \frac{dz^\nu_A}{dt} \frac{dz^\sigma_A}{dt} + \frac{1}{c}\Gamma^0_{\nu\sigma} \frac{dz^\nu_A}{dt} \frac{dz^\sigma_A}{dt} \frac{dz^i_A}{dt}$$

with $z^0_A = ct; dz^i_A/dt = v^i_A$. Written out, this equation reads

$$\frac{d^2 z^i_A}{dt^2} = -c^2 \left\{ \Gamma^i_{00} + 2\Gamma^i_{0j} \frac{v^j_A}{c} + \Gamma^i_{jk} \frac{v^j_A}{c} \frac{v^k_A}{c} \right. \qquad (3.16)$$

$$\left. - \left[\Gamma^0_{00} + 2\Gamma^0_{0j} \frac{v^j_A}{c} + \Gamma^0_{jk} \frac{v^j_A}{c} \frac{v^k_A}{c}\right] \frac{v^i_A}{c} \right\}.$$

Note that the Christoffel symbols appearing in the last equations result from the external metric, i.e., that part of the metric that is induced by external bodies B ≠ A only. One sees that the coordinate acceleration of a body is dominated by the term

$$-c^2 \Gamma^i_{00} = -w_{,i} + \mathcal{O}_2. \qquad (3.17)$$

In this Newtonian approximation, the acceleration of a body is given by the gradient of the gravitational potential $w = U + \mathcal{O}_2$.

From the form of the metric potentials (3.7) and (3.8) for a system of mass monopoles in the BCRS, in this way one derives the well-known Einstein–Infeld–Hoffmann (EIH) equations of motion that are the basis of every modern solar system ephemeris

$$\frac{d^2 \mathbf{z}_A}{dt^2} = -\sum_{B \neq A} \frac{GM_B}{r_{AB}^2} \mathbf{n}_{AB} \left\{ 1 + \frac{1}{c^2} \left[\mathbf{v}_A^2 + 2\mathbf{v}_B^2 - 4\mathbf{v}_A \cdot \mathbf{v}_B - \frac{3}{2}(\mathbf{n}_{AB} \cdot \mathbf{v}_B)^2 \right] \right.$$

$$- 4 \sum_{C \neq A} \frac{GM_C}{c^2 r_{AC}} - \sum_{C \neq B} \frac{GM_C}{c^2 r_{BC}} \left(1 + \frac{1}{2} \frac{r_{AB}}{r_{CB}} \mathbf{n}_{AB} \cdot \mathbf{n}_{CB} \right) \right\}$$

$$- \frac{7}{2} \sum_{B \neq A} \sum_{C \neq B} \mathbf{n}_{BC} \frac{G^2 M_B M_C}{c^2 r_{AB} r_{BC}^2}$$

$$+ \sum_{B \neq A} (\mathbf{v}_A - \mathbf{v}_B) \frac{GM_B}{c^2 r_{AB}^2} (4 \mathbf{n}_{AB} \cdot \mathbf{v}_A - 3 \mathbf{n}_{AB} \cdot \mathbf{v}_B)$$

(3.18)

with

$$r_{AB} \equiv |\mathbf{z}_A(t) - \mathbf{z}_B(t)|, \quad \mathbf{n}_{AB} \equiv [\mathbf{z}_A(t) - \mathbf{z}_B(t)] / r_{AB}.$$

Exercise 3.2: From the form of the metric potentials (3.7), (3.8), and (3.16), derive the EIH equations of motion (3.18).

Chapter 4
Barycentric Dynamical Reference System

4.1 Concepts

The barycentric dynamical reference system (BDRS) is a space–time reference system whose origin agrees with the solar system barycenter. Note that the BDRS is not yet an IAU-adopted name, in contrast to BCRS and GCRS; often, it is called *conventional dynamical realization of the ICRS*, a name that however lacks the reference to the barycenter.

Conceptually, the space–time metric of the system agrees with the canonical barycentric metric (3.2). Its timescale t is barycentric coordinate time (TCB) or, equivalently, TDB. Both its timescale and its spatial coordinates (x, y, z) are realized by observations of solar system objects like the Sun, the Moon, planets, asteroids, spacecrafts, etc. The BDRS is materialized by means of some solar system ephemeris.

4.2 Observational Methods

4.2.1 Optical Ground-Based Astrometry

Still today, ground-based optical astrometric data is used to determine the positions of Jupiter, Saturn, Uranus, and Pluto. Transit observations that are used for modern solar system ephemerides are from Washington, Herstmonceux, La Palma, Tokyo, El Leoncito, and Bordeaux. For Pluto photographic plates from Palomar, Pulkovo, Bordeaux, Valinhos, Asiago, Copenhagen, Lick, and Torino observatories are employed (Barbieri et al. 1972, 1975, 1979, 1988; Cohen et al. 1967; Gemmo and Barbieri 1994; Jensen 1979; Klemola and Harlan 1982, 1984, 1986; Rapaport et al. 2002; Rylkov et al. 1995; Sharai and Budnikova 1969; Zappala et al. 1980, 1983). Modern CCD astrometry from US Naval Observatory Flagstaff Station (USNOFS) (Stone et al. 2003, Fig. 4.1) and Table mountain

Fig. 4.1 The 1.55-m Kaj Strand Astrometric Reflector at USNO Flagstaff Station

Fig. 4.2 The Royal Greenwich Observatory's Automatic Transit Circle telescope at the La Palma Observatory in the Canary Islands. Courtesy of D.W. Evans

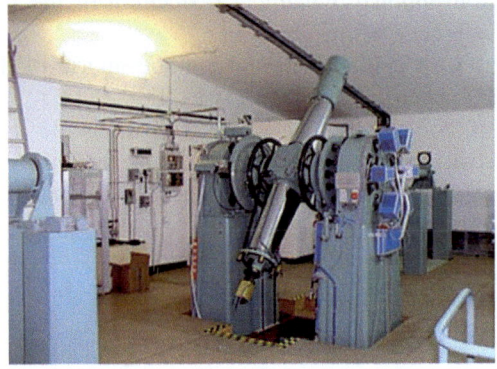

observatory (TMO) is also used. Fig. 4.2 shows the Greenwich Observatory's Automatic Transit Cirlce telescope at the La Palma Observatory that played an important role in the past.

4.2.2 Lunar Laser Ranging

Laser distance measurements to the Moon (lunar laser ranging, LLR) determine the light travel time, which can be converted into a distance, between a station on the Earth and a reflector on the Moon (Shelus 2001). The distance between the two centers of mass is about 384,400 km on average and varies due to the

4.2 Observational Methods

Fig. 4.3 The LLR retroreflector positions

elliptical orbit of the Moon by about 21,000 km in course of a month; the solar tidal forces lead to a variation of about 3,700 km. This highly variable distance can be measured by means of LLR with centimeter precision. This is achieved by means of laser pulses from the Earth to the Moon and back. During three US American Apollo missions (11, 14, and 15) and two unmanned Soviet missions (Luna 17 and Luna 21), retroreflectors were deployed near the landing sites between 1969 and 1973 (Fig. 4.3). In July 1969 the first laser reflector array was placed on the lunar surface in the *Mare Tranquillitatis* by Apollo 11 astronauts. Two further reflector arrays were installed 1971 in the *Fra Mauro* region and at the *Hadley Rille* by Apollo 14 and 15 astronauts. Finally, in 1973, a French reflector array was placed near the crater *Le Monnier* by means of Luna 21. The American reflectors consist of 100 (Apollo 11 and 14; Fig. 4.4) and about 300 (Apollo 15) triple prisms, respectively, of about 4 cm in diameter, mounted on an aluminum frame.

Within a few weeks after the first laser reflector array was installed by Apollo 11, LLR measurements were successfully performed by McDonald Observatory near Fort Davis, Texas. In 1984, two further LLR stations, one on Mt. Haleakala on Maui (Hawaii) and one station in Grasse (France), followed. Lunar activities on Mt. Haleakala were closed in 1990. LLR measurements with the 2.5-m telescope were finished in 1985 and continued with the *McDonald Laser Ranging System*, involving a dedicated 75-cm telescope. In July 2005, the new APOLLO (*Apache Point Observatory Lunar Laser-Ranging Operation*) system in the southern part of New Mexico (USA) started operation (Murphy et al. 2000).

The McDonald Observatory in Texas, USA; the Apache Point Observatory, New Mexico, USA; and the Observatoire de la Côte d'Azur (OCA), France, are the only currently operational LLR sites. The latter has undergone renovation since late 2004 and returned to action in September 2009. The McDonald Observatory has major problems to get further LLR tracking funded. Although no system upgrade could be made in the past years, lunar tracking could be continued at a certain level. The new

Fig. 4.4 The Apollo 14 laser retroreflector on the lunar surface. It consists of a total of 100 triple prisms of 4 cm in size mounted to an aluminum frame

APOLLO site is equipped with a 3.5-m telescope and is designed for millimeter accuracy ranging. A new set of data from APOLLO was released in 2011 with a total of about 940 normal points. The data are now available in the newly adopted ILRS CRD data format through a reformatting effort at the McDonald Observatory.

Also, other modern stations have demonstrated lunar capability, for example, the Matera Laser Ranging Station, Italy, in 2010, but all of them suffer from technical problems or funding restrictions. The Wettzell Observatory, Germany, plans to resume lunar tracking in the near future. The Australian station at Mt. Stromlo is expected to join this group in the future, and there are plans for establishing lunar capability at the South African site of Hartebeesthoek.

In LLR, one works with pulsed laser light with a pulse duration of about 150 ps containing about 10^{19} photons. Some of these photons hit one of the retroreflectors on the lunar surface, and typically one single photon per ten pulses finds its way back to the reception optics of the station. Then one faces the problem that the majority of received photons are just noise and the correct photons have to be identified in some way or another. To this end, one first works with sharp filtering in frequency and time domain. Only those photons with a CMO (calculated minus observed) difference of less than 10 m in space are taken into account. Then one constructs a histogram (Fig. 4.5), where the number of received photons is plotted that fall into a certain CMO time interval. For ideal measurements without noise and a perfect model, all photons would end up in a small interval around zero. This is clearly not the case, but one finds a large peak in the vicinity of zero containing about 100 photons from about 15 min of observation. These photons define a so-called *normal point* that presents a kind of pulse arrival time. Such normal points are the observational basis of LLR. Presently, the accuracy of LLR measurements lies in the range of a few centimeters (Fig. 4.6).

4.2 Observational Methods

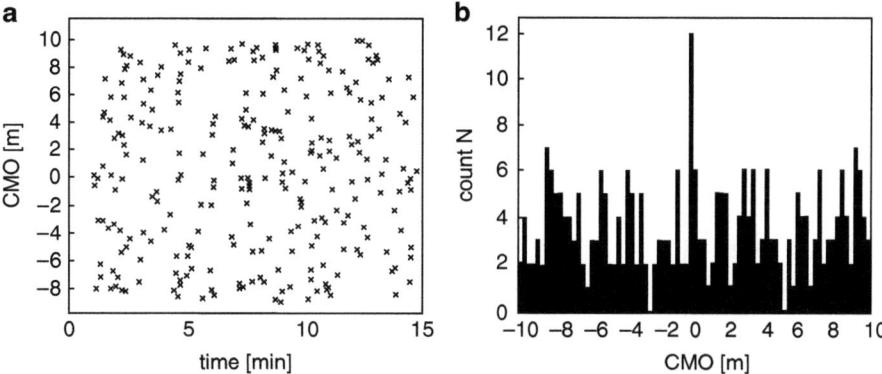

Fig. 4.5 (a) Difference of the Earth–Moon distance: "calculated minus observed" (CMO) as function of time. The image results from a block of measurements of 15 min duration. (b) Histogram of such differences. Indicated is the number of photons that fall into a certain CMO interval. For an ideal model and ideal observations, all photons would be found in the interval around zero

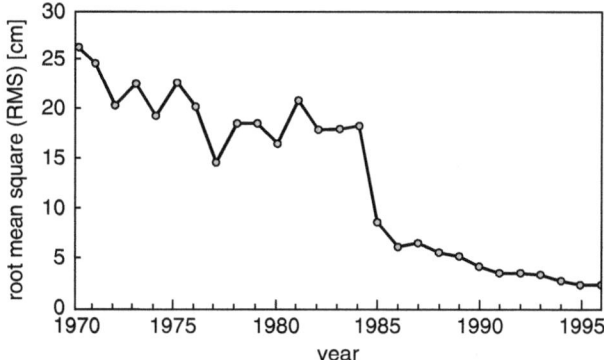

Fig. 4.6 Weighted mean residuals as function of time. Note the increase in accuracy in the 1980s

Because of the much larger signal-to-noise ratio, the problem of normal points is different for the APOLLO station that aims at millimeter precision. Figure 4.7 shows lunar photons observed with the APOLLO station; it shows the received photons during a 4-min run on 19 October 2005 corresponding to 5,000 shots at 20 shots per second, with about 700 lunar photons (website of Tom Murphy, http://physics.ucsd.edu/~tmurphy/); CMO values are on the x-axis. According to Murphy (2011), the construction of normal points at the APOLLO station involves the following: construct CMO values from the raw data, search for a signal in sliding 2 ns windows, fit a straight line to the residuals found within this window (taking out some background "hits"), find the centroid time of the *valid* hits, round to the nearest 5 s for the normal point epoch, and add to the prediction the linear fit to the CMO residuals evaluated at the normal point epoch.

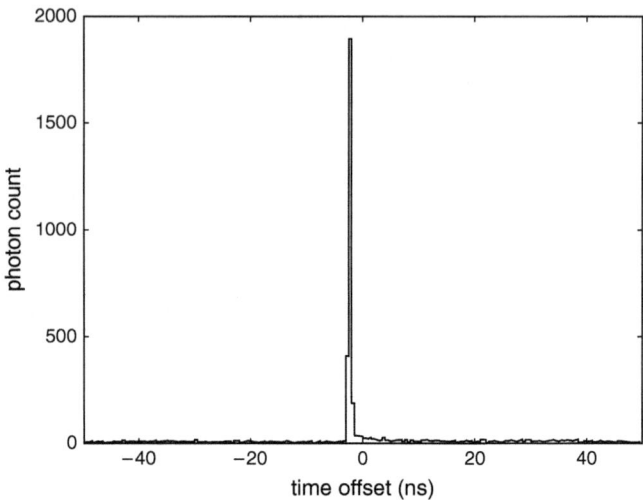

Fig. 4.7 APOLLO returns from the Apollo 14 array on 2008 October 17. Bins are 0.5 ns wide. Over 2400 photons are detected from the reflector in a 3000-shot run, spanning 150 seconds. Courtesy of Tom Murphy

The improvement in the SNR value of APOLLO in comparison, for example, with the OCA station results from using a bigger telescope and better astronomical seeing (APOLLO sees as many as ten photons in a single pulse). Only because of the large SNR value, the problem that the retroreflectors on the Moon introduce an error of more than 1 mm can be solved. These retroreflectors are not usually at an exact right angle to the incoming beam, so the different corner cubes of the retroreflectors, due to lunar libration, are at different distances from the transmitter. To determine the distance to the reflector to 1-mm precision, about 3,000 reflected photons are needed.

In April 2010, the APOLLO team announced that with the aid of photos from the Lunar Reconnaissance Orbiter, they had found the long-lost Lunokhod 1 rover and had received returns from the laser retroreflector. By the fall of 2010, the location of the rover had been determined to about a centimeter. The location near the limb of the moon, combined with the ability to range the rover even when it is in sunlight, promises to be particularly useful for determining aspects of the Earth–Moon system (http://en.wikipedia.org/, September 2011; Murphy et al. 2010).

The analysis of normal points involves substantial software packages. These contain an ephemeris part for the motion of solar system bodies and a part for parameter determinations via least-squares fit.

In the ephemeris part, the relativistic equations of motion for spherical bodies (the Einstein–Infeld–Hoffmann, EIH, equations of motion) are numerically integrated for Sun, Earth, Moon, planets, and a selected number of asteroids (Ceres, Pallas, Vesta, etc.). Deviations from spherical symmetry are considered for Earth, Moon, and Sun in Newtonian approximation with a usual expansion of the

4.2 Observational Methods

gravitational potential in terms of spherical harmonics. The rotational motion of the Moon is derived by numerical integration of the Euler gyroscope equations. From this, the libration angles of the Moon, describing the orientation of the lunar figure in space, are derived. The rotational motion of the Earth is divided into length of day (LOD), precession, nutation, and polar motion. Nutation is described by means of the usual nutation series. Finally, elastic effects in the Moon and tidal dissipation are described with heuristic parameters.

LLR measurements have provided essential information to diverse aspects of the Earth–Moon system. In the set of solve-for parameters are the geocentric station parameters that can be estimated with centimeter accuracy.

It is obvious that the measured distances depend upon the orientation of Earth and Moon in space. As far as the Moon is concerned, libration angles at some initial epoch can be determined. The least-squares fit also determines the initial Earth–Moon vector.

The large influence of the Sun on the lunar orbits leads to the possibility of a precise determination of the mass ratio $m_S/(m_E + m_M) = 328\,900.560 \pm 0.002$ with a precision of 10^{-8} according to the ratio of the LLR accuracy to the solar perturbation of the lunar orbit $\sim 3\,\text{cm}/3,000\,\text{km}$.

Besides the selenocentric reflector coordinates, LLR data provide important information about the lunar gravity field, for example, for certain combinations of lunar moments of inertia and mass multipole moments up to degree and order 3. Combined with Doppler data from lunar satellites, this leads to a precise determination of the polar moment of inertia, an important characteristic for the interior structure of the Moon.

Tidal deformation of the Moon is described with a certain constant of proportionality, the lunar Love number, that can also be inferred from LLR data. Tidal dissipation is accounted for in different ways. First of all, a dissipation parameter is in the set of solve-for parameters. Due to the tidal dissipation, the rotation of the Earth is slowed down, and the length of days is steadily increasing. Since the angular momentum in the Earth–Moon system is conserved, the Moon recedes from the Earth due to tidal dissipation. This might sound queer at first having in mind that artificial satellites due to atmospheric friction fall down to the Earth. However, due to the relatively large gravitational field of the Moon, it induces two tidal bulges that due to friction are moved away from the Earth–Moon line by the Earth's rotational motion. Then these tidal bulges gravitationally accelerate the Moon that thereby recedes from the Earth. According to Kepler's third law, the corresponding mean motion (mean angular velocity) of the Moon about the Earth, n_M, decreases with time. From LLR data, one finds for $\dot{n}_M$ a value of about $-26''/(100\,\text{year})^2$. This has the consequence that presently the Moon recedes from the Earth by about 3.8 cm/year.

The rate of precession and the nutation amplitude of the fundamental 18.6-year period, related with the motion of the lunar node, can be derived from LLR data together with many other Earth orientation parameters. Among them is an angle that describes the orientation of the Earth around its axis of rotation. This rotational phase called UT0 can be determined to be better than 0.1 ms. Polar motion is related with latitude variations that can be obtained with accuracies better than 1 mas.

Finally, LLR data has frequently been used to test various aspects of the gravitational interaction: the weak and the strong equivalence principle (Nordtvedt 1968, 1995; Williams et al. 1996; Müller and Nordtvedt 1998; Williams et al. 2009), the temporal variation of the gravitational constant G (Müller and Biskupek 2007; Williams et al. 2009), and the geodetic (or de Sitter) precession (e.g., Müller et al. 1991).

If we write Newton's law of gravity as

$$m_I \ddot{\mathbf{r}} = -G \frac{M m_G}{r^2} \frac{\mathbf{r}}{r} \qquad (4.1)$$

the *weak equivalence principle* asserts the equivalence of inertial mass m_I and gravitational mass m_G of some test body in an external gravitational field induced by some mass M, or more precisely that the world line of an uncharged test body is independent of its internal structure and composition (Will 1993). LLR yields an upper limit for the difference in the ratio of the gravitational and inertial masses for the Earth and Moon of the order 10^{-13} (Turyshev et al. 2004; Williams et al. 2004). The strong equivalence principle extends the weak form to self-gravitating bodies with nonnegligible gravitational self-energy. For example if

$$(m_G/m_I)_M - (m_G/m_I)_E \equiv \eta_N (\Delta_E - \Delta_M) \neq 0$$

with

$$\Delta \equiv -\frac{m_{G-\text{energy}}}{m_I}$$

the Earth (E) and Moon (M) would fall at different rates towards the Sun leading to an anomalous oscillation of the Earth–Moon distance with an amplitude of $(9.2 \, \eta_N)$ m. In Einstein's theory of gravity, the Nordtvedt parameter η_N vanishes exactly. LLR leads to an upper limit $\eta_N = (-0.6 \pm 5.2) \times 10^{-4}$ (Hofmann et al. 2010).

Theoretically, the Newtonian gravitational constant G might vary with time (e.g., Will 1993). LLR data leads to an upper limit of $\dot{G}/G = (-0.7 \pm 3.9) \times 10^{-13}$ year^{-1} (Hofmann et al. 2010).

Finally, one of the dynamical consequences of Einstein's theory of gravity, general relativity, is a non-Newtonian precession of the Moon as it orbits the Earth in a system freely falling in the gravitational field of the Sun (Shapiro et al. 1988). This geodetic precession, first noted by de Sitter (1916), amounts to about $2''$ per century. With LLR data, this geodetic (de Sitter) precession has been measured with an accuracy of a few parts in 10^3 (Müller et al. 1991).

4.2.3 Radar Measurements to Planets

Radar has been used as a remote sensing tool for the exploration of our solar system since 1946 (Jurgens 1982) when the deflection of echoes from the moon

4.2 Observational Methods

Fig. 4.8 The world's largest radio telescope: the 1,000-ft (305 m) Arecibo radio telescope in Puerto Rico

was reported (Mofenson 1946; Webb 1946; Bay 1947). Since these early radar observations, ground-based radar studies have frequently been made of the planets Mercury, Venus, Mars, the Galilean satellites, the rings of Saturn, and a certain number of asteroids (Jurgens 1982; Black et al. 2010, and cited references therein). Radar measurements to Mercury and Venus have been executed from the Arecibo (Fig. 4.8), Goldstone, Haystack, and the Byrd 100-m Green Bank radio telescopes (Pettengill et al. 1967; Black et al. 2010).

One problem with the radar technique is the atmosphere; another difficulty is signal strength. The signal will start to spread out as soon as it is sent. For planetary radar observations, the received signal is will be very weak. The echo will be Doppler shifted by the relative velocities of the ground radio station and the various parts of the planetary surface.

4.2.4 Radar Tracking to Spacecraft

Radar measurements have been made to the spacecrafts Mariner 10 (Mercury), Venus Explorer, Viking 1 and 2, Mars Pathfinder, Mars Global Surveyor (MGS), Odyssey, Mars Express (MEX), Mars Reconnaissance Orbiter (MRO) (Mars), Pioneer 10 and 11, Voyager 1 and 2, Ulysses, and Cassini. Three types of tracking data are used for planetary astrometry. The range is the time delay between the navigation antenna and the spacecraft orbiting or flybying a planet. These data give stringent constrains (a few meters) on the geocentric distances of the inner planets and on Jupiter and Saturn during flyby phases (a few kilometers to tenth of meters). VLBI observations of a spacecraft flying in the angular vicinity of ICRF sources during its flyby or orbiting phases give the most accurate angular positions of planets directly related with the ICRF. They are crucial to link the dynamical reference frame of the planetary ephemerides to the ICRF.

4.3 Solar System Ephemerides

4.3.1 Numerical Ephemerides

The word ephemeris originates from the Greek language ("εφημερος", ephemeros stands for a day) and means a list, table, or computer program for the determination of positions and velocities of large solar system bodies.

4.3.1.1 DE and EPM

The post-Newtonian equations of motion for a set of *point masses* (the EIH equations) are the basis of the three state-of-the-art solar system ephemerides: the American one, DE (Development Ephemeris; JPL); the Russian one, EPM (Ephemerides of Planets and the Moon; IPA, St. Petersburg); and the French one, INPOP (Intégration Numérique Planétaire de l'Observatoire de Paris). There, the EIH equations of motion are integrated numerically for the whole solar system including a set of selected minor planets. Considered are figure effects for Earth and Moon, the solar oblateness, the rotational motion of Earth and Moon (librations), and the effect of tidal dissipation (via some lag angle). Different versions of these ephemerides differ slightly in modeling of lunar libration, reference frames in which the ephemerides are computed, adopted value of solar oblateness, modeling of perturbation from asteroids, and the set of observations to which ephemerides are adjusted. Table 4.1 (Pitjeva 2005) lists some characteristics of various DE and EPM ephemerides. In course of time, all three leading groups for numerical ephemerides created a whole series of ephemerides.

DE200 was a worldwide standard for several decades. In the actual 400 series, the time between 1410 BC and 3000 AD is covered by DE403; DE404 covers the period between 3000 BC and 3000 AD For the present epoch, DE405 and DE406 were developed. Meanwhile, DE414, DE418, DE421, and DE423 (just produced for use by the *Messenger* project, making use of some Messenger tracking data) have been produced. Data can be retrieved from the web interface http://ssd.jpl.nasa.gov/ named *Horizons*. The *AstroRef* package supplies the function GetDE405Ephemeris that returns DE405 ephemerides data (see appendix B). The reader is also referred to the websites ftp://ssd.jpl.nasa.gov/pub/eph/planets/ and http://naif.jpl.nasa.gov/ and the comprehensive literature (Newhall et al. 1983; Konopliv et al. 2006, 2011; Standish 1982, 1990a,b, 1998, 2003, 2006; Standish et al. 1995; Folkner et al. 2007, 2009). DE721 is very similar to DE421 but covers a longer time span (Table 4.2).

Let us look more closely into DE421 that presents a significant advance over earlier ephemerides. Compared with DE418, it includes additional data, especially range and VLBI measurements of Mars spacecraft; range measurements to ESA's Venus Express; and use of current best estimates of planetary masses in the integration process (Folkner et al. 2009). It presents a combined fit of LLR and planetary measurements. Table 4.3 gives an overview over the data used for DE421.

4.3 Solar System Ephemerides

Table 4.1 Overview over the American (DE) and Russian (EPM) solar system ephemerides; from Pitjeva (2005)

Ephemeris	Interval of integration	Reference frame	Mathematical model	Type of observations	Number of observations	Time interval
DE118 (1981) ⇒ DE200	1599…2169	FK4 ⇒ J2000.0 system	Integration: the Sun, the Moon, nine planets + perturbations from three asteroids (two-body problem)	Optical	44,755	1911–1979
				Radar	1,307	1964–1977
				Spacecraft and landers	1,408	1971–1980
				LLR (lunar laser ranging)	2,954	1970–1980
				Total	50,424	1911–1980
EPM87 (1987)	1700…2020	FK4	Integration: the Sun, the Moon, nine planets + perturbations from five asteroids (two-body problem)	Optical	48,709	1717–1980
				Radar	5,344	1961–1986
				Spacecraft and landers	–	–
				LLR (lunar laser ranging)	1,855	1972–1980
				Total	55,908	1717–1986
DE403 (1995) ⇒ DE404	−1410…3000 ⇒ −3000…3000	ICRF	Integration: the Sun, the Moon, nine planets + perturbations from 300 asteroids (mean elements)	Optical	26,209	1911–1995
				Radar	1,341	1964–1993
				Spacecraft and landers	1,935	1971–1994
				LLR (lunar laser ranging)	9,555	1970–1995
				Total	39,057	1911–1995
EPM98 (1998)	1886…2006	DE403	Integration: the Sun, the Moon, nine planets, five asteroids + perturbations from 295 asteroids (mean elements)	Optical	–	–
				Radar	55,959	1961–1995
				Spacecraft and landers	1,927	1971–1982
				LLR (lunar laser ranging)	10,000	1970–1995
				Total	67,886	1961–1995
DE405 (1997) ⇒ DE406	1600…2200 ⇒ −3000…3000	ICRF	Integration: the Sun, the Moon, nine planets + perturbations from 300 (integrated) asteroids	Optical	28,261	1911–1996
				Radar	955	1964–1993
				Spacecraft and landers	1,956	1971–1995
				LLR (lunar laser ranging)	11,218	1969–1996
				Total	42,410	1911–1996

(continued)

Table 4.1 (continued)

Ephemeris	Interval of integration	Reference frame	Mathematical model	Type of observations	Number of observations	Time interval
EPM2000 (2000)	1886…2011	DE405	Integration: the Sun, the Moon, nine planets, 300 asteroids	Optical	–	–
				Radar	58,076	1961–1997
				Spacecraft and landers	24,587	1971–1997
				LLR	13,500	1970–1999
				Total	96,163	1961–1999
DE410 (2003)	1901…2019	ICRF	Integration: the Sun, the Moon, nine planets + perturbations from 300 asteroids	Optical	39,159	1911–2003
				Radar	978	1964–1997
				Spacecraft and landers	154,685	1971–2003
				LLR	9,555	1970–1995
				Total	204,377	1911–2003
EPM2004 (2004)	1880…2020	ICRF	Integration: the Sun, the Moon, nine planets, 301 asteroids, and an asteroid ring	Optical	46,064	1913–2003
				Radar	58,116	1961–1997
				Spacecraft and landers	197,271	1971–2003
				LLR	15,590	1970–2003
				Total	317,041	1913–2003

4.3 Solar System Ephemerides

Table 4.2 Overview over the latest numerical solar system ephemerides

Ephemeris	Interval of integration	Reference
DE414	1600–2201	
DE418	1899–2051	Folkner et al. (2007)
DE421	1900–2050	Folkner et al. (2009)
EPM2008	1800–2198	Pitjeva (2008)
EPM2010	1800–2200	Pitjeva et al. (2010c)
INPOP05	1600–2200	Fienga et al. (2008)
INPOP06	1600–2200	Fienga et al. (2008)
INPOP08		Fienga et al. (2009)
INPOP10a		Fienga et al. (2010, 2011)

Table 4.3 Data used for the ephemeris DE421

Object	Measurement	Type	Observatory	Span	No. Meas.
Moon	LLR	Range	McDonald 2.7 m	1970–1985	3,451
			MLRS/Saddle	1984–1988	275
			MLRS/Mt. Fowlkes	1988–2007	2,746
			Haleakala	1984–1990	694
			CERGA	1984–2005	9,177
			Matera	2004	11
			Apache Pt.	2006–2007	247
Mercury	Radar	Range	Arecibo	1967–1982	242
			Goldstone	1972–1997	283
			Haystack	1966–1971	217
			Eupatoria	1980–1995	75
	Radar	Closure	Goldstone	1989–1997	40
	Spacecraft	Range	Mariner 10	1974–1975	2
Venus	Spacecraft	Range	Venus Express	2006–2007	14,304
	Spacecraft	VLBI	Venus Express	2007	1
	Spacecraft	VLBI	Magellan	1990–1994	18
	Spacecraft	3-D	Cassini	1998–1999	2
	Radar	Range	Arecibo	1967–1970	227
			Goldstone	1970–1990	512
			Haystack	1966–1971	229
			Millstone	1964–1967	101
			Eupatoria	1962–1995	1,134
Mars	Spacecraft	Range	Viking L1	1976–1982	1,178
			Viking L2	1976–1977	80
			Pathfinder	1997	90
			MGS	1999–2006	164,781
			Odyssey	2002–2007	251,999
			Mars Express	2005–2007	63,133
			MRO	2006–2007	7,972

(continued)

Table 4.3 (continued)

Object	Measurement	Type	Observatory	Span	No. Meas.
	Spacecraft	VLBI	MGS	2001–2003	14
			Odyssey	2002–2007	66
			MRO	2006–2007	14
Jupiter	Spacecraft	3-D	Pioneer 10	1973	1
			Pioneer 11	1974	1
			Voyager 1	1979	1
			Voyager 2	1979	1
			Ulysses	1992	1
			Cassini	2000	1
	CCD	RA/Dec	USNOFS	1998–2007	2,533
	Spacecraft	VLBI	Galileo	1996–1997	24
	Transit	RA/Dec	Washington	1914–1994	2,053
			Herstmonceux	1958–1982	468
			La Palma	1992–1997	658
			Tokyo	1986–1988	98
			El Leoncito	1998	11
Saturn	Spacecraft	3-D	Pioneer 11	1979	1
			Voyager 1	1980	1
			Voyager 2	1981	1
			Cassini	2004–2006	31
	CCD	RA/Dec	USNOFS	1998–2007	3,153
			TMO	2002–2005	778
	Transit	RA/Dec	Bordeaux	1987–1993	119
			Washington	1913–1982	1,422
			Herstmonceux	1958–1982	405
			La Palma	1992–1997	730
			Tokyo	1986–1988	62
			El Leoncito	1998	18
Uranus	Spacecraft	3-D	Voyager 2	1986	1
	CCD	RA/Dec	USNOFS	1998–2007	1,612
			TMO	1998–2007	347
	Transit	RA/Dec	Bordeaux	1985–1993	165
			Washington	1914–1993	2,043
			Herstmonceux	1957–1981	353
			La Palma	1984–1997	1,030
			Tokyo	1986–1988	44
			El Leoncito	1997–1998	8
Neptune	Spacecraft	3-D	Voyager 2	1989	1
	CCD	RA/Dec	USNOFS	1998–2007	1,588
			TMO	2001–2007	267
	Transit	RA/Dec	Bordeaux	1985–1993	348

(continued)

4.3 Solar System Ephemerides

Table 4.3 (continued)

Object	Measurement	Type	Observatory	Span	No. Meas.
			Washington	1913–1993	1,838
			Herstmonceux	1958–1981	316
			La Palma	1984–1998	1,106
			Tokyo	1986–1988	59
			El Leoncito	1998–1999	11
Pluto	CCD	RA/Dec	USNOFS	1998–2007	852
			TMO	2001–2007	118
	Photo	RA/Dec	Misc.	1914–1958	42
			Palomar	1963–1965	8
			Pulkovo	1930–1992	53
			Bord/Valin	1995–2001	97
			Asiago	1969–1989	193
			Copenhagen	1975–1978	15
			Lick	1980–1985	11
			Torino	1973–1982	37
	Transit	RA/Dec	La Palma	1989–1998	380
			El Leoncito	1999	33

CERGA Centre d'Études et de Recherches Géodynamique et Astronomiques, *MGS* Mars Global Surveyor, *MLRS* McDonald Laser Ranging Station, *MRO* Mars Reconnaissance Orbiter, *USNOFS* US Naval Observatory Flagstaff Station, *TMO* Table Mountain Observatory

The orbit of Mercury is currently determined by radar range observations. Since the last radar range point is in 1999, the estimated Mercury orbit has not changed significantly for the past decade. The current orbit accuracy is a few kilometers (Folkner et al. 2009).

The Venus orbit accuracy has been significantly improved by inclusion of range measurements to the Venus Express spacecraft. Combined with VLBI measurements of Magellan and one VLBI observation of Venus Express, the Venus orbit accuracy is now about 200 m.

The orbits of Earth and Mars are continually improved through measurements of spacecraft in orbit about Mars. DE421 incorporated range data through the end of 2007. VLBI data through December 2007 have been included in the DE421 estimate. The Earth and Mars orbit accuracies are expected to be better than 300 m through 2008 (Folkner et al. 2009).

The orbits of Jupiter and Saturn are determined to accuracies of tens of kilometers using spacecraft tracking and modern ground-based astrometry. The orbit of Saturn is more accurate than that of Jupiter since the Cassini tracking data are more complete and more accurate than previous spacecraft tracking at Jupiter. The orbits of Uranus, Neptune, and Pluto are not well determined (Folkner et al. 2009). More information about orbit accuracies can be obtained from Figs. 4.9–4.22.

Latest versions of EPM include EPM2008 and EPM2010. The reader is referred to the website ftp://quasar.ipa.nw.ru/incoming/EPM/ and the corresponding literature (Krasinsky et al. 1993; Pitjeva 2001, 2005, 2009a, 2010a,b; Pitjeva and Standish 2009b) (Fig. 4.9).

76   4 Barycentric Dynamical Reference System

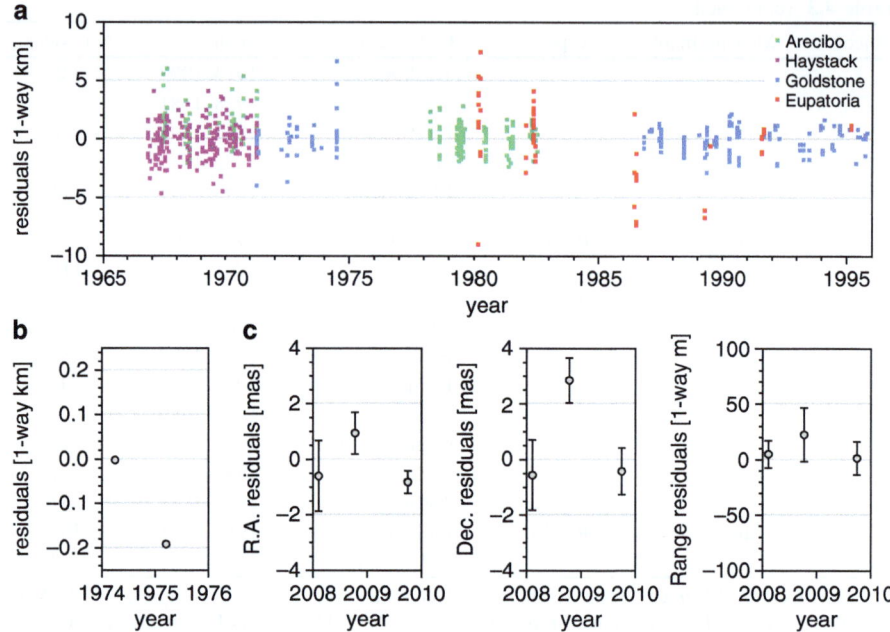

Fig. 4.9 Mercury orbit data. (**a**) Radar ranging, (**b**) Mariner 10 range, (**c**) Messenger encounters; from Folkner (2010). Courtesy of JPL, NASA

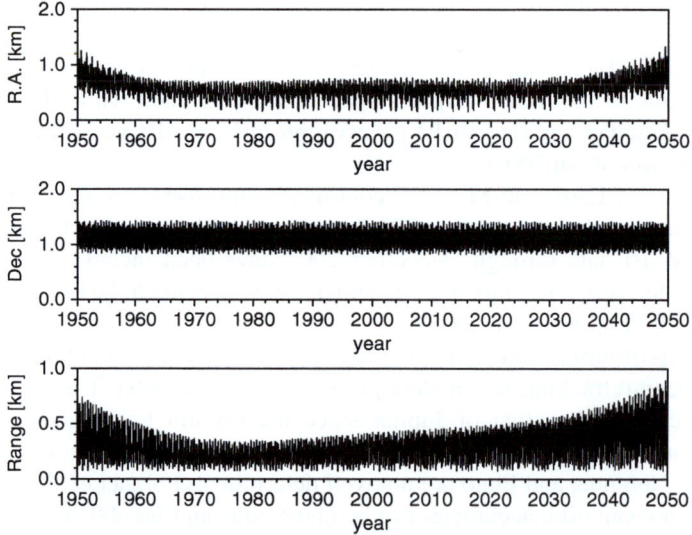

Fig. 4.10 Mercury orbit uncertainty with respect to the Earth; from Folkner (2010, 2011). Right ascension and declination have been converted into a distance by means of the corresponding range values. Courtesy of JPL, NASA

4.3 Solar System Ephemerides

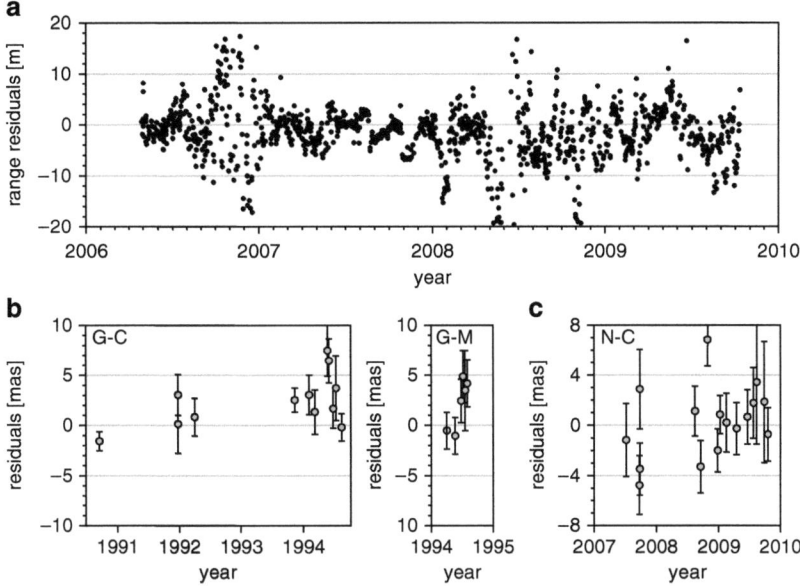

Fig. 4.11 Venus orbit data. (**a**) Venus Express Range, (**b**) Magellan VLBI with baseline Goldstone–Canberra (G–C) and Goldstone–Madrid (G–M), (**c**) Venus Express VLBI with base New Norcia–Ceberos; from Folkner (2010). Courtesy of JPL, NASA

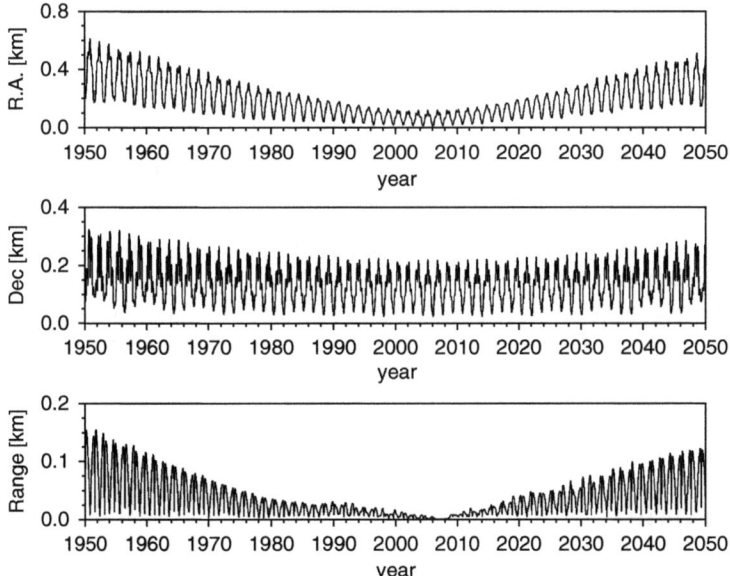

Fig. 4.12 Venus orbit uncertainty with respect to the Earth; from Folkner (2010, 2011). Courtesy of JPL, NASA

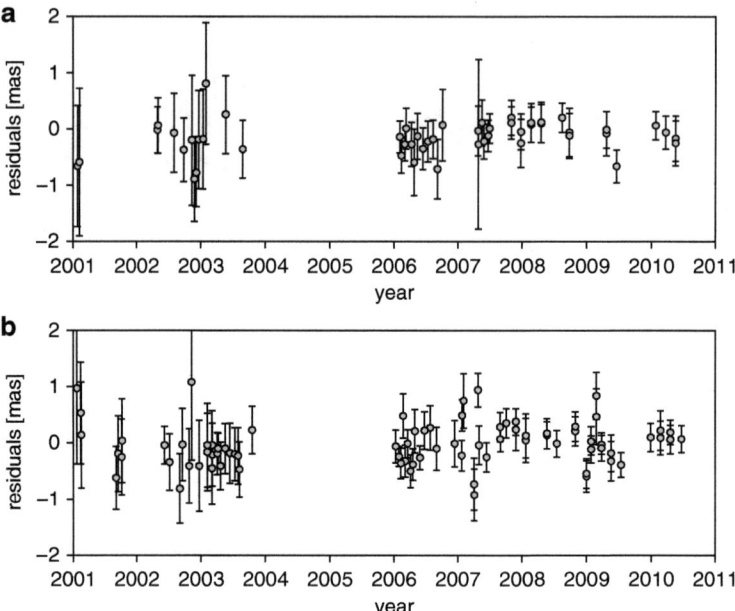

Fig. 4.13 Mars spacecraft VLBI data. (**a**) baseline Goldstone–Madrid, (**b**) baseline Goldstone–Canberra; from Folkner (2010). Courtesy of JPL, NASA

Exercise 4.1: Compute the geocentric position (AU) and velocity (AU/d) of the Sun in rectangular, equatorial coordinates for 1 January 2000, 0^h TT with INPOP10 and DE405 by making use of the IMCCE website.

Exercise 4.2: Compute the barycentric position (AU) and velocity (AU/year) of Mercury, Venus, Mars, Jupiter, and Saturn in rectangular, equatorial coordinates for 1 January 2000, 0^h TT with DE405. Use the function GetDE405Ephemeris of the *AstroRef* package.

Exercise 4.3: A solar light ray is reflected at Uranus on 23 July 2011, $14^h 32^m 44^s$ TT. At which time (UTC, accuracy: seconds) will an observer on the Earth be able to detect this light ray? Consider only the centers of mass for Uranus and Earth! Neglect effects from gravitational fields on the propagation of light rays.

Exercise 4.4: The illumination phase of a reflecting celestial body is given by the angle α between the observer and the illumination source measured at the reflecting body. Sometimes, this phase angle is converted into percent by $100 \cos^2(\alpha/2)$. Determine the phase of Venus for a terrestrial observation at 24 October 2011!

4.3.1.2 INPOP

INPOPxx are the first European planetary ephemerides built independently from the US JPL DExx ephemerides.

4.3 Solar System Ephemerides

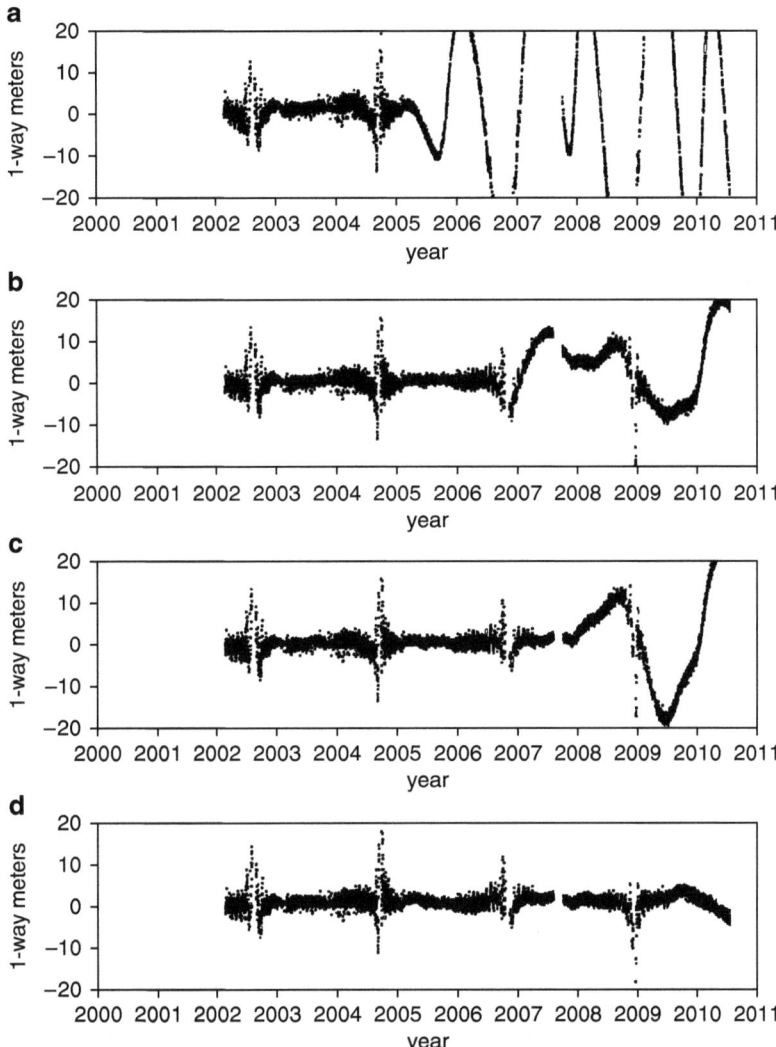

Fig. 4.14 Mars spacecraft range residuals (Mars Global Surveyor, Mars Odyssey, Mars Reconnaissance Orbiter, Mars Express) (**a**) DE414, (**b**) DE418, (**c**) DE421, (**d**) DE423; from Folkner (2010). Courtesy of JPL, NASA

The project began in 2003, and three versions of INPOP are presently available for users: INPOP06 (Fienga et al. 2008), INPOP08 (Fienga et al. 2009), and INPOP10a (Fienga et al. 2011). On the website www.imcce.fr/inpop, users have access to positions and velocities of the major planets of our solar system and of the Moon, the libration angles of the Moon, and also to the differences between terrestrial time TT (timescale used to date for the observations) and the barycentric

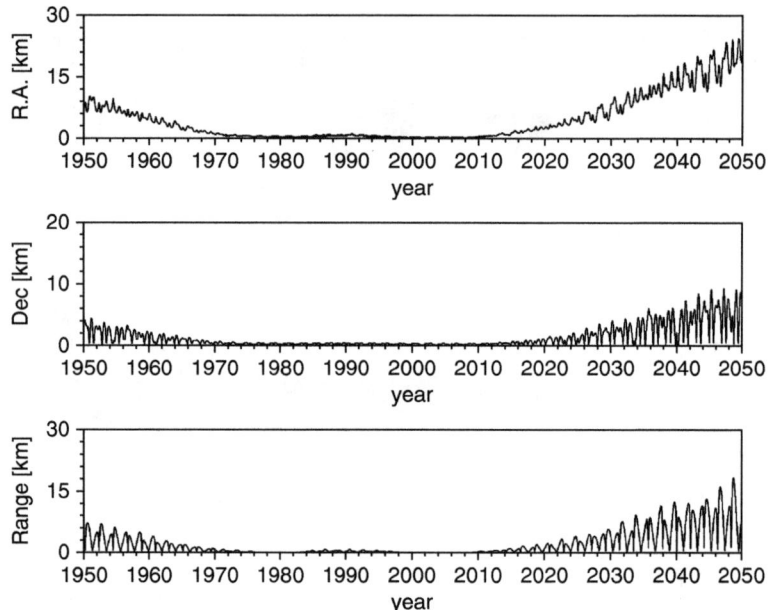

Fig. 4.15 Mars orbit uncertainties with respect to Earth; from Folkner (2010, 2011). Courtesy of JPL, NASA

times TDB or TCB (timescales used in the equations of motion). This difference, important for the accurate timing procedures (pulsar timing, Gaia), is estimated together with the planetary ephemerides. A perfect consistency between the timing of the observations and the motion of the solar system bodies is then obtained. Versions of INPOP computed in TCB are also provided together with a TCB–TCG (instead of TT–TDB) estimation. Following IAU recommendations, INPOP08 was the first planetary ephemeris solving for the mass of the Sun for a given fixed value of the astronomical unit. New estimations of the solar mass are regularly obtained since INPOP08 together with the oblateness of the Sun.

In the modeling of INPOP, developed in the PPN framework, included are the solar oblateness and the perturbations induced by more than 24, 000 asteroids as well as the tidal effects of the Earth and Moon. More than 136, 000 planetary observations (Table 4.4) including the latest Messenger, Mars Express, Venus Express, and Cassini data points and LLR normal points are used for the construction of INPOP. A fit of up to 200 parameters was needed for INPOP10a (Fienga et al. 2011). About 70 parameters related with the lunar orbit and initial rotation angles have been fitted separately by Manche et al. (2010). An iterative process between planetary and lunar adjustments is employed to keep consistency between the two parts of the ephemerides.

Based on INPOP, long-term planetary ephemerides are built for a period of validity of several million years. They have been aimed mainly at the paleoclimate reconstruction and geological timescale calibrations (Laskar et al. 2004, 2011).

4.3 Solar System Ephemerides

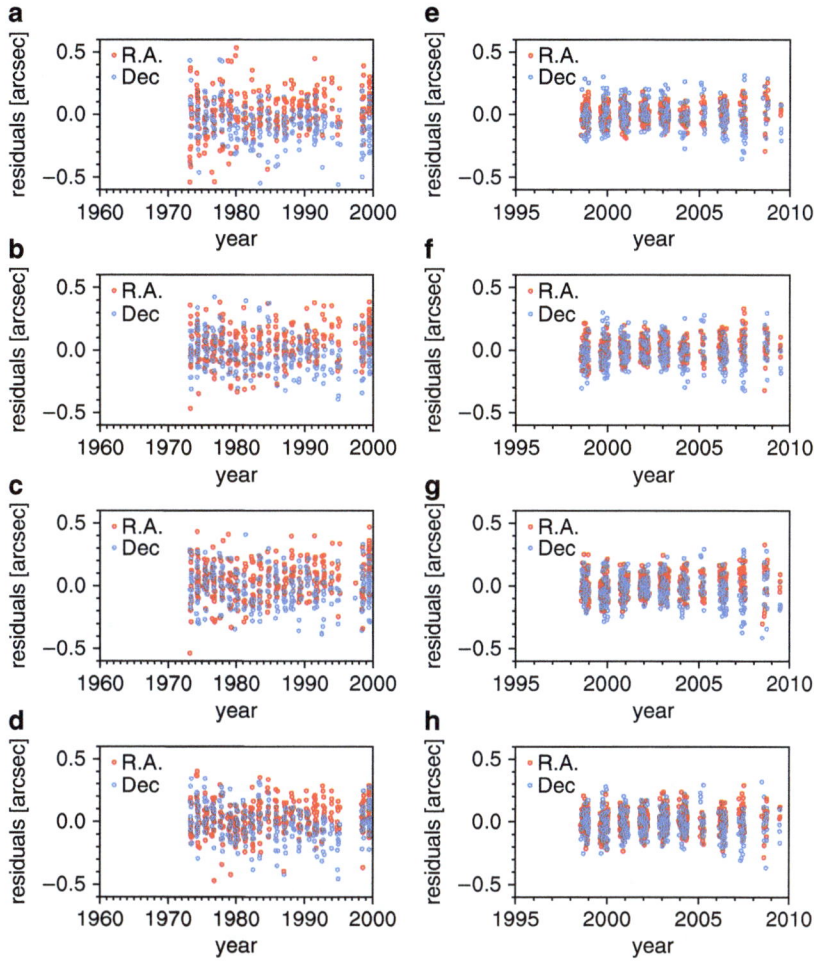

Fig. 4.16 Jupiter astrometry by position determination of Galilean moons. (**a**) Io by Nikolaev Observatory, (**b**) Europa by Nikolaev Observatory, (**c**) Ganymede by Nikolaev Observatory, (**d**) Callisto by Nikolaev Observatory, (**e**) Io by USNO Flagstaff, (**f**) Europa by USNO Flagstaff, (**g**) Ganymede by USNO Flagstaff, (**h**) Callisto by USNO Flagstaff; from Folkner (2010). Courtesy of JPL, NASA

The three charts of Fig. 4.23 show that the evolution of planetary ephemerides becomes more and more spacecraft dependent. INPOP06 was mainly driven by optical observations of outer planets and Mars tracking data. Thanks to new collaborations with ESA, Venus Express and Mars Express navigation data have been introduced in INPOP since INPOP08 (Fienga et al. 2009). With INPOP10a, positions of Mercury, Saturn, and Jupiter deduced from flybys of Messenger, Saturn, and other spacecrafts are included in the adjustment. As one can see, INPOP10a is then mainly driven by tracking the Mars spacecraft and the Venus Express

82   4 Barycentric Dynamical Reference System

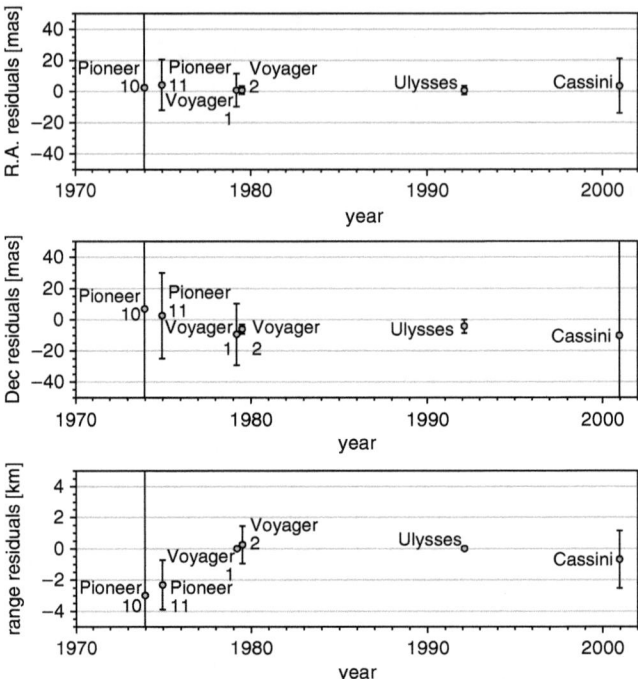

Fig. 4.17 Jupiter spacecraft data; from Folkner (2010). Courtesy of JPL, NASA

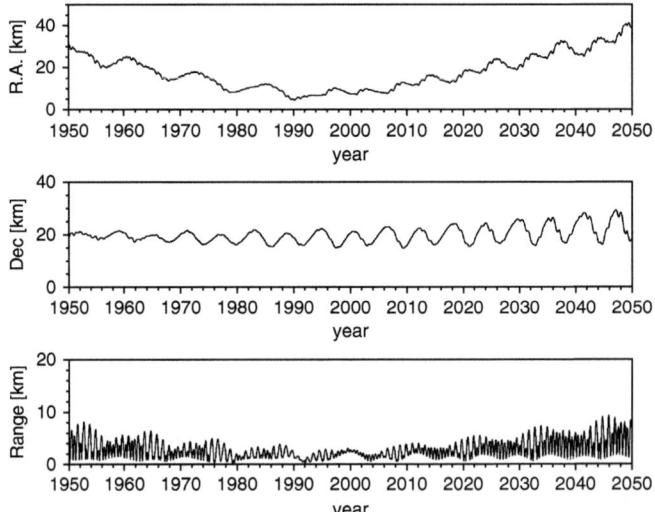

Fig. 4.18 Jupiter orbit uncertainties with respect to Earth; from Folkner (2010, 2011). Courtesy of JPL, NASA

4.3 Solar System Ephemerides

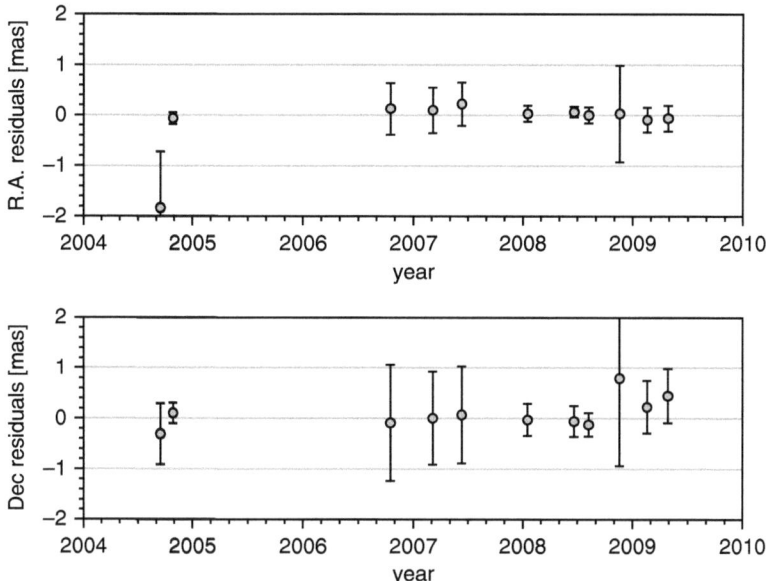

Fig. 4.19 VLBA observations of the Cassini spacecraft (Saturn); from Folkner (2010) and Jones et al. (2011). Courtesy of JPL, NASA

(VEX) data. The Jupiter and Saturn positions deduced from several flybys are also important for the fit of the ephemerides. The fact that Mars observations are now numerous and very accurate explains why the modeling of the asteroid perturbations on the inner planetary orbits is so crucial in the development of modern planetary ephemerides. Table 4.5 gives the main characteristics of the INPOP versions with an emphasis in the development of asteroid modeling. The aim here is to obtain good postfit residuals but also good extrapolation capabilities. INPOP10b (Fienga et al. 2011), the latest INPOP version, satisfies these two requirements as one can see in Fig. 4.24. This figure shows comparisons between Mars Express (MEX) observed one-way range in meters and Earth-Mars distance estimations using INPOP08, INPOP10a, INPOP10b, and DE423. Since these MEX data are not included in the fit of the planetary ephemerides, the plotted residuals give a good estimate of the extrapolation capabilities of each planetary ephemeris.

The internal accuracy reached by INPOP10a is less than 1 km for Venus and the Earth–Moon barycenter. For Mercury, Jupiter, and Saturn, the accuracy is about a few kilometers. For Mars, due to the asteroid perturbations, the accuracy is degraded to 10 km over one century and is more than 1,000 km for the other outer planets. Detailed estimations of uncertainties are given in Table 4.6 for three intervals of time.

Besides asteroid mass determinations obtained with the construction of INPOP, tests of fundamental physics are made since INPOP08 (Fienga et al. 2010, 2011).

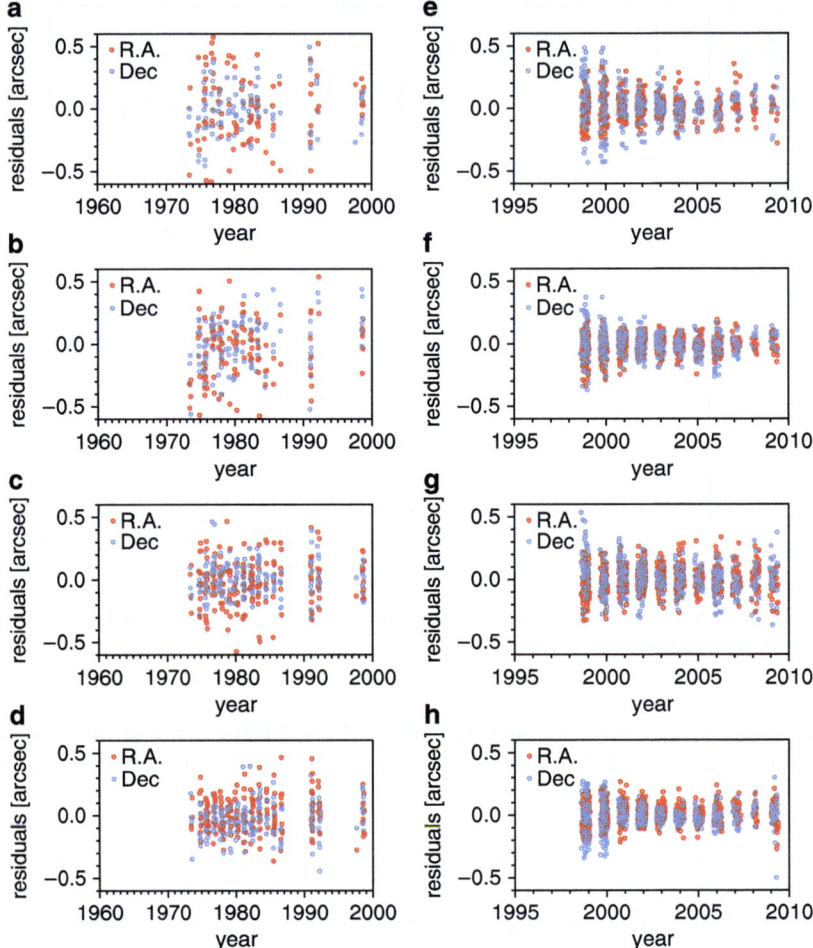

Fig. 4.20 Saturn astrometry by position determination of moons. (**a**) Tethys by Nikolaev Observatory, (**b**) Dione by Nikolaev Observatory, (**c**) Rhea by Nikolaev Observatory, (**d**) Titan by Nikolaev Observatory, (**e**) Dione by USNO Flagstaff, (**f**) Rhea by USNO Flagstaff, (**g**) Titan by USNO Flagstaff, (**h**) Iapetus by USNO Flagstaff; from Folkner (2010). Courtesy of JPL, NASA

Estimations of acceptable intervals of nonzero values of PPN parameters (β, γ) as well as possible supplementary advances of perihelia and nodes of planetary orbits give new constraints for alternative theories. Figure 4.25 illustrates the latest determinations obtained with INPOP10a and gives constraining values for (β, γ) violations of general relativity.

Finally, we would like to mention that DE, EPM, and INPOP ephemerides have been compared in the literature (see, e.g., Hilton and Hohenkerk 2011).

4.3 Solar System Ephemerides

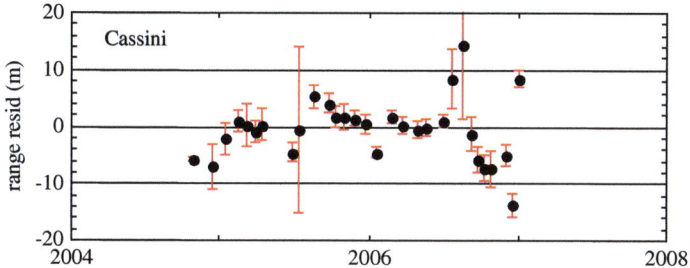

Fig. 4.21 Cassini range to Saturn (Folkner et al. 2009). Shown are residuals for DE421. Courtesy of JPL, NASA

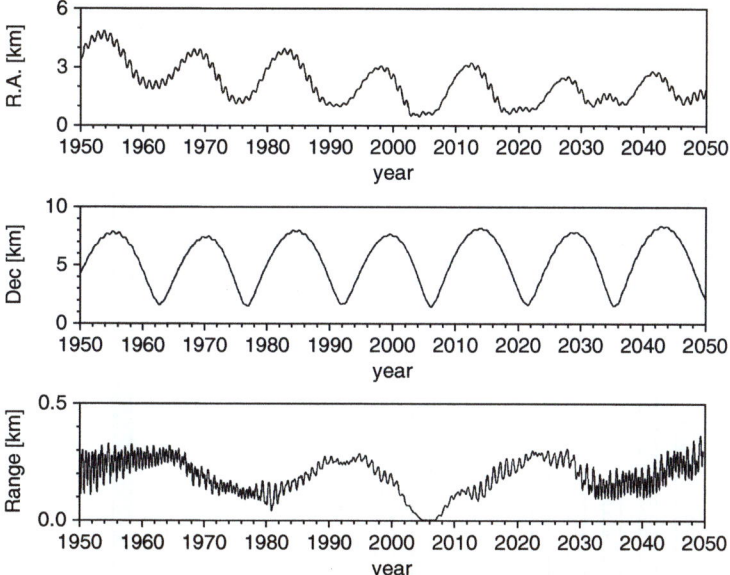

Fig. 4.22 Saturn orbit uncertainties with respect to Earth; from Folkner (2010, 2011). Courtesy of JPL, NASA

4.3.2 Semianalytical Ephemerides

For several applications, semianalytical ephemerides are of great value. Bretagnon and Francou (1988) theoretically derived series (polynomials in the time variable) for the motion of the planets named VSOP (*Variations Séculaire des Orbites Planétaires*); constants and initial conditions have been taken from DE200. Developed in 1982, VSOP82 (Bretagnon 1982; see also Bretagnon 1984) only yielded mean elements of planetary orbits. The extended theory, VSOP87 from 1987, also allowed the calculation of heliocentric and barycentric

Table 4.4 Planetary ephemerides data sets given by data types (column 1), by objects (column 2), and by mean accuracy (columns 3, 4, and 5) in right ascension α, declination δ, and geocentric distances R

		Mean accuracy			
Data type	Bodies	α (mas)	δ (mas)	R (m)	Number of observations
VLBI	Venus, Mars, Jupiter, Saturn	1–10	1–10		152
Flybys	Mercury, Jupiter, Saturn, Uranus, Neptune	0.1–1	0.1–1	1–30	43
Range tracking	Venus, Mars			2–30	56,881
Direct range	Mercury, Venus			1,000	951
Optical	Jupiter, Saturn, Uranus, Neptune, Pluto	300	300		24,800
LLR	Moon			0.05	xxx

The numbers of each type of observations used in the INPOP10a adjustment are given in column 6

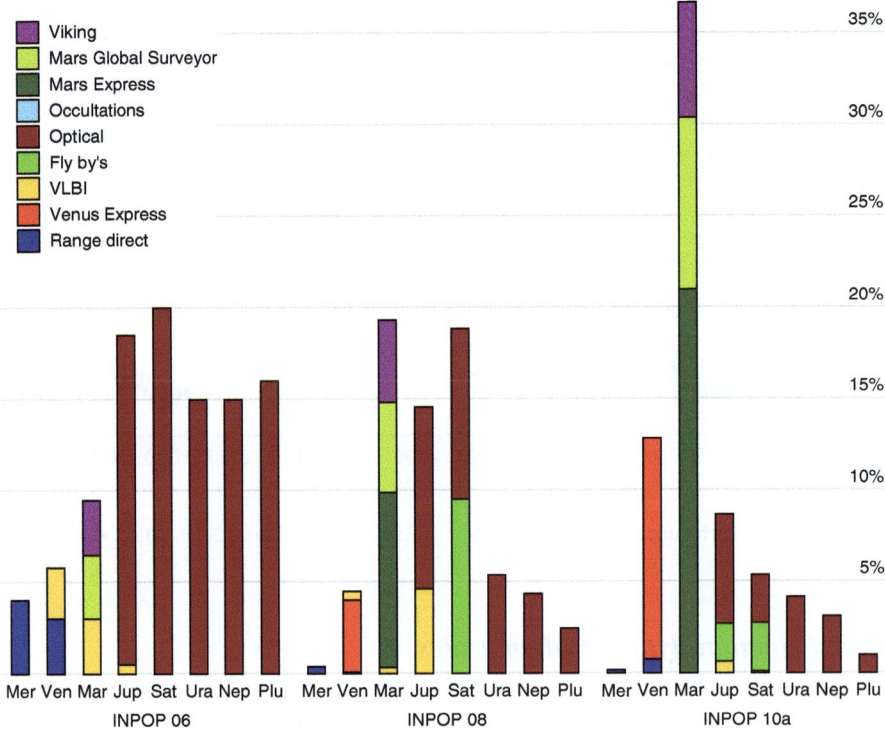

Fig. 4.23 Input data to INPOP06, INPOP08, and INPOP10a. Courtesy of Agnes Fienga

positions and velocities in various reference systems and with respect to different epochs. Different versions of VSOP87 are described in Table 4.7. Corresponding

4.3 Solar System Ephemerides

Table 4.5 Solve-for parameters for different ephemerides

Parameters	INPOP06	INPOP08	INPOP10a	INPOP10b	DE405	DE421
Asteroids	300	303	161	192	300	343
Fitted masses	5	34	145	192	3	8
Fixed masses	0	5	16	0	0	59
Asteroid ring	S	F	F	S	x	x
Densities	295	261	x	x	267	276
AU	F	S	F	F	S	S
M_E/M_M	F	S	S	F	S	S
Sun GM	F	F	S	S	F	F
Sun J2	S	S	S	F	F	F
Fit ends	2005.5	2008.5	2010.0	2010.0	1998.0	2008.0

Not modeled parameters are marked with "x," "S" marks the solved-for parameters, and "F" states fixed parameters

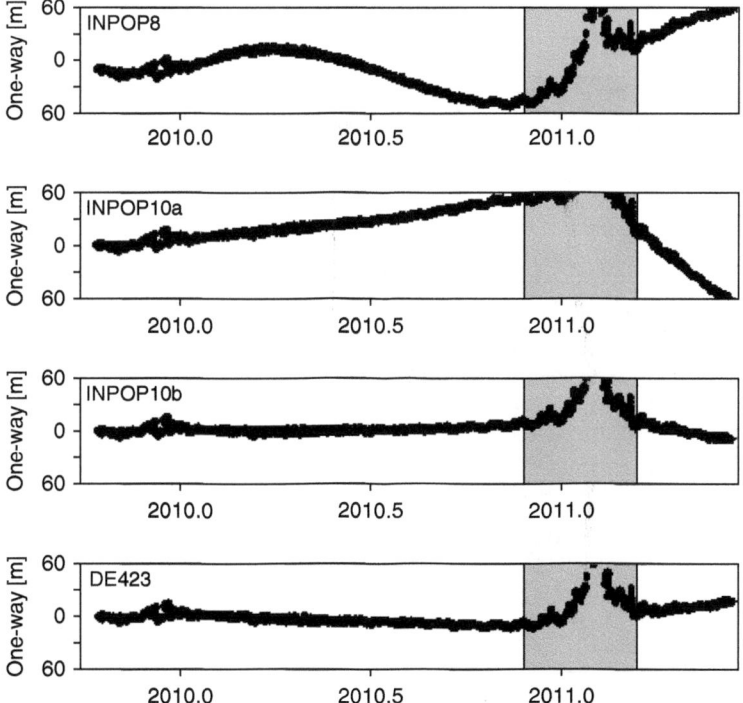

Fig. 4.24 Difference between MEX one-way range data and Earth–Mars distance estimates from various ephemerides. The *gray region* is strongly affected by the plasma of the solar corona. Courtesy of Agnes Fienga

software can be downloaded from ftp://ftp.imcce.fr/pub/ephem/planets/vsop87/. The *AstroRef* function `GetVSOPEphemeris` returns VSOP87 data.

Table 4.6 INPOP10a uncertainties over three intervals: J2000.0 ± 1 year, J2000.0 ± 10 year, and J2000.0 ± 100 year

	±1 year		±10 year		±100 year	
Body	σ (mas)	R (km)	σ (mas)	R (km)	σ (mas)	R (km)
Mercury	5	0.125	5	0.25	5	1
Venus	0.5	0.002	1	0.5	5	0.3
Mars	0.5	0.02	5	0.1	40	10
Jupiter	10	1	30	1	30	2
Saturn	0.05	0.02	0.2	0.5	1	3
Uranus	10	10	50	500	300	1,200
Neptune	7	60	25	800	300	3,000
Pluto	150	1,000	500	10,000	3,000	80,000
EMB	0.2	0.005	0.25	0.005	3	0.04

The uncertainties are given as angles σ as well as in form of geocentric (for the planets) or barycentric (for the Earth–Moon Barycenter, EMB) distance errors R

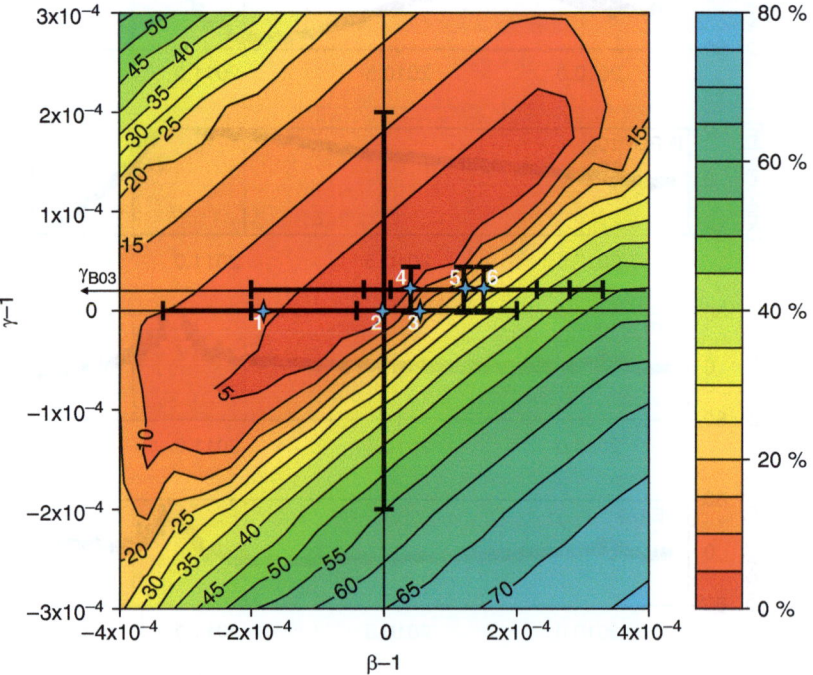

Fig. 4.25 Variations of postfit residuals obtained for different values of PPN β (x-axis) and γ (y-axis). Description of marks: PPN β values (1) obtained by Manche et al. (2010) using LLR observations with $\gamma - 1 = 0$; (2) obtained by Pitjeva (2010a) from a global fit of EPM planetary ephemerides; (3) deduced from INPOP08 (Fienga et al. 2010); (4) based mainly on Mars data analysis by Konopliv et al. (2011); (5) deduced from LLR for a fixed value of γ (Müller et al. 2008); (6) deduced from LLR for a fixed value of γ (Williams et al. 2009). The cross in the center (γ_{B03}) indicates a determination of γ during a solar conjunction in the frame of the Cassini mission by Bertotti et al. (2003). Courtesy of Agnes Fienga

4.3 Solar System Ephemerides

Table 4.7 Different versions of VSOP87

VSOP87	elliptic orbital elements, equinox and ecliptic J2000.0,
VSOP87A	heliocentric Cartesian coordinates, equinox and ecliptic J2000.0,
VSOP87B	heliocentric spherical coordinates, equinox and ecliptic J2000.0,
VSOP87C	heliocentric Cartesian coordinates, equinox and ecliptic of date
VSOP87D	heliocentric spherical coordinates, equinox and ecliptic of date
VSOP87E	barycentric Cartesian coordinates, equinox and ecliptic J2000.0.

In recent years, improved versions of VSOP appeared in the literature. VSOP2000 (Moisson and Bretagnon 2001) fitted to DE403 is 10 times better than the previous solutions VSOP82 and VSOP87, both fitted to DE200. Francou and Simon (2011) introduced further modifications and improvements, i.e., fits to the numerical integrations DE405 (JPL) and INPOP08A (IMCCE), for VSOP2010A and VSOP2010B, respectively, introduction of the perturbations coming from 298 asteroids, analytical theory of Pluto, perturbations due to Pluto, and better accuracy for the orbits of Jupiter and Saturn. In the time interval 1890–2000, the discrepancies between theories and numerical integrations are <0.04 mas for Mercury, Venus, and Earth–Moon barycenter; <0.07 mas for Jupiter; <0.3 mas for Saturn and Neptune; and <1.5 mas for Mars and Uranus. The precision is 3–10 times better than that of VSOP2000. In the time interval [−4000, +8000], the precision is a few $0''.1$ for Mercury, Venus, and Earth–Moon barycenter and a few arcseconds for the other planets. Here, the gain in precision is 5 times better for the telluric planets and 10–50 times better for the major planets in comparison to VSOP2000.

Chapter 5
Classical Astronomical Coordinates

5.1 Apparent Motion of Stars and Sun

Classical astronomical coordinates still play an important role for celestial reference systems (CRS). For their introduction we will study the apparent motion of stars and Sun. We start from an earthbound observer and will introduce some elementary facts about the apparent motion of stars and the Sun about the Earth. Figure 5.1 shows a long-exposure photograph of the night sky. There, all stars seem to revolve around the north celestial pole (NCP). This apparent motion simply results from the daily rotational motion of the Earth in space. The celestial pole is simply determined by the orientation of the Earth's rotation vector (angular velocity vector). Often astronomical directions are visualized by points on a unit sphere that we call directional or celestial sphere. The two piercing points of the rotational axis with the celestial sphere define the north and the south celestial poles.

The position of the celestial pole at the sky is determined by the celestial latitude of the observer. At the north pole the NCP stands practically in zenith direction, and the stars move parallel to the horizon that in this case coincides with the celestial equator (Fig. 5.2, upper panel). If, however, the observer is located at the equator, then the stars rise in the east vertically to the horizon, and they reach their highest point (where they culminate) in the meridian that runs through the zenith of the observer and the NCP, and they set in the west again perpendicular to the horizon plane. In the temperate zone (Fig. 5.2, lower panel), the NCP appears with an angle above the horizon that corresponds to the observer's latitude. The stars move accordingly about the pole (Fig. 5.3). Some are circumpolar and can always be seen; others rise in the east, culminate in the meridian, and set in the west.

Next we will consider the apparent path of the Sun on the celestial sphere. This path results from the motion of the Earth about the Sun in the ecliptic and the orientation of the Earth in space. As seen from Earth the Sun moves along the ecliptic in course of the year. The celestial equator is inclined by about 23.5° with respect to the ecliptic (Fig. 5.4). This angle is called obliquity of the ecliptic. The seasons essentially result from the obliquity of the ecliptic (Fig. 5.5). If we consider

Fig. 5.1 A long exposure shot of the night sky. All stars seem to revolve around the north celestial pole. Picture taken during the *11. Herzberger Teleskop Treffen* (http://www.herzberger-teleskoptreffen.de) Courtesy of U. Müller

the points of the Earth where the Sun stands in zenith direction in course of a year, then these points move between the tropic of Capricorn at the latitude of 23.5° south, across the equator to the tropic of Cancer at 23.5° north. Exactly two times a year the Sun stands exactly above the equator indicating the beginning of spring (around 21 March) and autumn (around 23 September) in the northern hemisphere. These are the vernal and autumn *equinoxes* (from Latin "aequus" and "nox" meaning "equal night" [...and day]). At the two equinoxes the Sun rises exactly in the east and sets exactly in the west (Fig. 5.6). After the vernal equinox the Sun moves towards the north implying that it rises in northeasterly direction and sets in the northwest. This motion continues until summer solstice when the Sun rises and sets maximally in northern direction; half a year later at winter solstice, the Sun rises and sets maximally in the south.

Figure 5.7 shows how the phases of the Moon come about. At new Moon, the Moon stands in the direction of the Sun and at full Moon in opposite direction so that we directly look at the illuminated lunar surface. During solar eclipses the Moon stands directly in the connecting line from the Earth towards the Sun so that Moon eclipses the image of the Sun. For the occurrence of a solar eclipse the lunar orbit in space is of relevance. It is inclined at about 5.2° with respect to the ecliptic, and the line of intersection between the lunar orbital plane and the ecliptic, called the nodal line, moves with a period of 18.6 years for a complete revolution. This motion of the lunar node has direct consequences for the occurrence of solar eclipses. As Fig. 5.8 shows, conditions for a solar eclipse are unfortunate if the Moon is not in the vicinity of the lunar node and if the node does not point towards the Sun. Without the space motion of the lunar node, favorable conditions for a solar eclipse would occur exactly two times a year. Due to the retrograde motion of the lunar node, however, the length of an eclipse year (with two favorable situations) is only 346.6 days. Nineteen eclipse years lead to a time span of 6585.8 days that is comparable with the duration of 223 months of 29.5 days. This implies that every 18 years and 11 days, the geometry of Sun, Earth, and Moon in space repeats leading to equivalent conditions for a solar eclipse. This time span is called Saros Cycle and

5.1 Apparent Motion of Stars and Sun

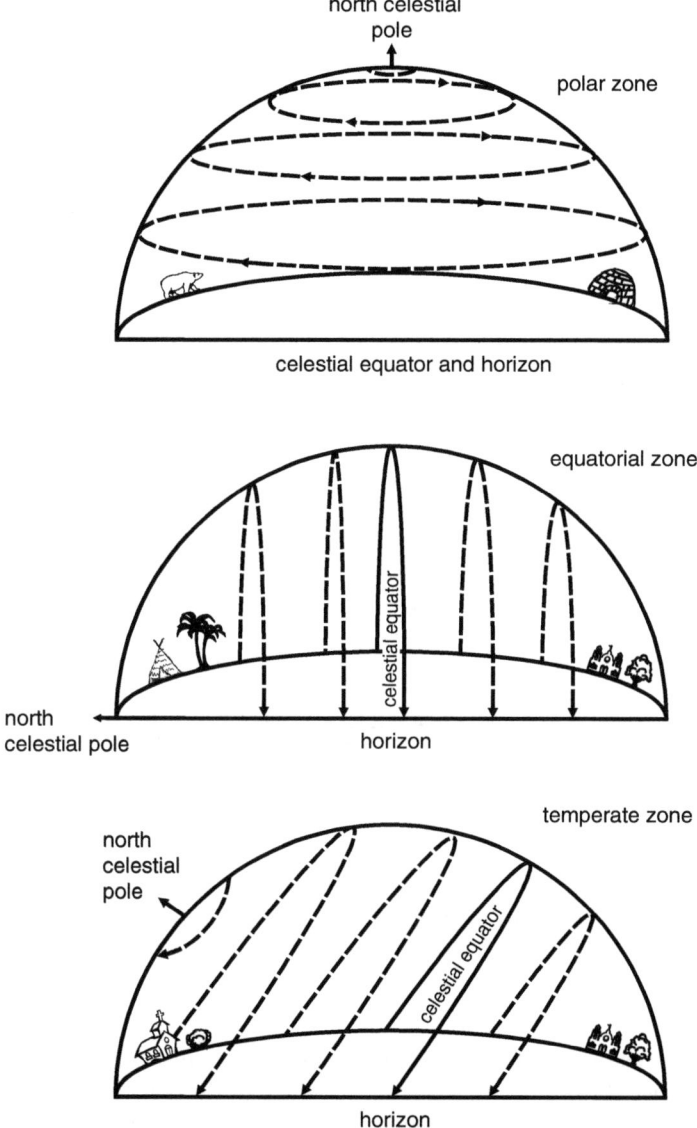

Fig. 5.2 The apparent motion of stars depends upon the observer's latitude. *Upper panel*, observer at the north pole; *middle panel*, observer at the equator; *lower panel*, observer in the temperate zone

was already known to the old Chaldeans. A certain solar eclipse very likely will occur again after one Saros Cycle.

Finally we would like to mention that the position of the (north) celestial pole varies slowly with time due to the precessional motion of the Earth's rotation axis. The whole Earth represents a kind of astronomical gyroscope that, due to its flatten-

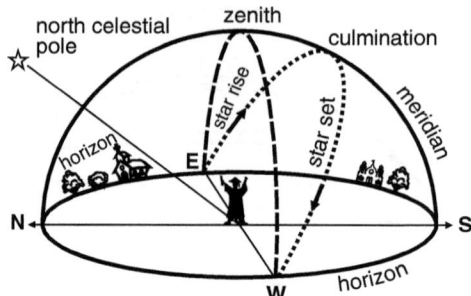

Fig. 5.3 For an observer on the ground, a position on the celestial sphere can be described with a few simple coordinates. Besides the north celestial pole (NCP), the zenith of the observer, defined by the direction of the plumb line, is an important label on the celestial sphere

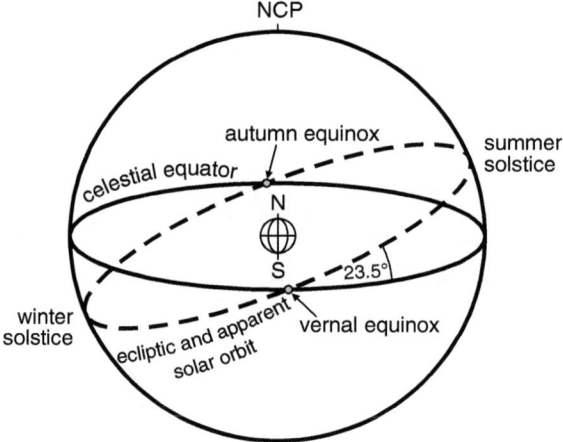

Fig. 5.4 The Earth's rotation axis is tilted with respect to the normal to the ecliptic by about 23.5°. On the celestial sphere ecliptic and celestial equators intersect in two points: the vernal and autumn equinoxes

ing, experiences torques from Sun, Moon, and planets that cause the precessional motion in space (Fig. 5.9). Due to precession the Earth's rotation axis moves around the normal to the ecliptic with a period of 25,800 years (the Platonic year). The Earth's rotation axis that presently points towards our pole star Polaris (α Ursae Minoris) rambles through the constellations. At 3000 BC, Thuban (α Draconis) was pole star; in the year 14000, Vega (α Lyrae) will be close to the NCP.

5.2 Spatial Coordinates

The elementary classical astronomical coordinate systems are the horizon system, the equatorial coordinates of the first and second kind, and the ecliptic system

5.2 Spatial Coordinates

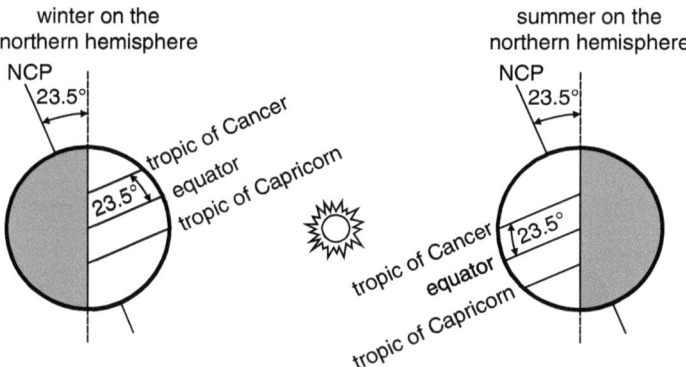

Fig. 5.5 The tilt of the Earth's rotation axis with respect to the ecliptic is the main reason for the seasons. Between the beginning of spring and the beginning of autumn, the solar rays fall perpendicular on points on the northern hemisphere; during the remaining part of the year, they fall perpendicular on points on the southern hemisphere. In this way the seasons are related with the angle between the incident solar rays and the zenith direction

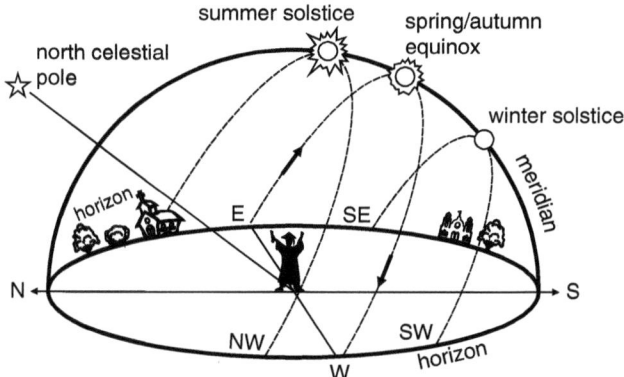

Fig. 5.6 The locations of sunrise and sunset vary in course of a year. At summer solstice the Sun rises in the maximal northerly direction of the eastern horizon and sets maximally in the north of the western horizon. At winter solstice the Sun rises deep in the southeast, remains deep at the southern sky during noonday, and sets maximally in the southern direction at the western horizon

(Tab. 5.1). Though each of these systems is described by means of two angles, it is important to realize that these are just polar angles of some Cartesian coordinate system. Such a Cartesian system has an origin and a handedness. The handedness can be right or left handed according to the possibility to present the x, y and z-axes with thumb, index, and middle finger of the right or left hand. Usually the z-axis of such a system is defined by its pole on the celestial sphere; the x–y-plane is called

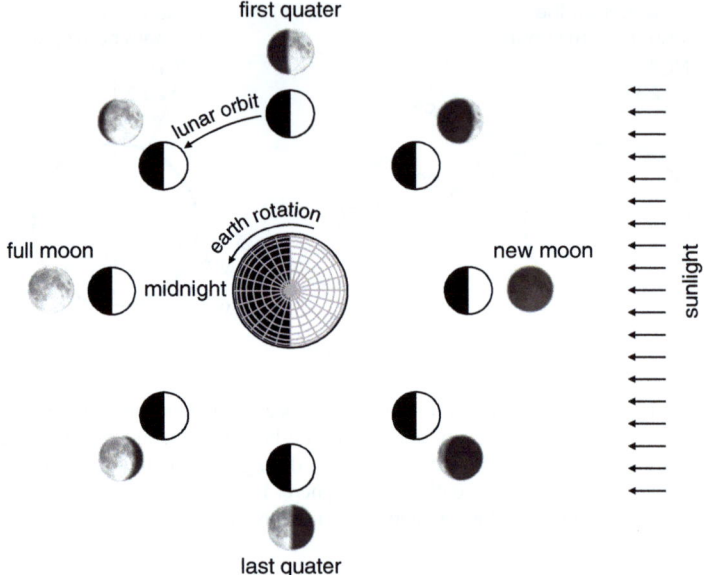

Fig. 5.7 Phases of the Moon including views for the northern hemisphere

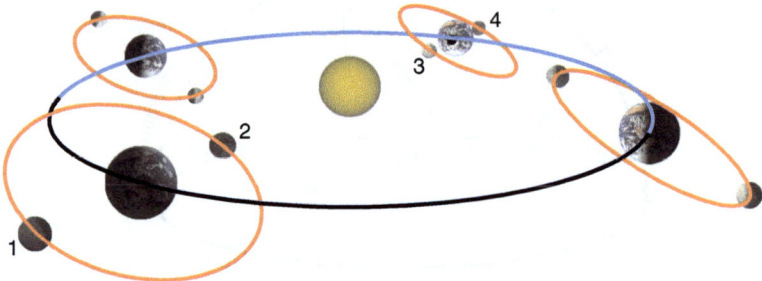

Fig. 5.8 The lunar orbit is inclined by about 5.2° with respect to the ecliptic. If the Moon is found outside the ecliptic, conditions for a solar eclipse are poor. A solar eclipse will only occur if the lunar node points towards the Sun. Only at the points 2 and 3 a solar eclipse can occur, a lunar eclipse at the points 1 and 4

equator. To visualize such a coordinate system, often, the concept of a directional (unit) sphere is used that might be called celestial sphere.

In astronomical spatial reference systems the direction of spatial coordinates is determined by the directions to selected astronomical objects (planets, stars, quasars, etc.). Usually only direction angles play a role (SLR and LLR are exceptions) and not the distance. In that case the position is well represented by a point on our celestial sphere.

5.2 Spatial Coordinates

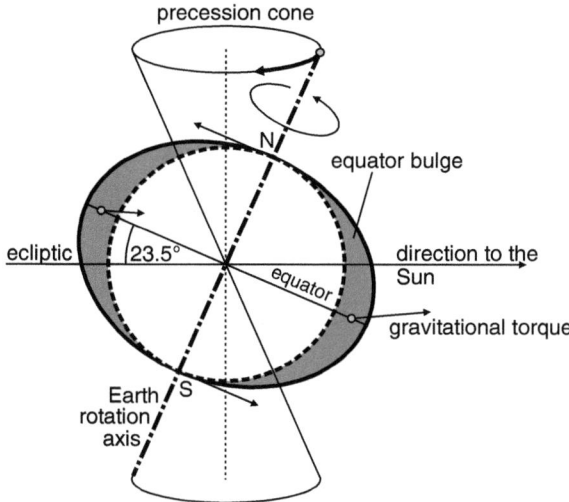

Fig. 5.9 Precessional motion of the Earth's rotation axis. As a result the (north) celestial pole moves on the celestial sphere. Presently Polaris (αUMi) is our pole star; in the year 14000, NCP will be close to Vega (αLyr) due to precession

Table 5.1 Overview over the various astronomical reference systems

System	Reference plane		Parameters measured from the reference plane	
	Primary	Secondary	Primary	Secondary
Horizon system	Celestial horizon	Celestial meridian (half containing north pole)	Elevation $-90° \leq a \leq +90°$ (+zenith)	Azimuth $0° \leq A < 360°$ (+east)
Equatorial system (1st kind)	Celestial equator	Hour circle of observer's zenith (half containing zenith)	Declination $-90° \leq \delta \leq +90°$ (+north)	Hour angle $0^h \leq h < 24^h$ $0° \leq h < 360°$ (+west)
Equatorial system (2nd kind)	Celestial equator	Equinoctial colure (half containing vernal equinox)	Declination $-90° \leq \delta \leq +90°$ (+north)	Right ascension $0^h \leq \alpha < 24^h$ $0° \leq \alpha < 360°$ (+east)
Ecliptic system	Ecliptic	Ecliptic meridian of the equinox (half containing vernal equinox)	(Ecliptic) latitude $-90° \leq \beta \leq +90°$ (+north)	(Ecliptic) longitude $0° \leq \lambda < 360°$ (+east)

5.2.1 Labels on the Celestial Sphere

Think about an observer on the Earth's surface observing the night sky. As simple tools for his orientation, he has the direction of the local plumb line and the celestial

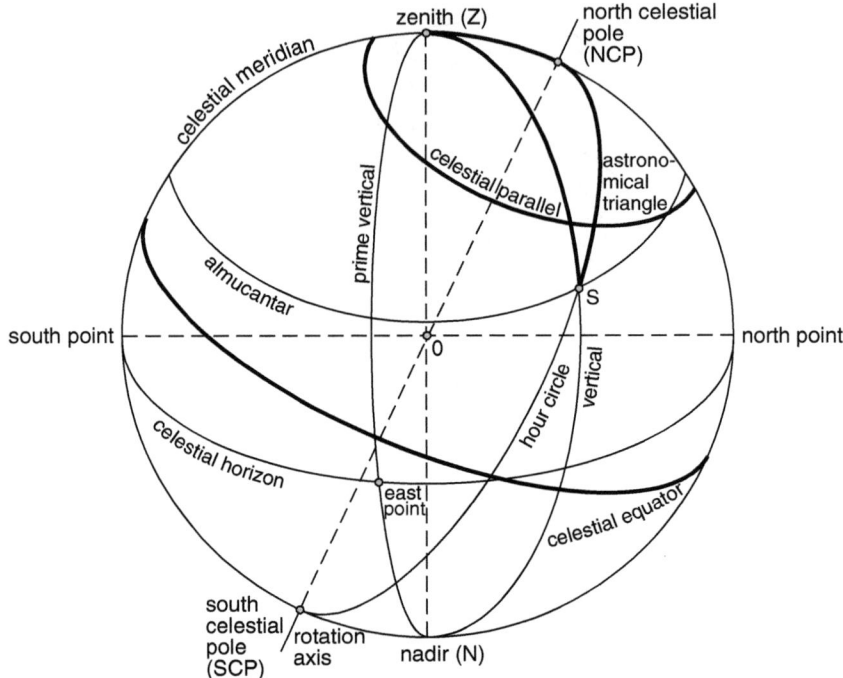

Fig. 5.10 Labels on the celestial sphere defined by the plumb line and the rotation axis of the Earth

pole about which the stars seem to move. In simple instruments the direction of the local plumb line is realized by means of a surveyor's level, a mercury level, or some compensator, e.g., in a theodolite. The Earth rotation axis defines the two poles on the celestial sphere, i.e., the north and south celestial pole. The plane perpendicular to the Earth rotation axis running through the geocenter is called the celestial equator. A great circle through the poles is called hour circle. A small circle parallel to the celestial equator is called celestial parallel (Fig. 5.10).

The direction of the local plumb line defines the zenith (upwards) and the nadir (downwards). The plane perpendicular to the direction of the plumb line through the observer (or through the geocenter) is called celestial horizon. A plane perpendicular to the celestial horizon is called vertical plane. A small circle parallel to the celestial horizon is called almucantar. The very vertical plane that contains the pole is called celestial meridian; it is also the hour circle through the zenith. The north and south points in the celestial horizon are defined by the two intersections of the celestial meridian with the horizon. The vertical plane perpendicular to the meridian is the prime vertical plane. The intersections of the prime vertical plane with the horizon are the east and the west points. These are also the intersection points of horizon and celestial equator.

5.2 Spatial Coordinates

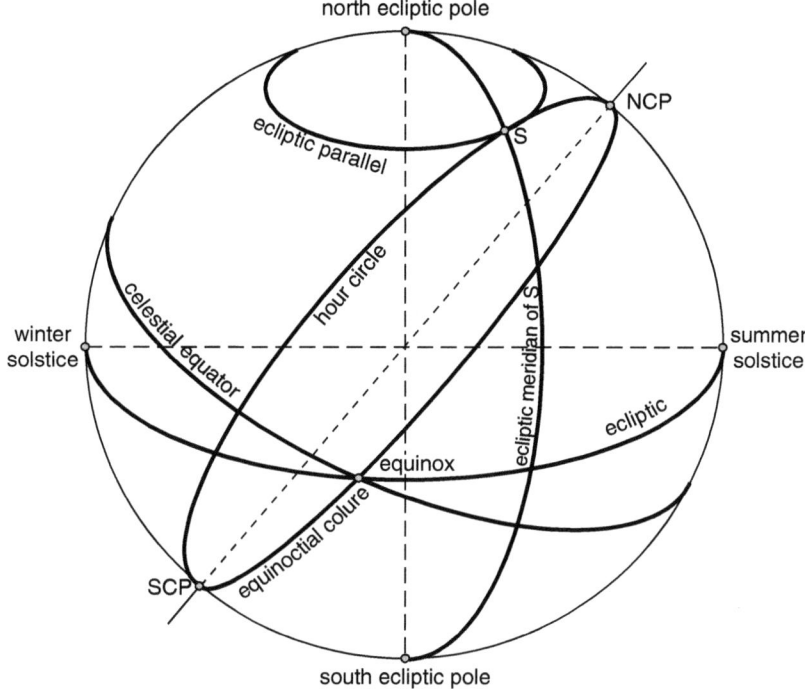

Fig. 5.11 Ecliptic labels on the celestial sphere

Let S be some point on the celestial sphere determined, e.g., by a star. Segments of the meridian, the hour circle, and the vertical circle running through S then define a spherical triangle: the astronomical or nautical triangle (Fig. 5.17) of S. The vertices of the astronomical triangle are the zenith Z of the observer, the NCP, and the point S.

The ecliptic plane is defined by the heliocentric position and velocity vectors of the Earth–Moon barycenter after corrections for certain periodic perturbations due to Venus and Jupiter. It always stays within $2''$ from the apparent path of the Sun about the Earth. The ecliptic normal through the geocenter defines the two ecliptic poles on the celestial sphere. A plane perpendicular to the ecliptic defines some ecliptic parallel on the celestial sphere. A great circle running through the ecliptic pole and the point S is called ecliptic meridian. The ecliptic intersects the celestial equator in a line that connects the two equinoctial points. One of the equinoxes, the vernal equinox, is of paramount importance in classical astronomy since it defines the classical astronomical x-axis of some equatorial or ecliptical coordinate system. The hour circle running through the vernal equinox is called equinoctial colure. The angle between celestial equator and the ecliptic is the obliquity of the ecliptic (Fig. 5.11).

Fig. 5.12 A right-handed Cartesian coordinate system and corresponding spherical angles

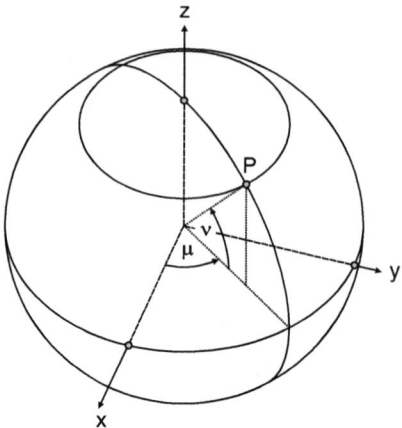

5.2.2 Cartesian and Spherical Coordinates

In the following, several astronomical spatial coordinate systems will be introduced. Each of these systems basically is a Cartesian coordinate system (x, y, z). Such a coordinate system has an origin ($\mathbf{x} = \mathbf{0}$) first of all. For the horizon coordinates the origin is primarily given by the location of the observer, the topocenter. However, when we discuss the relations between different astronomical coordinates, we will often assume the origin of the horizon system to coincide with the geocenter. Note that in that case, corresponding corrections have to be applied to a stellar position. The ecliptic system primarily has its origin in the barycenter of the solar system. However, for the sake of comparisons between coordinates, we will shift the origin into the geocenter.

Besides an origin every Cartesian coordinate system has its handedness. The intersection of the z-axis of some Cartesian coordinate system with some unit sphere often is called its pole, while the x–y-plane is called its equator or more common the primary reference plane.

Usually the position of a point P on the celestial sphere is indicated by two angles, say μ and ν (Fig. 5.12). The first angle μ is defined in the primary reference plane reckoned from some x-axis up to a secondary reference plane perpendicular to the primary reference plane and containing P. The second angle ν is defined in that secondary reference plane and reckoned from the primary reference plane. Such two angles are nothing but the spherical coordinates of a Cartesian system. If the two spherical angles are given, the Cartesian coordinates (the directional cosines) of the corresponding point on the unit sphere are given by

$$\begin{pmatrix} x \\ y \\ z \end{pmatrix} = \begin{pmatrix} \cos \nu \cos \mu \\ \cos \nu \sin \mu \\ \sin \nu \end{pmatrix}.$$

5.2 Spatial Coordinates

The *AstroRef* functions `VectorToSpherical` and `SphericalToVector` provide the transformation between the Cartesian and the spherical coordinate system (see Appendix B).

5.2.3 Horizon Coordinates

Basically the origin of the horizon system is the topocenter of some observer. Conceptually it is usually displaced into the geocenter by applying certain corrections to the positions of astronomical objects. The pole is given by the direction of the local plumb line, the z-axis pointing in upward direction. Conceptually the plumb line is realized by some level or some compensator in the observing instrument. The equator of the horizon system is called celestial horizon (the primary reference plane). The point where the z-axis hits the celestial sphere is called zenith. Every great circle emerging from the zenith hitting the celestial horizon perpendicularly is called a vertical circle. A vertical circle of special importance is the celestial meridian; it runs through the zenith and the NCP. The NCP is the point on the celestial sphere that is given by the rotation axis of the Earth on the northern hemisphere. The meridian hits the celestial horizon in two points: the north and the south points. The angle between the zenith and a star along the corresponding vertical circle is called *zenith distance* z; its complement counted upwards from the horizon is called *elevation a*

$$a = 90° - z. \tag{5.1}$$

The elevation is positive (negative) for positions above (below) the horizon. The second horizon angle counted in the horizon is the *azimuth* A; it counts from the north point in eastward direction. This implies that the x-axis of the horizon system points towards the north point and the y-axis to the east (i.e., in the direction where the equatorial angle takes the value of $+90°$); it is therefore a left-handed system. In Fig. 5.13, the horizon coordinates are depicted.

5.2.4 Equatorial Coordinates of the First Kind

The pole of both equatorial systems (of first and second kind) is the NCP; the corresponding equatorial plane is the celestial equator. Any great circle emerging from the NCP and hitting perpendicularly the celestial equator is called *hour circle*. The angle between the celestial equator and a star along the hour circle through the star is called *declination*, the first equatorial coordinate. The meridian hits the

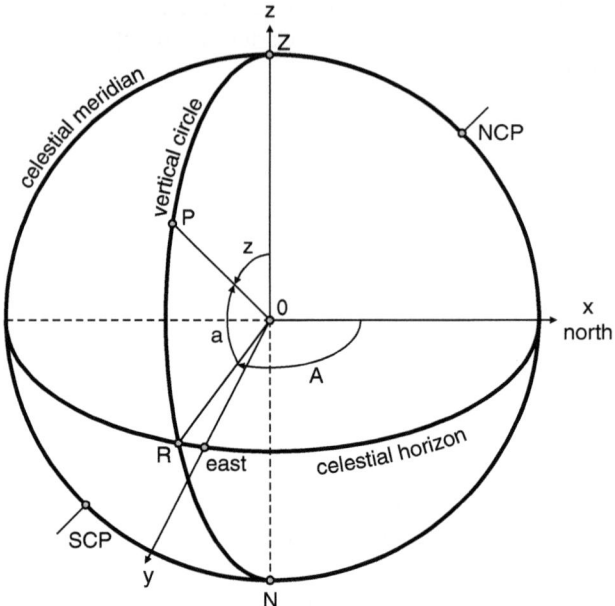

Fig. 5.13 Horizon coordinates; after Mueller (1969)

celestial equator in the corresponding north and south point. The second angle of the equatorial system of the 1st kind is the *hour angle h*; it is counted in the celestial equator from the south point in westward direction. This system, therefore, is left handed. Usually the hour angle is formally counted in (angle) hours, minutes, and seconds, where by definition 24 (angle) hours correspond to an angle of 360°, i.e.,

$$24^h \cong 360° \qquad 1^h \cong 15° \qquad 1^m \cong 15' \qquad 1^s \cong 15''.$$

In the equatorial coordinate system of the first kind (see Fig. 5.14), the z-axis points towards the NCP, the x-axis towards the south in the celestial equator, and the y-axis towards the west.

Note that in order to improve readability, we use a non-SI standard syntax for the unit of measurement symbols of hours, minutes, and seconds with a superscripted "h," "m," and "s," respectively. This is especially useful for the notation of angles in sexagesimal time format, like, e.g. $12^h 13^m 59^s$.

5.2.5 Equatorial Coordinates of the Second Kind

If we think about the problem of a stellar catalog, there clearly one faces the need for astronomical coordinates that are independent of the observer's location. This is true

5.2 Spatial Coordinates

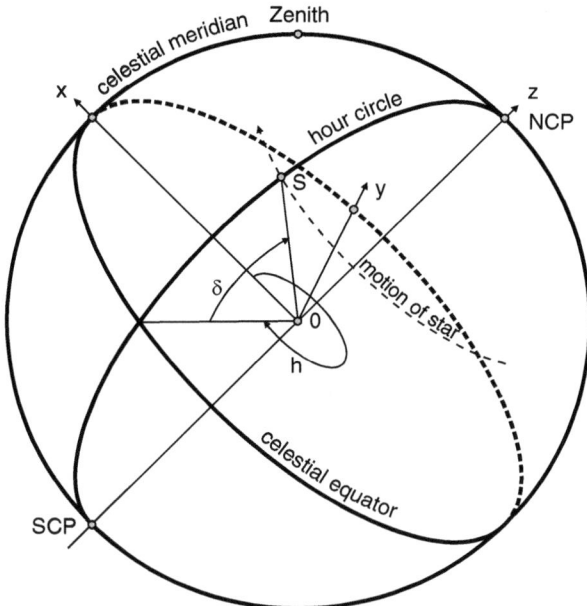

Fig. 5.14 Equatorial coordinates of the first kind; after Mueller (1969)

for the declination but not for the hour angle since it depends upon the observer's meridian. For that reason, the x-axis of the equatorial system of the second kind is defined by relation of the apparent position of the Sun.

The x-axis of the equatorial coordinates of the second kind points in the direction of the *vernal equinox*. The corresponding angle in the celestial equator is called *right ascension* α and is counted in eastward direction. The (α, δ) coordinate system therefore is right handed. The angle α is usually counted in (angle) hours, minutes, and seconds. The equatorial coordinates of the second kind are depicted in Fig. 5.15.

5.2.6 Ecliptic Coordinates

Ecliptic coordinates, i.e., *ecliptic longitude* λ and *ecliptic latitude* β, are constructed in analogy to (α, δ). The pole now is the ecliptic pole, and the primary reference plane is the ecliptic. Again, the x-axis points towards the vernal equinox, and it is a right-handed system. Ecliptic coordinates are depicted in Fig. 5.16.

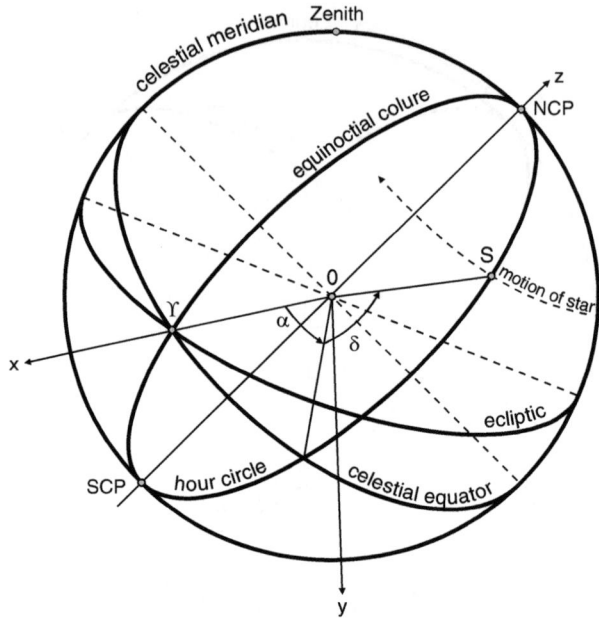

Fig. 5.15 Equatorial coordinates of the second kind; after Mueller (1969)

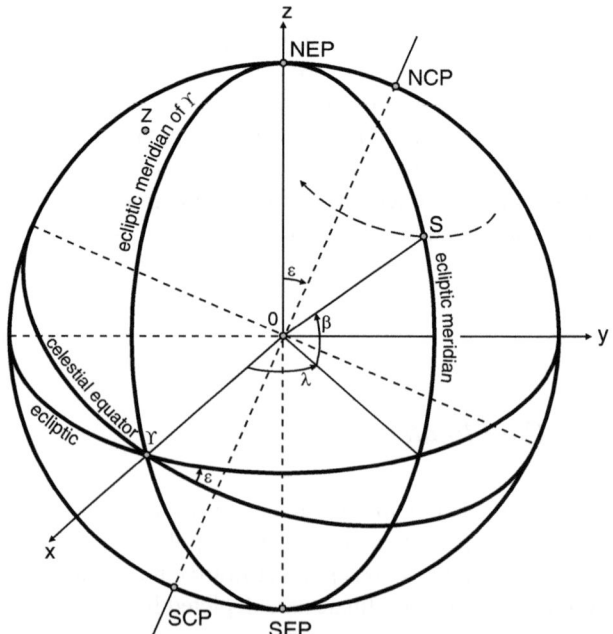

Fig. 5.16 Ecliptic coordinates; after Mueller (1969)

5.3 Relations Between Astronomical Coordinates

The relations between the various astronomical coordinates can be derived with the methods of spherical trigonometry on the celestial (unit) sphere or by means of rotational matrices applied to the corresponding Cartesian coordinates.

5.3.1 Relations Between (A, z) and (h, δ)

Let us consider the astronomical (or nautical) triangle in (Fig. 5.17). Since the azimuth A is reckoned from the north towards the east and the hour angle h from the south towards the west, the inner angles at Z (zenith) and NCP depend if the star is found west or east of the meridian. If the star is found west of the meridian, then the inner angles are h at NCP, $360° - A$ at Z, and the so-called parallactic angle p at S. If, however, the star is found east of the meridian, the inner angles are $24^h - h$ at NCP, A at Z, and p at S.

We now come to the sides of the astronomical triangle. The astronomical latitude is reckoned from the celestial equator along an hour circle to the zenith of the observer, hence the side between NCP and Z equals $90° - \Phi$. The declination of a star is reckoned from the celestial equator along an hour circle through S, hence the side between NCP and S equals $90° - \delta$. Finally the side between Z and S is given by the zenith distance z.

Let us now consider a star west of the meridian. From the law of sines for spherical triangles, we get

$$\frac{\sin(360° - A)}{\sin h} = \frac{\sin(90° - \delta)}{\sin z}$$

or since $\sin(360° - A) = -\sin A$ and $\sin(90° - \delta) = \cos \delta$

$$\sin z \sin A = -\cos \delta \sin h. \qquad (5.2)$$

Application of the law of cosines yields

$$\cos z = \cos(90° - \Phi)\cos(90° - \delta) + \sin(90° - \Phi)\sin(90° - \delta)\cos h$$

or

$$\cos z = \sin \Phi \sin \delta + \cos \Phi \cos \delta \cos h. \qquad (5.3)$$

Finally, the law of sine-cosine leads to

$$\sin z \cos(360° - A) = \cos(90° - \delta)\sin(90° - \Phi)$$
$$- \sin(90° - \delta)\cos(90° - \Phi)\cos h$$

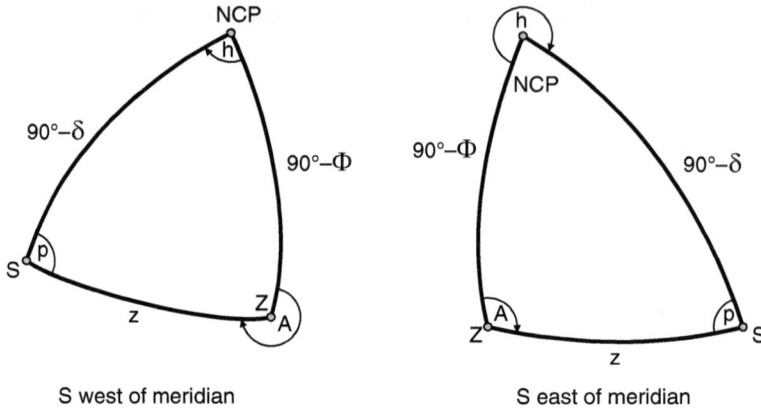

Fig. 5.17 The astronomical triangle. *Left*: star S west of meridian. *Right*: star S east of meridian

or

$$\sin z \cos A = \sin \delta \cos \Phi - \cos \delta \cos h \sin \Phi. \qquad (5.4)$$

The observer's (astronomical) latitude Φ is assumed to be known. Then relations (5.2)–(5.4) provide the transformation from (h, δ) to (A, z). Relation (5.3) provides the zenith distance z, and the azimuth A can be derived from (5.2) and (5.4) according to

$$\tan A = \frac{-\cos \delta \sin h}{\sin \delta \cos \Phi - \cos \delta \sin \Phi \cos h}.$$

For the inverse relations we write (5.2) in the form

$$\cos \delta \sin h = -\sin z \sin A. \qquad (5.5)$$

If (5.3) is multiplied by $\sin \Phi$ and (5.4) by $\cos \Phi$ and the results are added, one finds

$$\cos z \sin \Phi + \sin z \cos A \cos \Phi = \sin^2 \Phi \sin \delta + \sin \Phi \cos \Phi \cos \delta \cos h$$
$$+ \cos^2 \Phi \sin \delta - \sin \Phi \cos \Phi \cos \delta \cos h$$
$$= \sin \delta,$$

i.e.,

$$\sin \delta = \cos z \sin \Phi + \sin z \cos A \cos \Phi. \qquad (5.6)$$

Finally, multiplying (5.4) by $\sin \Phi$ and subtracting from (5.3) multiplied by $\cos \Phi$ leads to

$$\cos z \cos \Phi - \sin z \cos A \sin \Phi = \sin \Phi \cos \Phi \sin \delta + \cos^2 \Phi \cos \delta \cos h$$
$$- \sin \Phi \cos \Phi \sin \delta + \sin^2 \Phi \cos \delta \cos h$$

5.3 Relations Between Astronomical Coordinates

or
$$\cos \delta \cos h = \cos z \cos \Phi - \sin z \cos A \sin \Phi. \tag{5.7}$$

From relations (5.5) to (5.7), we get (δ, h) for given values of (A, z).

From these relations we can learn a lot about the apparent motion of the stars. Due to the rotational motion of the Earth every star apparently moves from east to west. The zenith distance or elevation depends upon the hour angle according to relation (5.3) for given values of δ and φ. For that reason the apparent stellar trajectory is symmetric about the meridian that is given by $h = 0$. Differentiation of (5.3) with respect to h leads to

$$-\sin z \frac{dz}{dh} = -\cos \Phi \cos \delta \sin h$$

or
$$\frac{dz}{dh} = \frac{\cos \Phi \cos \delta \sin h}{\sin z}. \tag{5.8}$$

Now, $\cos \Phi$, $\cos \delta$, and $\sin z$ are always positive, and we get:

- $h < 180° \cong 12^h$: zenith distance increases
- $h > 180° \cong 12^h$: zenith distance decreases

In the first case the star is found in the west on the setting side and in the second case, in the east on the rising side. For $h = 0$, when the star crosses the meridian, it attains the largest elevation above the horizon. This is its upper *culmination* (Fig. 5.18). For a meridian passage ($h = 0$), we have

$$\cos z = \sin \Phi \sin \delta + \cos \Phi \cos \delta = \cos(\Phi - \delta)$$

or
$$z = \Phi - \delta, \tag{5.9}$$

if the star culminates south of the zenith. If it culminates north of the zenith,

$$z = \delta - \Phi. \tag{5.10}$$

5.3.2 Relations Between (h, δ) and (α, δ)

The relation between the hour angle h and right ascension α is determined by an angle called *local apparent sidereal time* (LAST). LAST is defined by the hour angle of the actual true equinox Eqx:

$$\text{LAST} = h(\text{Eqx}).$$

From Fig. 5.19, we see that hour angle h and right ascension α of an arbitrary object at the celestial sphere are linked to LAST by

$$\alpha = \text{LAST} - h.$$

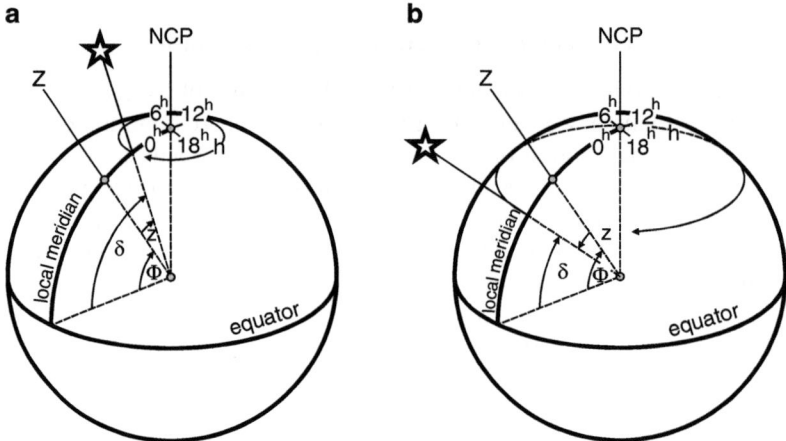

Fig. 5.18 (a) Upper culmination north of the zenith ($\delta > \Phi$). (b) Upper culmination south of the zenith ($\delta < \Phi$)

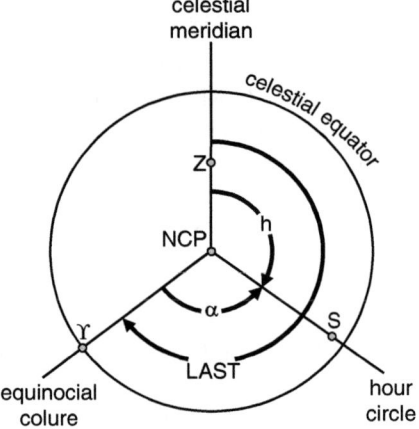

Fig. 5.19 Sidereal time

5.3.3 Rotational Matrices

To relate different Cartesian coordinates with the same origin and handedness, one may effectively use rotational matrices. Let us consider two right-handed Cartesian coordinate systems with the same origin. First we will assume that the z-axes of the two systems coincide. This situation is depicted in Fig. 5.20.

A rotation with positive angle should be carried out counterclockwise if we look from positive z-values on the x–y-plane. In Fig. 5.20a, the situation is depicted for a point on the old x-axis. For such a point, $\cos\theta = x'/x$ and $\sin\theta = -y'/x$, i.e.,

5.3 Relations Between Astronomical Coordinates

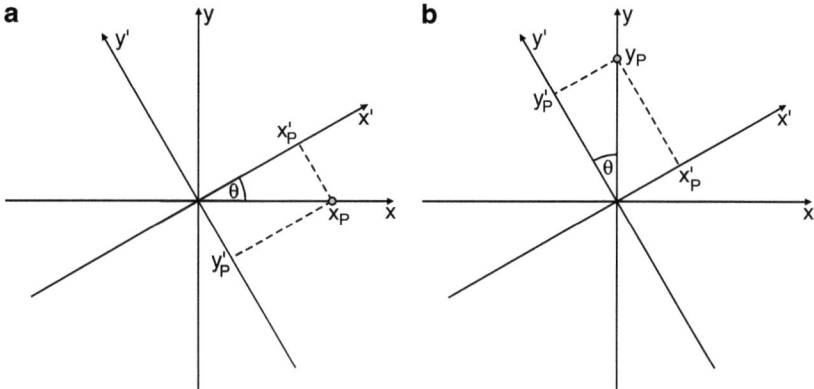

Fig. 5.20 Positive rotation of a fixed angle θ about the z-axis. (**a**) A point on the old x-axis; (**b**) a point on the old y-axis

$$x' = x \cos \theta ; \qquad y' = -x \sin \theta.$$

For a point on the old y-axis the situation is shown in Fig. 5.20b. We see that in this case, $\sin \theta = x'/y$ and $\cos \theta = y'/y$ so that

$$x' = y \sin \theta ; \qquad y' = y \cos \theta.$$

In the general case we get

$$\begin{aligned} x' &= x \cos \theta + y \sin \theta , \\ y' &= -x \sin \theta + y \cos \theta. \end{aligned} \qquad (5.11)$$

For a rotation about the common z-axis obviously $z' = z$ so that the 3-dimensional rotation of the position vector **x** about the z-axis can be described with the rotational matrix:

$$\mathcal{R}_z(\theta) = \begin{pmatrix} \cos \theta & \sin \theta & 0 \\ -\sin \theta & \cos \theta & 0 \\ 0 & 0 & 1 \end{pmatrix} \qquad (5.12)$$

by

$$\mathbf{x}' = \mathcal{R}_z(\theta)\,\mathbf{x}. \qquad (5.13)$$

The corresponding rotational matrices about the x- and y-axes read

$$\mathcal{R}_x(\theta) = \begin{pmatrix} 1 & 0 & 0 \\ 0 & \cos \theta & \sin \theta \\ 0 & -\sin \theta & \cos \theta \end{pmatrix} \qquad (5.14)$$

and
$$\mathcal{R}_y(\theta) = \begin{pmatrix} \cos\theta & 0 & -\sin\theta \\ 0 & 1 & 0 \\ \sin\theta & 0 & \cos\theta \end{pmatrix}. \tag{5.15}$$

Often, we write $\mathcal{R}_1, \mathcal{R}_2, \mathcal{R}_3$ instead of $\mathcal{R}_x, \mathcal{R}_y, \mathcal{R}_z$. Easy access on these matrices in Maple^{TM} is given by the function RotMatrix of the *AstroRef* package.

Exercise 5.1: Calculate the three rotational matrices $\mathcal{R}_1, \mathcal{R}_2$ and $\mathcal{R}_3$ for the angle $\theta = 2.7$ rad.

Exercise 5.2: Calculate the products $\mathcal{R}_1(2.7\,\text{rad}) \cdot \mathcal{R}_2(3.1\,\text{rad})$ and $\mathcal{R}_2(3.1\,\text{rad}) \cdot \mathcal{R}_1(2.7\,\text{rad})$! Compare your results!

Exercise 5.3: Calculate the rotational matrix for the reverse rotation about the x-axis $\mathcal{R}_1(-2.7\,\text{rad})$ and compare your result to the forward rotation from Exercise 5.1!

We will now derive the relation between $(a = 90° - z, A)$ and (h, δ) with the help of rotational matrices (the corresponding Cartesian coordinate systems are both left handed). To this end, we start from the (a, A) system. The z-axis points towards the zenith of the observer and the x-axis towards north. If we rotate the coordinates about the z-axis by $+180°$, then the x-axis will point towards the south from which the hour angle is reckoned. If then we rotate about the y-axis so that the z-axis points towards the NCP, we have reached the Cartesian coordinate system related with (h, δ). To this end, we have to rotate about the angle $-(90° - \Phi)$, i.e.,

$$\begin{pmatrix} x \\ y \\ z \end{pmatrix}_{h,\delta} = \mathcal{R}_2(\Phi - 90°)\,\mathcal{R}_3(180°) \begin{pmatrix} x \\ y \\ z \end{pmatrix}_{a,A}. \tag{5.16}$$

Note that in a series of rotational matrices, the one standing rightmost operates first. We now want to verify this relation. Written out, we get

$$\begin{pmatrix} \cos\delta\cos h \\ \cos\delta\sin h \\ \sin\delta \end{pmatrix} = \begin{pmatrix} \sin\Phi & 0 & \cos\Phi \\ 0 & 1 & 0 \\ -\cos\Phi & 0 & \sin\Phi \end{pmatrix} \begin{pmatrix} -1 & 0 & 0 \\ 0 & -1 & 0 \\ 0 & 0 & 1 \end{pmatrix} \begin{pmatrix} \cos a\cos A \\ \cos a\sin A \\ \sin a \end{pmatrix}$$

$$= \begin{pmatrix} \sin\Phi & 0 & \cos\Phi \\ 0 & 1 & 0 \\ -\cos\Phi & 0 & \sin\Phi \end{pmatrix} \begin{pmatrix} -\cos a\cos A \\ -\cos a\sin A \\ \sin a \end{pmatrix}$$

$$= \begin{pmatrix} -\cos a\sin\Phi\cos A + \sin a\cos\Phi \\ -\cos a\sin A \\ +\cos a\cos\Phi\cos A + \sin a\sin\Phi \end{pmatrix}$$

$$= \begin{pmatrix} \cos z\cos\Phi - \sin z\sin\Phi\cos A \\ -\sin z\sin A \\ \cos z\sin\Phi + \sin z\cos\Phi\cos A \end{pmatrix}$$

in agreement with (5.2)–(5.4).

5.4 Sidereal Times

Exercise 5.4: Check the relations between (a, A) and (h, δ) with Maple™ using the *AstroRef* function `RotMatrix`.

If we want to switch from the (h, δ) system to the (α, δ) system, we can first go to a right-handed system with a reflection of the y-axis, i.e., $y \to -y$. A rotation about the z-axis with angle −LAST then leads to the desired result:

$$\begin{pmatrix} x \\ y \\ z \end{pmatrix}_{\alpha,\delta} = \mathcal{R}_3(-\text{LAST}) \begin{pmatrix} x \\ -y \\ z \end{pmatrix}_{h,\delta} . \quad (5.17)$$

5.3.4 Relations Between (α, δ) and (λ, β)

The equatorial coordinates of the second kind and the ecliptic coordinates differ from each other only by a rotation about their common x-axis by the obliquity of the ecliptic ϵ. One finds

$$\begin{pmatrix} \cos\beta \cos\lambda \\ \cos\beta \sin\lambda \\ \sin\beta \end{pmatrix} = \mathcal{R}_1(\epsilon) \begin{pmatrix} \cos\delta \cos\alpha \\ \cos\delta \sin\alpha \\ \sin\delta \end{pmatrix} = \begin{pmatrix} \cos\delta \cos\alpha \\ \cos\delta \sin\alpha \cos\epsilon + \sin\delta \sin\epsilon \\ -\cos\delta \sin\alpha \sin\epsilon + \sin\delta \cos\epsilon \end{pmatrix} \quad (5.18)$$

and for the inverse transformation

$$\begin{pmatrix} \cos\delta \cos\alpha \\ \cos\delta \sin\alpha \\ \sin\delta \end{pmatrix} = \begin{pmatrix} \cos\beta \cos\lambda \\ \cos\beta \sin\lambda \cos\epsilon - \sin\beta \sin\epsilon \\ \cos\beta \sin\lambda \sin\epsilon + \sin\beta \cos\epsilon \end{pmatrix} . \quad (5.19)$$

After such a transformation, one obtains the new angles, e.g., β and λ from

$$\lambda = \arctan\frac{y}{x}$$

$$\beta = \arctan\frac{z}{\sqrt{x^2 + y^2}} = \arcsin z.$$

Exercise 5.5: Use *AstroRef* to find the conversion between (a, A) and (λ, β).

5.4 Sidereal Times

For many applications, LAST and Greenwich apparent sidereal time (GAST) play an important role. If the symbol "Eqx" again stands for the actual true equinox, LAST and GAST are given by

$$\text{LAST} = h(\text{Eqx})$$
$$\text{GAST} = h_{\text{Gr}}(\text{Eqx}),$$

where $h(\text{Eqx})$ is the local and $h_{\text{Gr}}(\text{Eqx})$ is the Greenwich hour angle of the true vernal equinox respectively. By means of a mean equinox of date, mEqx, that differs from the actual true equinox by effects from nutation (see below), one introduces local mean sidereal time (LMST) and Greenwich mean sidereal time (GMST):

$$\text{LMST} = h(\text{mEqx})$$
$$\text{GMST} = h_{\text{Gr}}(\text{mEqx}).$$

The difference between the hour angles of true and mean equinox is called *equation of equinoxes* (Eq.E.):

$$\text{Eq.E.} = h(\text{Eqx}) - h(\text{mEqx}) = h_{\text{Gr}}(\text{Eqx}) - h_{\text{Gr}}(\text{mEqx}), \quad (5.20)$$

so that

$$\text{LAST} = \text{LMST} + \text{Eq.E.}, \quad \text{GAST} = \text{GMST} + \text{Eq.E.} \quad (5.21)$$

The equation of equinoxes results from the nutational motion of the Earth that will be discussed below. The classical formula reads

$$\text{Eq.E.} = \Delta\psi \cos\epsilon. \quad (5.22)$$

Here, $\Delta\psi$ is the nutation in longitude, and the factor $\cos\epsilon$ provides the projection from the ecliptic onto the equator. At the XXIInd General Assembly in 1994, IAU Resolution C7 Recommendation 3 indicated that the equation of the equinoxes should be amended by the addition of new complementary terms related with precession–nutation cross terms. This change was needed in order to refine the definition of the origin of coordinates from which earth rotation was measured to give accuracy better than a milliarcsecond (Capitaine and Gontier 1993). Also, for higher accuracies, one distinguishes a dynamical from a kinematical equation of equinoxes (Capitaine and Gontier 1993).

Sidereal times are derived from a local time measurement. The observed zone time like CET is first converted to coordinated universal time (UTC). By adding the offset $\Delta\text{UT1} = \text{UT1} - \text{UTC}$, universal time UT1 is determined from which GAST can be derived. The calculation of GAST proceeds in two steps: first, GMST is computed and then the equation of equinoxes (Eq.E) is added (Fig. 5.21).

An approximate value for GMST in degrees can be derived from

$$\text{GMST} = 280.46061837° + 360.98564736629° \times (\text{JD} - \text{J2000.0})$$
$$+ 0.000387933° t^2 - 0.00000002583° t^3 \quad (5.23)$$

5.4 Sidereal Times

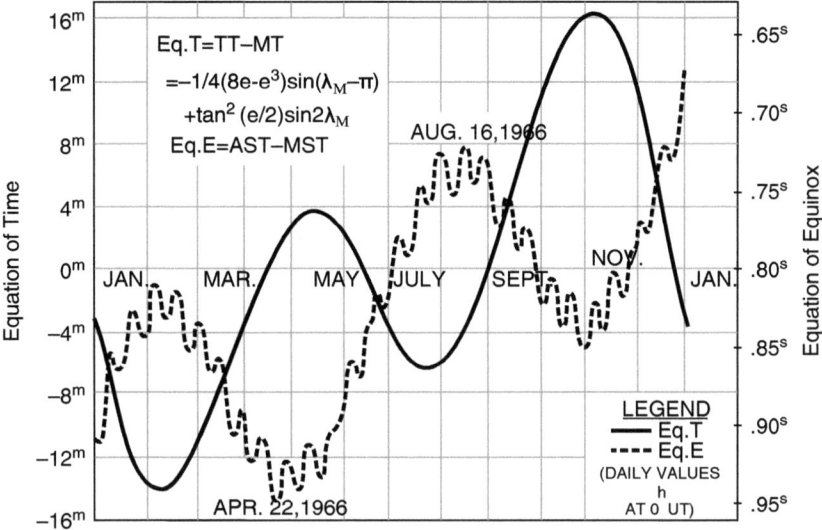

Fig. 5.21 The equation of equinoxes (Eq.E) (*dashed line*) for the year 1966. The *solid line* shows the equation of time (Eq.T) that will be discussed below

with $t = (\text{JD} - \text{J2000.0})/36{,}525$ and $\text{J2000.0} = 2\,451\,545.0$ (e.g., Meeus 1999). This approximation neglects the offset ΔUT1. Therefore, UTC is used directly in the Julian date value JD. A more precise definition of GMST will be discussed below in relation with universal time UT1.

If Λ denotes the eastward astronomical longitude of an observer, its LAST is given by

$$\text{LAST} = \text{GAST} + \Lambda. \qquad (5.24)$$

Exercise 5.6: For 1 January 2000, $12^h00^m00\overset{s}{.}00$ TT, compute GMST, GAST, and LAST for an observer with eastward longitude $\Lambda = 13.730417°$ using the *AstroRef* functions GAST and GMST.

Exercise 5.7: Determine the equation of equinoxes for 3 July 2005, $14^h30^m51^s$ CEST, by comparison of GAST and GMST!

Exercise 5.8: Every year there is exactly 1 day when GMST becomes 0^h two times. Find out which day it was in 2010!

Chapter 6
Astrometry

If a stellar position is measured from the Earth, it will vary because of refraction in the Earth's atmosphere, parallax, aberration, proper motion and precession, nutation, and polar motion. Parallax and aberration require the selection of a reference point that usually is chosen as the barycenter of the solar system. For that reason, the computation of parallax and aberration requires solar system ephemerides.

6.1 Refraction

The astronomical refraction results from the varying index of refraction in the Earth's atmosphere. If a light ray passes from one optically homogenous medium with refractive index n into another with index n', then Snell's law (Fig. 6.1).

$$n \sin \alpha = n' \sin \alpha' \tag{6.1}$$

is valid, where the angles of incidence and reflection are defined with respect to the interface normal. Since the density of the Earth's atmosphere decreases with height, incident light rays are deflected towards the plumb line, i.e., a star appears at smaller zenith distance z. It should be noted here that the index of refraction also depends upon the wavelength of light. For very precise modeling of refraction this has to be taken into account.

For an approximate treatment of refraction we will first consider a small region near the zenith of the observer. We will assume the atmosphere in hydrostatic equilibrium in the Earth's gravity field to be spherically symmetric, i.e., consisting of a large number of homogeneous spherical shells whose density increases exponentially with height. For the case of very small zenith distances we will neglect the curvature of these shells and consider a system of plane-parallel layers (Fig. 6.2). Starting outside the atmosphere with refractive index equal to unity and the zenith distance z at each separating layer interface, we encounter refraction according to

Fig. 6.1 Geometry in Snell's law of refraction

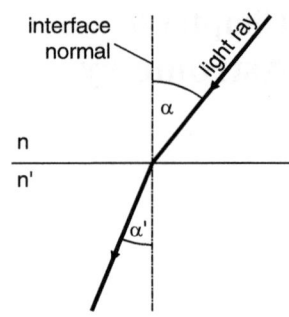

Fig. 6.2 Refraction in the model of plane-parallel layers

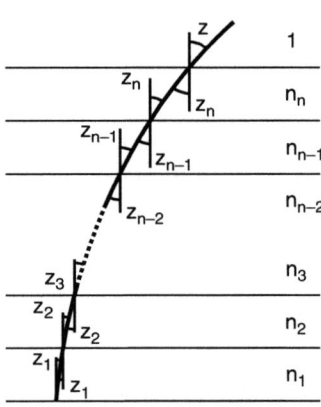

Snell's law. Let n_n be the index of refraction of the outermost atmospheric layer; we then have

$$\sin z = n_n \sin z_n,$$

while z_n denotes the zenith distance of the light ray within the outermost layer. For the next layer we get

$$n_n \sin z_n = n_{n-1} \sin z_{n-1}$$

and so forth. For the transition to the bottom layer we finally get

$$n_2 \sin z_2 = n_1 \sin z_1,$$

so that altogether we have

$$\sin z = n_1 \sin z_1. \tag{6.2}$$

We will call $z_1 \equiv z'$ the apparent zenith distance, z the true zenith distance, and $\zeta \equiv z - z'$ the correction of refraction. Because of $\cos \zeta \simeq 1, \sin \zeta \simeq \zeta$ we get

$$\sin(z' + \zeta) = n_1 \sin z'$$

$$\sin z' + \zeta \cos z' = n_1 \sin z'$$

or

$$\zeta = (n_1 - 1) \tan z' = \eta \tan z'. \tag{6.3}$$

6.1 Refraction

The quantity $\eta = (n_1 - 1)$ is called the coefficient of refraction. For $n_1 = 1.00029$, this coefficient will be of order $\eta \simeq 1'$. For our simple model of plane-parallel layers, we thus found the remarkable result that the true zenith distance z can be computed from the apparent directly measurable one, z', solely from the index of refraction at the bottom of the atmosphere. For this model a detailed knowledge about the physical conditions of the atmosphere in the various heights is unnecessary. We also found that the correction of refraction depends upon the tangent of the apparent zenith distance. This is true as long as the curvature of the atmospheric layers can be neglected.

The curvature of the atmospheric layers can be considered in different ways. A simple possibility consists of considering an expansion for ζ in terms of powers of $\tan z'$

$$\zeta = \eta_1 \tan z' + \eta_2 \tan^3 z' + \cdots . \tag{6.4}$$

The $\tan^2$-term will be negligibly small. For the constants η_1, η_2, etc., appearing in such an expansion, one finds from observations at meteorological standard conditions (air pressure: $p_0 = 1,013$ hPa, air temperature: $T_0 = 0°C$)

$$\eta_1 = 60''\!.1; \qquad \eta_2 = -0''\!.072.$$

By means of a simple formula the actual air pressure p (in hPa) and temperature T (in degrees Celsius) on the ground can be taken into account:

$$\zeta = \frac{p}{1013} \frac{273.15}{273.15 + T} (60''\!.1 \tan z' - 0''\!.072 \tan^3 z'). \tag{6.5}$$

6.1.1 The Model of Saastamoinen

An improvement of (6.5) is given by the model of Saastamoinen (1972). Besides the air pressure p in hPa and temperature T in degrees Celsius, the relative humidity $H_{\rm rel}$ is another input quantity. First one computes the water vapor pressure $p_{\rm w}$ according to

$$p_{\rm w} = H_{\rm rel} \left(\frac{273.15 + T}{247.1} \right)^{18.36} \tag{6.6}$$

With the auxiliary quantity

$$Q = \frac{p - 0.156 \, p_{\rm w}}{273.15 + T}$$

one gets (amplitudes in arcseconds)

$$\eta_1 = 16.271 \, Q - 7.49 \times 10^{-5} p$$
$$\eta_2 = 6.410774 \times 10^{-4} Q^2 - 7.49 \times 10^{-5} p$$

Fig. 6.3 Refraction in a model of spherical layers

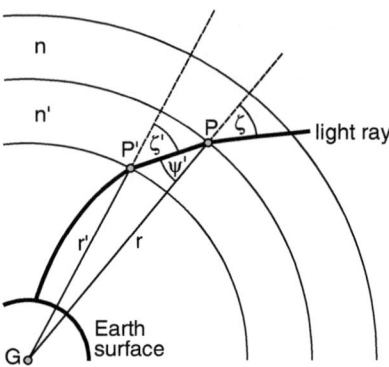

or

$$\zeta = 16\overset{''}{.}271\, Q(\tan z' + 0.0000394\, Q \tan^3 z') - 0\overset{''}{.}0000749\, p\, (\tan z' + \tan^3 z').$$

Saastamoinen's model yields reasonable results for zenith distances $z' < 70°$. It is based on a mean wavelength for visible light (574 nm). The function `RefrSaas` of the *AstroRef* package calculates refraction corrections using this model.

6.1.2 Refraction Corrections as Integrals

We will now consider the curvature of atmospheric layers in more detail (Fig. 6.3). For two adjacent layers let n and n' be the two indices of refraction. With the notation from Fig. 6.3, we get from Snell's law

$$n \sin \zeta = n' \sin \psi'.$$

We now consider the triangle $GP'P$, with $GP = r$, $GP' = r'$ (G: geocenter), and the inner angle at $P' = 180° - \zeta'$. From the law of sines, we get

$$\frac{\sin(180° - \zeta')}{r} = \frac{\sin \psi'}{r'}$$

or

$$r' \sin \zeta' = r \sin \psi'.$$

Together with the law of refraction, we infer that

$$nr \sin \zeta = n'(r \sin \psi') = n'r' \sin \zeta' = \text{const.} \tag{6.7}$$

Let us denote the constant appearing here by κ; we get

$$\sin \zeta = \frac{\kappa}{nr} \tag{6.8}$$

6.1 Refraction

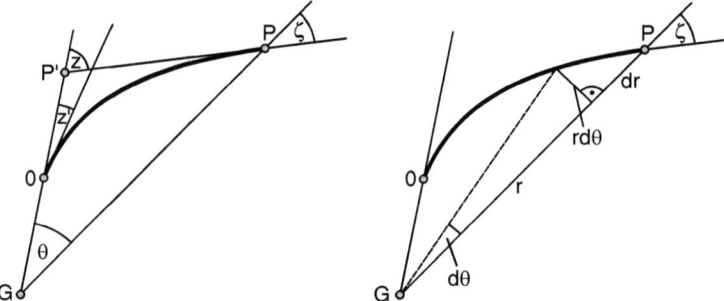

Fig. 6.4 Zenith distances in the spherical model of refraction; *right*: details for the computation of dθ

or

$$\tan \zeta = \frac{\sin \zeta}{\sqrt{1 - \sin^2 \zeta}} = \frac{\kappa}{\sqrt{n^2 r^2 - \kappa^2}}. \tag{6.9}$$

We will now introduce the zenith distance of a light ray with respect to an observer. In Fig. 6.4 the inner angle at P' is $180° - z$, i.e., $(180° - z) + \zeta + \theta = 180°$, or

$$z = \zeta + \theta.$$

We will now analyze the differential relation

$$dz = d\zeta + d\theta \tag{6.10}$$

in more detail. The small angle $d\zeta$ can be obtained from the law of refraction. By differentiation of

$$nr \sin \zeta = \text{const.}$$

we get

$$d(nr) \sin \zeta + nr \cos \zeta \, d\zeta = 0,$$

or

$$d\zeta = -\tan \zeta \frac{d(nr)}{nr}. \tag{6.11}$$

Another consequence is the relation

$$\frac{d(nr)}{dr} \tan \zeta + (nr) \frac{d\zeta}{dr} = 0$$

or

$$\frac{d\zeta}{dr} = -\frac{(n + r \, dn/dr) \tan \zeta}{nr}. \tag{6.12}$$

The small angle $d\theta$ can be obtained from Fig. 6.4 (right). One sees that $\tan \zeta = rd\theta/dr$ or

$$d\theta = \tan \zeta \frac{dr}{r}. \tag{6.13}$$

Thus we finally obtain

$$\begin{aligned} dz &= d\theta + d\zeta \\ &= \tan \zeta \frac{dr}{r} - \tan \zeta \frac{d(nr)}{nr} \\ &= -\tan \zeta \frac{dn}{n} = -\frac{\kappa}{\sqrt{n^2 r^2 - \kappa^2}} \frac{dn}{n}. \end{aligned}$$

The refraction correction $\zeta \equiv z - z'$ is therefore given by

$$\zeta = -\int_{n_0}^{1} \frac{\kappa}{\sqrt{n^2 r^2 - \kappa^2}} \frac{dn}{n}. \tag{6.14}$$

The index of refraction above the Earth's atmosphere was chosen to be unity. The quantity κ can be expressed in terms of quantity at the location of the observer:

$$\kappa = n_0 r_0 \sin \zeta_0$$

where ζ_0 equals the observed (apparent) zenith distance z', i.e.,

$$\zeta = n_0 r_0 \sin z' \int_{1}^{n_0} \left(n^2 r^2 - n_0^2 r_0^2 \sin^2 z'\right)^{-1/2} \frac{dn}{n}.$$

Another representation of ζ results from

$$\begin{aligned} \zeta &= \int_{\zeta_0}^{\zeta_\infty} \left[d\zeta + \tan \zeta \frac{dr}{r}\right] = \int_{\zeta_0}^{\zeta_\infty} \left[1 + \frac{\tan \zeta}{r} \left(\frac{d\zeta}{dr}\right)^{-1}\right] d\zeta \\ &= \int_{\zeta_0}^{\zeta_\infty} \frac{r \, dn/dr}{n + r \, dn/dr} d\zeta. \end{aligned} \tag{6.15}$$

Sophisticated laws of astronomical refraction can be derived from this integral expression for the refraction correction. For a precise treatment of atmospheric refraction, the integral appearing in (6.15) can be integrated numerically by means of an atmospheric model. Often a so-called polytropic model is employed (Seidelmann 1992). This means that the model atmosphere consists of two layers, the troposphere and the stratosphere. The troposphere extends from the Earth's surface up to a height of $h_t = 11$ km. In the troposphere the temperature reduces with increasing height by a constant rate of $\alpha = 0.0065$ (deg per km), and humidity is assumed to agree with that measured by the observer. Temperature in the

6.1 Refraction

stratosphere is assumed to be constant with a value determined at the tropopause. Moreover, it is assumed that in the stratosphere the water vapor pressure vanishes. The stratosphere extends from the troposphere up to a height of $h_s = 80$ km. The refraction integral in (6.15) is divided into a tropospheric and a stratospheric part

$$\zeta = \zeta_t + \zeta_s.$$

As input quantities, we have:

- z': observed zenith distance,
- p: measured air pressure in hPa,
- T_0: absolute air temperature in K (i.e., measured temperature in degrees Celsius plus 273.15 K),
- H_{rel}: relative humidity in fractions of 1,
- λ: wavelength of light in μm,
- Φ: latitude of observer, and
- h_0: height above the geoid in m.

Starting with the parameters for the Earth and the atmosphere model $S = 8314.36$, $M_d = 28.966$, $M_w = 18.016$, $\delta = 18.36$, $R_E = 6378120$, $h_t = 11{,}000$, $h_s = 80{,}000$, and $\alpha = 0.0065$, one first computes several quantities:

$$p_w = H_{\text{rel}} \left(\frac{T_0}{247.1} \right)^\delta$$

$$g = 9.784 \left(1 - 26 \times 10^{-4} \cos 2\Phi - 28 \times 10^{-8} h_0 \right)$$

$$A = \left(287.604 + \frac{1.6288}{\lambda^2} + \frac{0.0136}{\lambda^4} \right) \frac{273.15}{1013.25} \times 10^{-6}$$

$$C_1 = \alpha$$

$$C_2 = g M_d / S$$

$$C_3 = C_2 / C_1 \equiv \gamma$$

$$C_4 = \delta$$

$$C_5 = \frac{p_w (1 - M_d/M_w) \gamma}{(\delta - \gamma)}$$

$$C_6 = \frac{A(p + C_5)}{T_0}$$

$$C_7 = \frac{(A C_5 + 11.2684 \times 10^{-6} p_w)}{T_0}$$

$$C_8 = \frac{\alpha(\gamma - 1) C_6}{T_0}$$

$$C_9 = \frac{\alpha(\delta - 1) C_7}{T_0}.$$

That method requires the value of distance r as function of the angle ζ, which can be determined from Snell's law by iteration:

$$r_{i+1} = r_i - \left[\frac{n_i r_i - n_0 r_0 \sin z'/\sin z_r}{n_i + r_i (dn/dr)_i}\right] \qquad i = 1, 2, \ldots.$$

Calculations for n and dn/dr differ from troposphere and stratosphere. In the troposphere with $r \leq r_t \equiv R_E + h_t$, one has

$$n = 1 + \left[C_6 \left(\frac{T}{T_0}\right)^{\gamma-2} - C_7 \left(\frac{T}{T_0}\right)^{\delta-2}\right]\frac{T}{T_0}$$

$$\frac{dn}{dr} = -C_8 \left(\frac{T}{T_0}\right)^{\gamma-2} + C_9 \left(\frac{T}{T_0}\right)^{\delta-2}$$

with

$$T = T_0 - \alpha(r - r_0).$$

In the stratosphere with $r_t \leq r \leq r_s \equiv R_E + h_s$ and a temperature of T_t:

$$T_t = T_0 - \alpha(r_t - r_0)$$

one has

$$n = 1 + (n_t - 1)e^{-C_2(r-r_t)/T_t}$$

$$\frac{dn}{dr} = -\frac{C_2}{T_t}(n_t - 1)e^{-C_2(r-r_t)/T_t}.$$

The values for ζ at the integration boundaries, z_t and z_s, can be derived from Snell's law, e.g.,

$$z_t = \arcsin \frac{n_0 r_0 \sin z'}{n_t r_t}.$$

With *AstroRef*, the refraction correction using the numerical integral method may be obtained by `RefrNumInt`.

6.1.3 Applying Refraction Corrections

Refraction causes variations in the coordinates of an astronomical object. The changes in right ascension α and declination δ can be derived from (e.g., Mueller 1969)

$$(\Delta\alpha)_R = \alpha' - \alpha = \zeta \sin h \operatorname{cosec} z' \cos \Phi \sec \delta,$$

$$(\Delta\delta)_R = \delta' - \delta = \zeta(\sin \Phi \operatorname{cosec} z' \sec \delta - \tan \delta \cot z').$$

Exercise 6.1: For values of zenith distance $z = 10°, 20°, \ldots, 70°$, temperature of $0°C$, pressure of $1,000$ hPa, humidity of 50%, altitude of 0 m, latitude of $50°$, and wavelength of $0.574\,\mu m$, compute values ζ_i for the refraction corrections with the Maple™ functions `RefrSaas` and `RefrNumInt` and compare the results.

Exercise 6.2: By the help of Saastamoinen's refraction model of (`RefrSaas`), determine the influence of meteorological parameters pressure, humidity, and temperature on the refraction correction and sort them by importance!

Exercise 6.3: Sitting at the sea, you are watching the sunset. Right now, the lower rim of the Sun hits the water level which marks the geometrical horizon. The angular diameter of the Sun is $30'$. Find out how this scene would look like if there was no atmosphere, i.e., no refraction.

6.2 Parallax

An object at finite distance appears in different directions if seen from two different observational points. This phenomenon is called *parallax*. Usually astronomical parallaxes are related with a chosen reference point. This reference point usually is the barycenter of the solar system (the heliocenter in some approximation). The reduction from the geocenter to the barycenter is called the annual parallax because the induced period of angular changes is 1 year. If topocentric positions are reduced to the geocenter, one speaks about the geocentric or diurnal parallax since those angular changes have a period of 1 day.

In 1838, Friedrich Wilhelm Bessel has been the first who succeeded in measuring an annual parallax of 61 Cygni. During the following year, Thomas James Henderson in South Africa determined the parallax of α Centauri. In 1840 Friedrich Georg Wilhelm Struve measured the parallax of Vega.

The next stellar system, α Centauri, has a parallax of $0\rlap{.}''76$. According to their distances the parallaxes of all other stars are correspondingly smaller.

6.2.1 Annual Parallax

Due to the annual motion of the Earth about the Sun, the geocentric direction towards a star changes with respect to quasi-inertial axes. For that reason one reduces the geocentric positions to the barycenter of the solar system. The geometry of annual parallax is depicted in Fig. 6.5. Here θ and θ' are the directional angles to the light source with respect to the Earth-Sun line. If the triangle defined by Earth, Sun, and light source shows a right angle (Fig. 6.5), the angle Π at the star is called *annual parallax*. Neglecting variations in Earth-Sun distance, one has

$$\sin \Pi \simeq \Pi = \frac{1\,\text{AU}}{d}, \qquad (6.16)$$

Fig. 6.5 Geometry in the problem of annual parallax

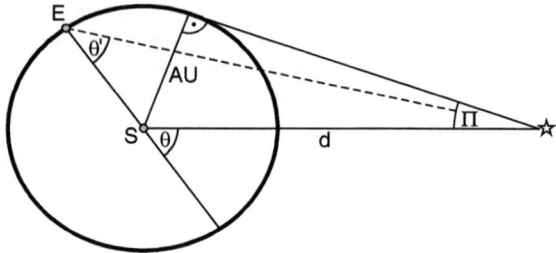

where the astronomical unit (AU) denotes the mean distance of the Earth from the Sun. It is approximately

$$1 \text{ AU} \cong 1.495985 \times 10^8 \text{ km}. \tag{6.17}$$

One sees that the annual parallax is nothing but the inverse distance d of a star from the barycenter (Sun) and

$$\Pi \text{ [rad]} \simeq \frac{1}{d \text{ [AU]}}. \tag{6.18}$$

From this we can derive the usual astronomical distance scale called *parallax second* or short *parsec* (pc): 1 parsec is that distance for which the annual parallax equals 1 arcsec, i.e.,

$$1 \text{ pc} \cong \frac{1 \text{ AU}}{(1'')_{\text{rad}}} = 1 \text{ AU} \left(\frac{3600 \cdot 180}{\pi}\right) = 206264.8 \text{ AU}$$

$$\cong 3.0856 \times 10^{13} \text{ km}$$

$$\cong 3.2615 \text{ ly}, \quad \text{(ly : light} - \text{year)}.$$

where the unit ly means light-years. With the *AstroRef* package, conversions between the various length scales may be executed by the function `ConvertLength`.

The geometry of annual parallax is shown in Fig. 6.6. Here, x_Q denotes a vector from the barycenter B to the light source Q, x_E a vector from the barycenter to the geocenter E, and x'_Q a vector from the geocenter to the light source. The equation for annual parallax simply reads

$$x_Q = x_E + x'_Q. \tag{6.19}$$

From this simple equation, we can easily derive first-order corrections for right ascension and declination due to annual parallax. To this end we define Euclidean unit vectors $\mathbf{m}$ and $\mathbf{m}'$ from barycenter and geocenter to the light source, i.e., we write $x_Q = d \cdot \mathbf{m}$ and $x'_Q = d' \cdot \mathbf{m}'$. Therefore,

$$\mathbf{m}' = \frac{1}{d'}(x_Q - x_E) = \frac{d}{d'}\left(\mathbf{m} - \frac{x_E}{d}\right).$$

6.2 Parallax

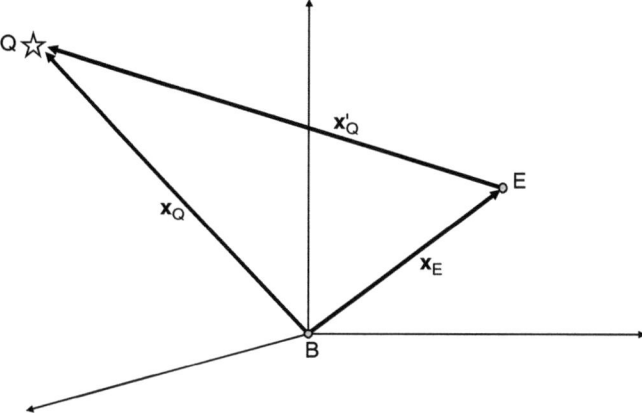

Fig. 6.6 Vectors in the problem of annual parallax

For the problem of annual parallax we will now neglect the variations in the barycentric distance of the geocenter and write

$$\mathbf{x}_S = -\mathbf{x}_E = 1\,\text{AU} \cdot \hat{\mathbf{x}}, \tag{6.20}$$

where $\mathbf{x}_S$ denotes the vector from the geocenter to the barycenter and $\hat{\mathbf{x}}$ the corresponding unit vector. Together with $\eta = d/d'$ and $\Pi = 1\,\text{AU}/d$, we end up with the relation

$$\mathbf{m}' = \eta(\mathbf{m} + \Pi \cdot \hat{\mathbf{x}}). \tag{6.21}$$

We will encounter an equation of that form more often; we will generally write it as

$$\mathbf{m}' = \eta(\mathbf{m} + \boldsymbol{\delta}) \tag{6.22}$$

with $|\boldsymbol{\delta}| \ll 1$. In terms of right ascension and declination, the unit vector $\mathbf{m}$ is given by

$$\mathbf{m} = \begin{pmatrix} \cos\delta\cos\alpha \\ \cos\delta\sin\alpha \\ \sin\delta \end{pmatrix}$$

and a similar relation holds for the unit vector $\mathbf{m}'$. For that reason relation (6.22) reads

$$\cos\delta'\cos\alpha' = \eta(\cos\delta\cos\alpha + \delta_x)$$
$$\cos\delta'\sin\alpha' = \eta(\cos\delta\sin\alpha + \delta_y) \tag{6.23}$$
$$\sin\delta' = \eta(\sin\delta + \delta_z).$$

If we add the first equation, multiplied by $\cos\alpha$, and the second one multiplied by $\sin\alpha$, we get
$$\cos\delta' \cos\Delta\alpha = \eta(\cos\delta + \delta_x \cos\alpha + \delta_y \sin\alpha)$$
with $\Delta\alpha = \alpha' - \alpha$. If, however, we multiply the first equation by $\sin\alpha$ and subtract the second one multiplied by $\cos\alpha$, we get
$$-\cos\delta' \sin\Delta\alpha = \eta(\delta_x \sin\alpha - \delta_y \cos\alpha).$$
Therefore,
$$\tan\Delta\alpha = \frac{\delta_y \cos\alpha - \delta_x \sin\alpha}{\cos\delta + \delta_x \cos\alpha + \delta_y \sin\alpha}. \tag{6.24}$$
Neglecting terms of second order in $|\delta|$, (i.e., $\tan\Delta\alpha \simeq \Delta\alpha$) we get
$$\Delta\alpha \simeq \delta_y \cos\alpha \sec\delta - \delta_x \sin\alpha \sec\delta \tag{6.25}$$
and by taking into account also the third relation in (6.23):
$$\Delta\delta \simeq \delta_z \cos\delta - \delta_y \sin\delta \sin\alpha - \delta_x \cos\alpha \sin\delta. \tag{6.26}$$
In the case of the annual parallax, we can express $\delta_z = \Pi\hat{z}$ by δ_y. From
$$\mathbf{x}_{\lambda,\beta} = \mathcal{R}_1(\epsilon)\, \mathbf{x}_{\alpha,\delta}$$
we infer for the solar orbit about the Earth
$$z_{\lambda,\beta} = -y_{\alpha,\delta} \sin\epsilon + z_{\alpha,\delta} \cos\epsilon = 0,$$
i.e., $z_{\alpha,\delta} = \tan\epsilon \cdot y_{\alpha,\delta}$. For the case of the annual parallax we therefore get
$$\delta_z = \tan\epsilon \cdot \delta_y. \tag{6.27}$$

The first-order corrections for α and δ due to annual parallax can therefore be written as
$$(\Delta\alpha)_P = \Pi(Yc - Xd); \qquad (\Delta\delta)_P = \Pi(Yc' - Xd'). \tag{6.28}$$
Here, $X = \hat{x}$ and $Y = \hat{y}$ are the Cartesian equatorial components of the unit vector pointing from the geocenter to the barycenter (heliocenter). The *stellar constants* c, c', d, and d' are given by
$$\begin{aligned} c &= \cos\alpha \sec\delta, \\ d &= \sin\alpha \sec\delta, \\ c' &= \tan\epsilon \cos\delta - \sin\alpha \sin\delta, \\ d' &= \cos\alpha \sin\delta. \end{aligned} \tag{6.29}$$

Fig. 6.7 Geometry in the problem of geocentric parallax

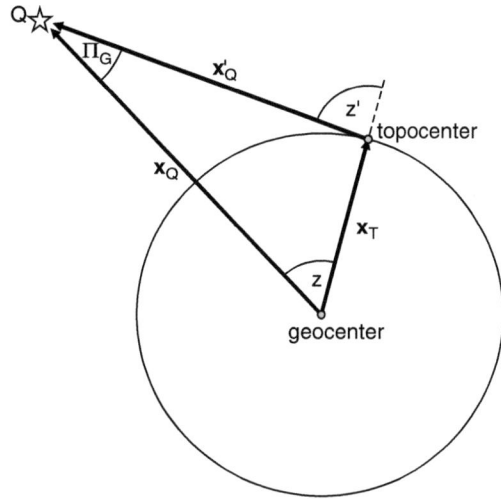

6.2.2 Geocentric Parallax

Figure 6.7 shows the geometry for the problem of geocentric or diurnal parallax. The equation of geocentric parallax simply reads

$$\mathbf{x}_Q = \mathbf{x}_T + \mathbf{x}'_Q. \tag{6.30}$$

Here, $\mathbf{x}_Q$ is a vector from the geocenter to the light source Q, $\mathbf{x}'_Q$ the corresponding vector from the topocenter, and $\mathbf{x}_T$ denotes the geocentric vector of the observer's topocenter. From the triangle in Fig. 6.7, we obtain ($r = |\mathbf{x}_Q|$; $R = |\mathbf{x}_T|$)

$$\frac{r}{\sin(180° - z')} = \frac{R}{\sin \Pi_G}$$

or

$$\sin \Pi_G \simeq \Pi_G \simeq \frac{R}{r} \sin z' \simeq \frac{R}{r'} \sin z. \tag{6.31}$$

From the equation of geocentric parallax, we recover our basic relation (6.22) with

$$\boldsymbol{\delta} \simeq -\frac{R}{r'} \mathbf{m}_T \simeq -\Pi_G \operatorname{cosec} z \cdot \mathbf{m}_T, \tag{6.32}$$

where $\mathbf{m}_T$ denotes the unit vector from the geocenter to the topocenter. We then get

$$\Delta \alpha \simeq \delta_y \cos \alpha \sec \delta - \delta_x \sin \alpha \sec \delta$$
$$\simeq -\Pi_G \operatorname{cosec} z \sec \delta \, (\hat{y}_T \cos \alpha - \hat{x}_T \sin \alpha)$$
$$\simeq -\Pi_G \operatorname{cosec} z \sec \delta \cos \delta_T \sin(\alpha_T - \alpha).$$

Now, the declination of the topocenter, δ_T, equals the astronomical latitude Φ. The right ascension of the topocenter equals the local apparent sidereal time LAST of the observer. With $\alpha_T - \alpha = \text{LAST} - \alpha = h$ (h: hour angle), we finally obtain

$$(\Delta \alpha)_{\text{GP}} = -\Pi_G \sin h \, \text{cosec} \, z \cos \Phi \sec \delta. \tag{6.33}$$

In an analogous way, one finds

$$(\Delta \delta)_{\text{GP}} = -\Pi_G (\sin \Phi \, \text{cosec} \, z \sec \delta - \tan \delta \cot z). \tag{6.34}$$

Exercise 6.4: Sirius is the brightest star on the night sky when viewed from Earth. Its distance is about 8.3 ly.
- Determine the annual parallax Π and the geocentric parallax Π_G!
- Calculate the stellar constants (6.29) for the equatorial position of Sirius $\alpha = 06^h 45^m 08\overset{s}{.}917, \delta = -16° 42' 58''\!.02$!
- For 1 January 2000, 12:00 UTC, calculate the first-order corrections for right ascension and declination due to annual parallax.

6.3 Aberration

Aberration results from the velocity of the observer with respect to some reference point and the finite value for the vacuum speed of light. The following situation is illustrative: if someone stands still and it is raining just from above, the rain will come from the front if he is moving (Fig. 6.8, left). It is obvious that a description of aberration requires a nonmoving reference (rest) point that usually is chosen as the barycenter of our solar system.

Figure 6.8 (right) shows a simple vector diagram for the illustration of aberration. The vector $\mathbf{c}_Q$ points from the observer at rest to the light source. We will assume the length of this vector to be given by the vacuum speed of light, c. The vector $\mathbf{v}$ denotes the moving observer's velocity with respect to the rest frame pointing at some fictitious point at infinity: the *apex* of motion. Due to aberration, the image of some remote body will be displaced towards the apex. If we add the vectors $\mathbf{c}_Q$ and $\mathbf{v}$, we obtain a vector $\mathbf{c}'_Q$ that for the moving observer points towards the light source:

$$\mathbf{c}'_Q = \mathbf{c}_Q + \mathbf{v},$$

or expressed in terms of unit vectors

$$\mathbf{m}' = \frac{\mathbf{m} + \mathbf{v}/c}{|\mathbf{m} + \mathbf{v}/c|}. \tag{6.35}$$

Let θ be the apex angle of a star for an observer at rest, θ' the corresponding apex angle for a moving observer. Then the law of sines yields ($\Delta \theta = \theta - \theta'$)

$$\frac{\sin \Delta \theta}{\sin \theta'} = \frac{v}{c}$$

6.3 Aberration

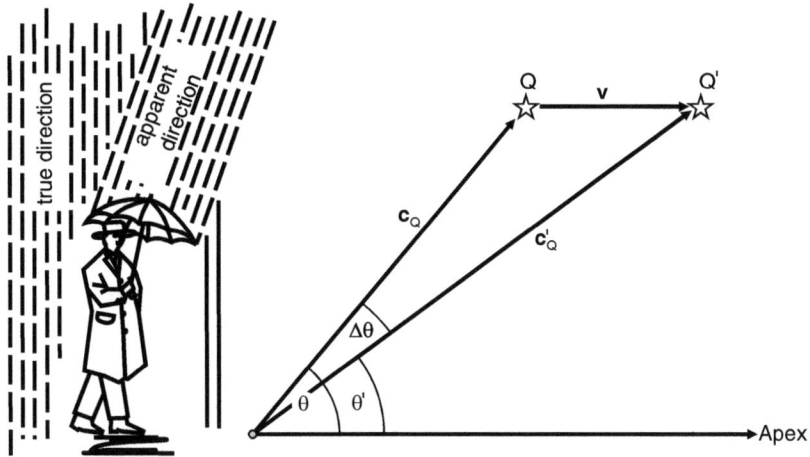

Fig. 6.8 The problem of aberration. *Left*: illustration of the effect with raindrops instead of photons; *right*: a vector diagram for a quasi-Newtonian description of aberration

or since $\sin \Delta\theta \simeq \Delta\theta$; $\sin \theta' \simeq \sin \theta$

$$\Delta\theta \simeq \frac{v}{c} \sin\theta \equiv \kappa \sin\theta. \qquad (6.36)$$

The quantity $\kappa = v/c$ is called the *constant of aberration*.

Exercise 6.5: Referring to Fig. 6.8 (left), imagine a car riding through the rain with a constant velocity of 30 km/h. The velocity of the raindrops of 5 m/s should also be constant. Now consider a wind influencing the direction θ of the raindrops with respect to the ground. Let the angle θ vary from 0° (opposite to the car's moving direction) over 90° (perpendicular to the ground) to 180° (in the car's moving direction).

- Find out from which directions θ' the driver sees the raindrops coming from (plot θ' against θ).
- How do the corrections $\Delta\theta = \theta - \theta'$ look like (plot $\Delta\theta'$ against θ)?

Now increase the velocity of the raindrops to the unrealistic value of 500 m/s and see how the plots change. From the plot $\Delta\theta$ against θ for the fast raindrops, determine the constant of aberration.

Relation (6.35) is correct in a quasi-Newtonian framework but not in relativity. In the framework of special relativity, it has to be replaced by

$$\mathbf{m}' = \left[\mathbf{m} + \gamma\boldsymbol{\beta} + (\gamma - 1)(\boldsymbol{\beta}\cdot\mathbf{m})\boldsymbol{\beta}/\beta^2\right] \frac{1}{\gamma(1 + \boldsymbol{\beta}\cdot\mathbf{m})} \qquad (6.37)$$

with $\boldsymbol{\beta} \equiv \mathbf{v}/c$ and

$$\gamma \equiv \frac{1}{\sqrt{1 - \beta^2}}. \qquad (6.38)$$

From this, one derives

$$\sin \Delta\theta = \frac{(v/c)\sin\theta + (v/c)^2 \sin 2\theta/[2(1+\gamma^{-1})]}{1+(v/c)\cos\theta}$$

$$= \frac{v}{c}\sin\theta - \frac{1}{4}\left(\frac{v}{c}\right)^2 \sin 2\theta + \cdots .$$

The aberration effect on the position of an observed celestial body may be taken into account with the function `Aberration` of the *AstroRef* package.

Exercise 6.6: Proof the special relativistic aberration formula (6.37) in the following way: in some inertial coordinate system $\mathcal{I} = (t, \mathbf{x})$, consider the following set of four (labeled by α) vector fields:

$$e^\mu_{(\alpha)} = \delta_{\alpha\mu},$$

i.e.,

$$e^\mu_{(0)} = (1, \mathbf{0})^T \; ; \quad e^\mu_{(i)} = \delta_{\mu i}.$$

With respect to the metric tensor $\eta_{\mu\mu} = \text{diag}(-1, +1, +1, +1)$, these vectors obey the following orthogonality relations:

$$\eta_{\mu\nu} e^\mu_{(0)} e^\nu_{(0)} = -1$$

$$\eta_{\mu\nu} e^\mu_{(i)} e^\nu_{(j)} = \delta_{ij}$$

$$\eta_{\mu\nu} e^\mu_{(0)} e^\nu_{(i)} = 0.$$

For that reason the set of such four vectors is called a *tetrad field*. For an observer that is moving with coordinate velocity $\mathbf{v}$ in $\mathcal{I}$, the corresponding tetrad field is given by

$$\bar{e}^\mu_{(\alpha)} = \Lambda^\mu_\nu(-\boldsymbol{\beta}) e^\nu_{(\alpha)},$$

so that

$$\bar{e}^\mu_{(0)} = \Lambda^\mu_0(-\boldsymbol{\beta}) = \gamma(1; \boldsymbol{\beta})$$

and

$$\bar{e}^\mu_{(i)} = \Lambda^\mu_i(-\boldsymbol{\beta}) = \left(\gamma\beta_i; \delta_{ij} + (\gamma - 1)\frac{\beta^i\beta^j}{\beta^2}\right).$$

Note that the orthogonality relations remain valid so that, e.g.,

$$\bar{e}^\mu_{(0)} = \frac{1}{c}\frac{dx^\mu_{\text{obs}}}{d\tau},$$

where τ is the observer's proper time, since

$$\eta_{\mu\nu} \bar{e}^\mu_{(0)} \bar{e}^\nu_{(0)} = \frac{1}{c^2}\frac{1}{d\tau^2}\left(\eta_{\mu\nu} dx^\mu_{\text{obs}} dx^\nu_{\text{obs}}\right) = \frac{1}{c^2}\frac{1}{d\tau^2} ds^2_{\text{obs}} = -1.$$

This shows that $\bar{e}^\mu_{(0)}$ (the 4-velocity of the observer divided by c) is a mathematical representation of the observer's clock. The three fields $\bar{e}^\mu_{(i)}$ represent the three spatial reference directions of the observer. We now consider a light ray in $\mathcal{I}$ with tangent vector

$$k^\mu = (1, -\mathbf{m})$$

6.3 Aberration

with $\mathbf{m} \cdot \mathbf{m} = 1$. Since $ds^2 = 0$ for light rays, k^μ is a null vector (i.e., a vector of zero length), i.e.,

$$\eta_{\mu\nu} k^\mu k^\nu = 0.$$

We can now project k^μ into the observer's (rest) space, orthogonal to $\bar{e}^\mu_{(0)}$, by means of the projection operator

$$P^\mu_\nu = \delta^\mu_\nu + \eta_{\nu\alpha} \bar{e}^\mu_{(0)} \bar{e}^\alpha_{(0)}.$$

Proof that $P^\mu_\nu \bar{e}^\nu_{(0)} = 0$. Let us define

$$\bar{k}^\mu = P^\mu_\nu k^\nu.$$

Proof that

$$\bar{k}^\mu = k^\mu - \gamma \sigma \bar{e}^\mu_{(0)}$$

and

$$|\bar{k}^\mu| \equiv (\eta_{\mu\nu} \bar{k}^\mu \bar{k}^\nu)^{1/2} = \gamma\sigma,$$

where

$$\sigma \equiv 1 + \boldsymbol{\beta} \cdot \mathbf{m}.$$

The moving observer will then see the incident light ray to come from the direction

$$m'_i = -\eta_{\mu\nu} \bar{e}^\mu_{(i)} \frac{\bar{k}^\nu}{|\bar{k}^\nu|}.$$

Show that the answer is given by relation (6.37).

6.3.1 Annual Aberration

The annual aberration results from the orbital motion of the Earth about the solar system barycenter with a velocity of about 30 km/s. From this one gets a constant of annual aberration of

$$\kappa = \frac{30 \text{ km/s}}{300000 \text{ km/s}} = 10^{-4} \text{ rad} = 20\overset{''}{.}5.$$

To derive the corresponding corrections for right ascension and declination, we start from relation (6.35) that agrees with our basic relation (6.22) with

$$\boldsymbol{\delta} = \boldsymbol{\beta} \equiv \frac{\mathbf{v}}{c}. \tag{6.39}$$

Similar to the case of the annual parallax, one derives for the velocity of the Earth about the solar system barycenter.

$$\beta_z = \tan \epsilon \cdot \beta_y. \tag{6.40}$$

To first order one finds ($\boldsymbol{\beta} \equiv \mathbf{v}/c$)

$$(\Delta\alpha)_A = \beta_y \cos\alpha \sec\delta - \beta_x \sin\alpha \sec\delta$$
$$(\Delta\delta)_A = \beta_y (\tan\epsilon \cos\delta - \sin\alpha \sin\delta) - \beta_x \cos\alpha \sin\delta. \tag{6.41}$$

This result is usually written in the form

$$(\Delta\alpha)_A = Cc + Dd, \qquad (\Delta\delta)_A = Cc' + Dd'. \tag{6.42}$$

The quantities c, c', d, and d' are again the stellar constants (6.29), and C and D are the *Besselian day numbers* given by

$$C \equiv \beta_y, \qquad D \equiv -\beta_x. \tag{6.43}$$

These relations for $(\Delta\alpha)_A$ and $(\Delta\delta)_A$ are correct only to first order in $(1/c)$. Including quadratic terms in $1/c$, one gets

$$(\Delta\alpha)_A = \beta_y \cos\alpha \sec\delta - \beta_x \sin\alpha \sec\delta$$
$$+ (\beta_x \sin\alpha - \beta_y \cos\alpha)(\beta_x \cos\alpha + \beta_y \sin\alpha) \sec^2\delta + \cdots$$

$$(\Delta\delta)_A = -\beta_y \sin\alpha \sin\delta + \beta_z \cos\delta - \beta_x \cos\alpha \sin\delta \tag{6.44}$$
$$-\frac{1}{2}(\beta_x \sin\alpha - \beta_y \cos\alpha)^2 \tan\delta$$
$$+ (\beta_x \cos\delta \cos\alpha + \beta_y \cos\delta \sin\alpha + \beta_z \sin\delta)$$
$$\times (\beta_x \sin\delta \cos\alpha + \beta_y \sin\delta \sin\alpha - \beta_z \cos\delta) + \cdots.$$

Exercise 6.7: Use the simple formulae (6.43) to compute the Besselian day numbers C and D for 1 February 1994, 0^h TT, by using the JPL ephemeris for the barycentric velocity vector of the Earth.

6.3.2 Diurnal Aberration

Diurnal aberration results from the diurnal rotation of the Earth leading to a topocentric velocity $\mathbf{v}_T$ with respect to the geocenter. If the rotation of the Earth is described with an angular velocity $\mathbf{\Omega}_E$, the velocity of some earthbound observer in geocentric coordinates $\mathbf{x}_T$ is given by

$$\mathbf{v}_T = \mathbf{\Omega}_E \times \mathbf{x}_T. \tag{6.45}$$

Due to Earth's rotation every earthbound observer therefore has a velocity in eastward direction of

$$v = \Omega_E R \cos\Phi.$$

Here, Φ is the observer's latitude,

$$\Omega_E = \frac{2\pi}{86164} \mathrm{s}^{-1}$$

and R denotes the observer's distance to the geocenter. The resulting constant of diurnal aberration therefore reads

$$k = \frac{v}{c} = 0\rlap{.}''319 \, \rho \cos\Phi \tag{6.46}$$

with $\rho = R/R_E$ and $R_E = 6,378$ km. First-order corrections for right ascension and declination can again be derived from our basic relation (6.22) with $\boldsymbol{\delta} = \boldsymbol{\beta}$. Now, the unit vector to the topocenter projected into the equatorial plane is given by $\hat{\mathbf{e}} = (\cos(\text{LAST}), \sin(\text{LAST}), 0)^T$, and the unit vector perpendicular to $\hat{\mathbf{e}}$ pointing in eastward direction reads $\hat{\mathbf{e}}_{\text{east}} = (-\sin(\text{LAST}), \cos(\text{LAST}), 0)^T$. The vector $\boldsymbol{\beta}$ is then given by

$$\boldsymbol{\delta} = \boldsymbol{\beta} = \frac{\mathbf{v}_T}{c} = k\,\hat{\mathbf{e}}_{\text{east}} = k \begin{pmatrix} -\sin(\text{LAST}) \\ +\cos(\text{LAST}) \\ 0 \end{pmatrix}. \tag{6.47}$$

From this, we infer:

$$(\Delta\alpha)_{\text{DA}} = k \sec\delta \, [\sin(\text{LAST})\sin\alpha + \cos(\text{LAST})\cos\alpha]$$
$$= k \sec\delta \cos(\text{LAST} - \alpha)$$
$$= k \sec\delta \cos h$$

or inserting the expression for k:

$$(\Delta\alpha)_{\text{DA}} = 0\rlap{.}''319 \, \rho \cos\Phi \cos h \sec\delta. \tag{6.48}$$

In an analogous way, we obtain:

$$(\Delta\delta)_{\text{DA}} = k \sin\delta \sin h$$

or

$$(\Delta\delta)_{\text{DA}} = 0\rlap{.}''319 \, \rho \cos\Phi \sin h \sin\delta. \tag{6.49}$$

6.4 Space Motion of Stars

Let us consider some barycentric Cartesian coordinate system oriented according to the geocentric equatorial coordinates of the second kind. In such a system the position vector of a star can be written as

$$\mathbf{x} = r \begin{pmatrix} \cos\alpha \cos\delta \\ \sin\alpha \cos\delta \\ \sin\delta \end{pmatrix}, \tag{6.50}$$

Fig. 6.9 Proper motion of stars

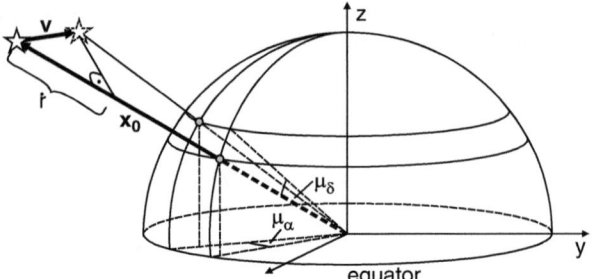

where the barycentric distance of the star can be derived from its annual parallax Π. Assuming the distance is measured in astronomical units, we can write

$$r = \frac{1}{\sin \Pi}. \qquad (6.51)$$

Let $\mu_\alpha = d\alpha/dt$ and $\mu_\delta = d\delta/dt$ be the the proper motion in right ascension and declination, measured in radians per Julian century and $\dot{r} = dr/dt$ the radial velocity, measured in astronomical units per Julian century (Fig. 6.9). By differentiation of (6.50), we get

$$\mathbf{v} = r \begin{pmatrix} -\cos\delta \sin\alpha & -\sin\delta \cos\alpha & \cos\delta \cos\alpha \\ \cos\delta \cos\alpha & -\sin\delta \sin\alpha & \cos\delta \sin\alpha \\ 0 & \cos\delta & \sin\delta \end{pmatrix} \begin{pmatrix} \mu_\alpha \\ \mu_\delta \\ \dot{r}/r \end{pmatrix}. \qquad (6.52)$$

If $\mathbf{x}_0$ denotes the star's position vector at epoch J2000.0, then its position vector at another time will be

$$\mathbf{x} = \mathbf{x}_0 + \mathbf{v}t, \qquad (6.53)$$

where

$$t = \frac{JD - J2000.0}{36{,}525}$$

is the time that has elapsed since J2000.0 in Julian centuries. Here we have assumed that μ_α, μ_δ, and $\dot{r}/r$ have been converted into radians per century. The final values for α and δ are obtained from the components (x, y, z) of vector $\mathbf{x}$ according to

$$\tan\alpha = \frac{y}{x}; \quad \tan\delta = \frac{z}{\sqrt{x^2 + y^2}}.$$

With *AstroRef*, proper motion or even space motion may be taken into account by `ProperMotion`.

Fig. 6.10 Precession and nutational motion of the Earth's rotation axis

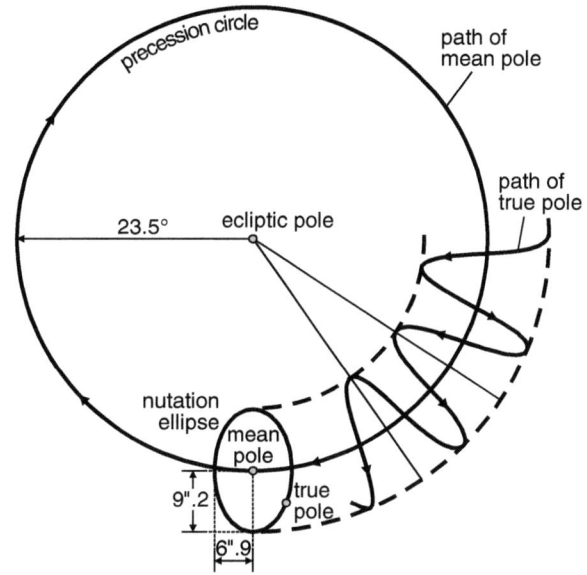

6.5 Astronomical Precession

The precession of the equator (formerly called lunisolar precession) results from the torques that the Moon, Sun, and planets exert on the equatorial bulge of the Earth. The period of precession is about 25, 780 years (the Platonic year) corresponding to an annual precession of about 50″. In the classical equinox-based reference frame, one faces also the precession of the ecliptic, resulting from gravitational perturbations of other planets onto the orbit of the Earth-Moon barycenter about the Sun (the ecliptic). Due to such perturbations the obliquity of the ecliptic varies between 21°55″ and 24°18″. Nutations are variations of the Earth's rotation axis with shorter periods caused mainly by the eccentricity of the Earth orbit and the inclination of the lunar orbit with respect to the ecliptic (see Fig. 6.10). Due to nutation the true celestial pole moves along a small ellipse about the mean celestial pole that is given solely by the precessional motion. This ellipse of nutation has a semimajor axis of 9″.2 and a semiminor axis of 6″.9 (Fig. 6.10).

The general precession (precession of equator plus precession of the ecliptic) of the classical astronomical reference system from some reference epoch T_0 to some other epoch T (here and in the following the epochs are given in terrestrial time, TT) is generally described by three precession angles ζ_A, z_A, and θ_A illustrated in Fig. 6.11. One considers two reference systems related with the two epochs T_0 and T. The z-axis of each system points towards the corresponding celestial poles P_0 (at epoch T_0) and P_T (at epoch T). The corresponding x-y-planes are denoted by Eq_0 (mean equatorial plane at epoch T_0) and Eq_T (mean equatorial plane at

Fig. 6.11 The precession angles $\zeta_A, z_A,$ and θ_A

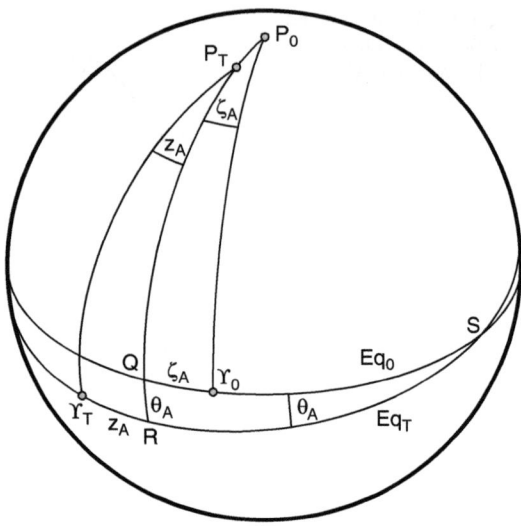

epoch T). Usually we will start with our fixed reference epoch (J2000.0 $= T_0$) and transform mean places at time T_0 to some epoch of date T. The direction of x-axes is given by Υ_0 on Eq$_0$ and Υ_T on Eq$_T$. We now consider the great circle running through P_0 and Υ_0.

We first rotate the Cartesian coordinate system at epoch T_0 about its z-axis by the angle $-\zeta_A$. After that our great circle is now running through the celestial pole P_T, it hits Eq$_0$ now in Q, and the y-axis is shifted towards S, the point of intersection of the two equatorial planes. The z-axis still points to P_0. We now rotate the coordinate system about the y-axis about the angle θ_A such that Eq$_T$ becomes the new x–y-plane. The x-axis now points at R, the y-axis to S, and the z-axis to P_T. By means of a third rotation with an angle $-z_A$ about the z-axis, we succeed that the x-axis points to the mean equinox at epoch T, Υ_T. This construction reveals the definitions of the three precession angles ζ_A, z_A, and θ_A. The right ascension of S in the T_0-system equals $90° - \zeta_A$. The right ascension of S in the T-system is $90° + z_A$, and the inclination of Eq$_T$ in the T_0-system is θ_A. From this follows that the transformation from $\mathbf{x}(\alpha_0, \delta_0)$ to $\mathbf{x}(\alpha, \delta)$ is described with the precession matrix $\mathcal{P}$:

$$\begin{pmatrix} x \\ y \\ z \end{pmatrix}_{\alpha,\delta} = \mathcal{P} \begin{pmatrix} x \\ y \\ z \end{pmatrix}_{\alpha_0,\delta_0} \tag{6.54}$$

with

$$\mathcal{P} = \mathcal{R}_3(-z_A)\mathcal{R}_2(+\theta_A)\mathcal{R}_3(-\zeta_A). \tag{6.55}$$

Explicitly the components of the precession matrix read

6.5 Astronomical Precession

$$p_{11} = \cos z_A \cos\theta_A \cos\zeta_A - \sin z_A \sin\zeta_A$$
$$p_{12} = -\cos z_A \cos\theta_A \sin\zeta_A - \sin z_A \cos\zeta_A$$
$$p_{13} = -\cos z_A \sin\theta_A$$
$$p_{21} = \sin z_A \cos\theta_A \cos\zeta_A + \cos z_A \sin\zeta_A$$
$$p_{22} = -\sin z_A \cos\theta_A \sin\zeta_A + \cos z_A \cos\zeta_A \quad (6.56)$$
$$p_{23} = -\sin z_A \sin\theta_A$$
$$p_{31} = \sin\theta_A \cos\zeta_A$$
$$p_{32} = -\sin\theta_A \sin\zeta_A$$
$$p_{33} = \cos\theta_A.$$

Without the last rotation, one has

$$\cos\delta \sin(\alpha - z_A) = \cos\delta_0 \sin(\alpha_0 + \zeta_A)$$
$$\cos\delta \cos(\alpha - z_A) = \cos\theta_A \cos\delta_0 \cos(\alpha_0 + \zeta_A) - \sin\theta_A \sin\delta_0 \quad (6.57)$$
$$\sin\delta = \sin\theta_A \cos\delta_0 \cos(\alpha_0 + \zeta_0) + \cos\theta_A \sin\delta_0$$

and

$$\sin(\alpha_0 + \zeta_A)\cos\delta_0 = +\sin(\alpha - z_A)\cos\delta$$
$$\cos(\alpha_0 + \zeta_A)\cos\delta_0 = +\cos(\alpha - z_A)\cos\theta_A \cos\delta + \sin\theta_A \sin\delta \quad (6.58)$$
$$\sin\delta_0 = -\cos(\alpha - z_A)\sin\theta_A \cos\delta + \cos\theta_A \sin\delta.$$

In the past the Lieske et al. (1977) expressions for the precession angles have been used:

$$\zeta_A = 2306''\!.2181\,t + 0''\!.30188\,t^2 + 0''\!.017998\,t^3,$$
$$z_A = 2306''\!.2181\,t + 1''\!.09468\,t^2 + 0''\!.018203\,t^3, \quad (6.59)$$
$$\theta_A = 2004''\!.3109\,t - 0''\!.42665\,t^2 - 0''\!.041833\,t^3$$

with

$$t = \frac{\text{JD}_{\text{TT}} - \text{J2000.0}}{36{,}525}.$$

These angles and the appropriate precession matrix may be calculated by calling the function `Precession1977` from the *AstroRef* package.

Often in the literature, additional precession angles are defined. Some of them are depicted in Fig. 6.12:

- π_A = angle at the intersection point R between the moving ecliptic (T) and the original ecliptic (T_0)

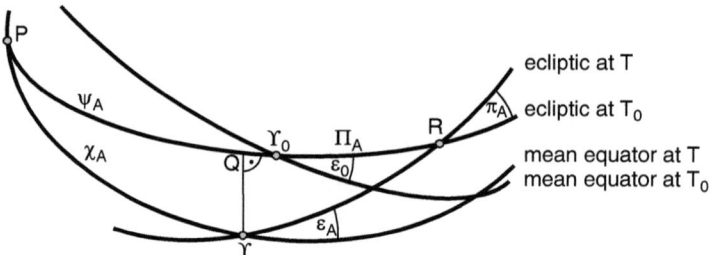

Fig. 6.12 Further precession angles

- Π_A = longitude of the ascending node of the moving ecliptic (T) with respect to the original ecliptic (T_0)
- ϵ_A = mean obliquity of the ecliptic (T); ϵ_0 = mean obliquity of the ecliptic (T_0)
- ψ_A = precession of the equator = angle in the original ecliptic reckoned from the original equinox (Υ_0) to the dynamical mean equator (P)
- χ_A = precession of the ecliptic = angle in the dynamical equator (T), reckoned from the original ecliptic (P) to the dynamical mean equator (Υ)

The expressions for these angles given by (Lieske et al. 1977) read

$$\pi_A = 47''.0029\,t - 0''.03302\,t^2 + 0''.000060\,t^3$$
$$\Pi_A = 174°52'34''.982 - 869''.8089\,t + 0''.03536\,t^2$$
$$\psi_A = 5038''.7784\,t - 1''.07259\,t^2 - 0''.001147\,t^3 \quad (6.60)$$
$$\chi_A = 10''.5526\,t - 2''.38064\,t^2 - 0''.001125\,t^3.$$

The true obliquity of the ecliptic ϵ_T at a certain instant of time results as a sum of the mean ecliptic of date, ϵ_A, and the nutation in obliquity.

Exercise 6.8: Around 300 BC, Egyptian astronomers divided the celestial zone along the ecliptic into 12 equally wide parts. The stellar constellations within these sections became famous as the zodiac. At their time the vernal equinox was located within the stellar constellation Aries. Find out how many constellations the vernal equinox passed by up to now because of precession. Use the *AstroRef* function `Precession` for your calculations.

6.5.1 Approximate Treatment of Precession

For short time spans around the reference epoch, the precession angles will be small and one may resort to various approximations. The mean equator of the reference epoch and that of the dynamical epoch are almost parallel to each other. The same holds for the two ecliptic planes. This situation is depicted in Fig. 6.13.

6.5 Astronomical Precession

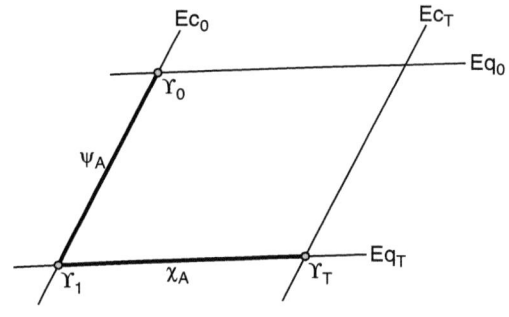

Fig. 6.13 Approximate treatment of precession with the angles ψ_A and χ_A. Ec$_0$ and Ec$_T$ mark the ecliptic planes at the reference and the dynamical epoch, while Eq$_0$ and Eq$_T$ mark the corresponding equators

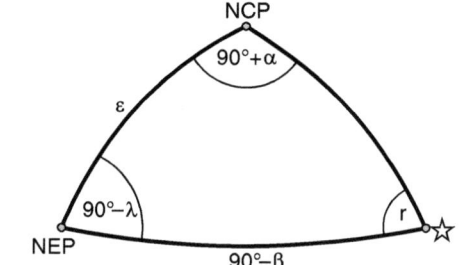

Fig. 6.14 Elimination of the parallactic angle r

Here it is convenient to work with the two angles *precession of the equator*, ψ_A, and *precession of the ecliptic*, χ_A.

Let us first consider the precession of the equator only. This will leave the ecliptic unchanged but the equator will move from Eq$_0$ to Eq$_T$. As a consequence, the equinox will move from Υ_0 to Υ_1 (Fig. 6.13). We see that the angle ψ_A can be interpreted as small change in the ecliptic longitude dλ. The corresponding changes in right ascension and declination can then be derived from the differential relations

$$d\alpha = (\cos r \cos \beta \, d\lambda - \sin r \, d\beta - \sin \delta \cos \alpha \, d\epsilon) \sec \delta$$
$$d\delta = \sin r \cos \beta \, d\lambda + \cos r \, d\beta + \sin \alpha \, d\epsilon$$

with

$$d\beta = d\epsilon = 0; \quad d\lambda = \psi_A$$

and the parallactic angle r. From the triangle in Fig. 6.14, we obtain

$$\sin r \cos \beta = \sin \epsilon \cos \alpha$$
$$\cos r \cos \beta = \cos \epsilon \cos \delta + \sin \epsilon \sin \alpha \sin \delta.$$

Insertion of these relations into

$$\Delta \alpha = \alpha - \alpha_0 = \psi_A \cos r \cos \beta \sec \delta$$
$$\Delta \delta = \delta - \delta_0 = \psi_A \sin r \cos \beta$$

we finally end up with

$$\Delta\alpha = \psi_A(\cos\epsilon + \sin\epsilon \sin\alpha \tan\delta)$$
$$\Delta\delta = \psi_A \sin\epsilon \cos\alpha.$$

The precession of the ecliptic only reduces the value for right ascension by χ_A, so that general precession approximately can be described by

$$(\Delta\alpha)_{\text{Pr}} = (\psi_A \cos\epsilon - \chi_A) + \psi_A \sin\epsilon \sin\alpha \tan\delta$$
$$(\Delta\delta)_{\text{Pr}} = \psi_A \sin\epsilon \cos\alpha. \qquad (6.61)$$

For very short time spans this can be written in the form

$$(\Delta\alpha)_{\text{Pr}} = (m + n \sin\alpha \tan\delta)(T - T_0)$$
$$(\Delta\delta)_{\text{Pr}} = n \cos\alpha (T - T_0) \qquad (6.62)$$

with

$$m \equiv \frac{d\psi_A}{dt} \cos\epsilon - \frac{d\chi_A}{dt}$$
$$n \equiv \frac{d\psi_A}{dt} \sin\epsilon. \qquad (6.63)$$

6.6 Astronomical Nutation

The dominant contribution to astronomical nutation results from the fact that the lunar orbit is inclined by about 5° with respect to the ecliptic and the motion of the lunar node. This leads to periodic variations of the celestial equator with respect to the ecliptic called *nutation in obliquity* and denoted by $\Delta\epsilon$. Another reason for astronomical nutation is the eccentricity of the Earth orbit about the Sun leading to *nutation in longitude*, $\Delta\psi$.

A purely precessing coordinate system defines *mean quantities*, a system undergoing precession, and nutation defines true quantities (true equinox, true celestial equator, etc.). The action of nutation is described with the nutation matrix. If the precessional motion has been taken into account, we face two planes: the mean ecliptic of date and the mean equator of date. Now nutation does not influence the location of the ecliptic but the location of the equator. By means of the nutation matrix, we get from the mean equator to the true equator of date. This is depicted in Fig. 6.15.

The nutation in obliquity, $\Delta\epsilon$, is defined by

$$\epsilon_T = \epsilon_A + \Delta\epsilon, \qquad (6.64)$$

6.6 Astronomical Nutation

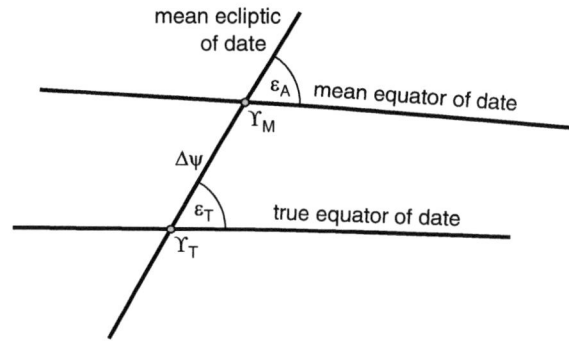

Fig. 6.15 The nutation angles $\Delta\psi$ and $\Delta\epsilon = \epsilon_T - \epsilon_A$

where the mean obliquity of the ecliptic is given by

$$\epsilon_A = 23°26'21''.448 - 46''.8150 t - 0''.00059 t^2 + 0''.001813 t^3 \tag{6.65}$$

with $t = (\text{JD}_{TT} - \text{J2000.0})/36{,}525$ (cf., *AstroRef* function `Obliquity`, appendix B). From Fig. 6.15, we can read off the nutation matrix $\mathcal{N}$:

$$\mathcal{N} = \mathcal{R}_1(-\epsilon_T)\mathcal{R}_3(-\Delta\psi)\mathcal{R}_1(\epsilon_A). \tag{6.66}$$

The following rotations are employed:

- A rotation about the x-axis with angle ϵ_A moves the x–y-plane of the mean equator of date to the mean ecliptic of date.
- A second rotation about the z-axis with angle $-\Delta\psi$ shifts the x-axis from the mean equinox of date (Υ_M) to the true equinox of date (Υ_T).
- A third rotation about the x-axis with angle $-\epsilon_T$ moves the x–y-plane into the true equator of date.

The nutation matrix, $\mathcal{N}$, transforms the mean system of date, $\mathbf{x}_M$, into the true system of date $\mathbf{x}_T$:

$$\mathbf{x}_T = \mathcal{N} \mathbf{x}_M. \tag{6.67}$$

The angles $\Delta\epsilon$ and $\Delta\psi$ are called nutation in obliquity and longitude (see Fig. 6.15). Explicitly the components of the nutation matrix read

$$n_{11} = \cos\Delta\psi$$
$$n_{12} = -\sin\Delta\psi\cos\epsilon_A$$
$$n_{13} = -\sin\Delta\psi\sin\epsilon_A$$
$$n_{21} = \sin\Delta\psi\cos\epsilon$$
$$n_{22} = \cos\Delta\psi\cos\epsilon\cos\epsilon_A + \sin\epsilon\sin\epsilon_A$$
$$n_{23} = \cos\Delta\psi\cos\epsilon\sin\epsilon_A - \sin\epsilon\cos\epsilon_A$$

$$n_{31} = \sin \Delta\psi \sin \epsilon$$
$$n_{32} = \cos \Delta\psi \sin \epsilon \cos \epsilon_A - \cos \epsilon \sin \epsilon_A$$
$$n_{33} = \cos \Delta\psi \sin \epsilon \sin \epsilon_A + \cos \epsilon \cos \epsilon_A \tag{6.68}$$

The nutation angles $\Delta\psi$ and $\Delta\epsilon$ are usually derived from a nutation series. One writes

$$\Delta\psi = \sum_i \left[(A_i + A'_i t) \sin(\arg_i) + (A''_i + A'''_i t) \cos(\arg_i)\right]$$
$$\Delta\epsilon = \sum_i \left[(B_i + B'_i t) \cos(\arg_i) + (B''_i + B'''_i t) \sin(\arg_i)\right]. \tag{6.69}$$

The sum extends over a whole series of partial nutations. The quantities $A_i, B_i, \ldots$ are the nutation amplitudes; $\arg_i$ indicate the arguments or phases. The partial nutations are dominated by the amplitudes A_i and B_i, i.e., the dominant contributions to $\Delta\psi$ go with $\sin(\arg_i)$ and those to $\Delta\epsilon$ with $\cos(\arg_i)$. These are the *inphase terms* of nutation; the A''_i and B''_i are the *out-of-phase* terms. In the IAU 1980 nutation series, only inphase terms were given, while the IAU 2000 series additionally includes out-of-phase terms. The arguments or phases are derived from so-called fundamental angles. Nutation series of low accuracy employ only the elementary fundamental angles l, l', F, D, Ω. Any argument is then given by

$$\arg_i = i_l l + i_{l'} l' + i_F F + i_D D + i_\Omega \Omega, \tag{6.70}$$

where the factors i_l, $i_{l'}$, etc. take integer values. For the elementary fundamental angles, one has

$$l = L - \varpi$$
$$l' = L' - \varpi'$$
$$F = L - \Omega \tag{6.71}$$
$$D = L - L'$$

with

L: longitude of the Moon,
L': longitude of the Sun,
ϖ: longitude of the lunar pericenter,
ϖ': longitude of the solar pericenter, and
Ω: longitude of the lunar ascending node.

More precise nutation series, like IAU 2000, do include additional fundamental angles describing the mean longitudes of the major planets. Definitions of the principle fundamental angles according to the IAU 2000 nutation theory are given in (9.50).

6.6 Astronomical Nutation

Table 6.1 Some important nutation terms of the IAU 2000 series

l	l'	F	D	Ω	Period	A_i	B_i
0	0	0	0	1	6798.383	−17206.4161	9205.2331
0	0	2	−2	2	182.621	−1317.0906	573.0336
0	0	2	0	2	13.661	−227.6413	97.8459
0	0	0	0	2	3399.192	207.4554	−89.7492
0	1	0	0	0	365.260	147.5877	7.3871
0	1	2	−2	2	121.749	−51.6821	22.4386
1	0	0	0	0	27.555	71.1159	−0.6750
0	0	2	0	1	13.633	−38.7298	20.0728
1	0	2	0	2	9.133	−30.1461	12.9025
0	−1	2	−2	2	365.225	21.5829	−9.5929

Periods are given in days, amplitudes in mas

The most important inphase nutation terms of the IAU 2000 series are given in Table 6.1. Let us first consider the first line in Table 6.1. It starts with integer numbers, presenting factors in front of the fundamental angles, that defines the argument of the first partial nutation, which reads $0 \times l + 0 \times l' + 0 \times F + 0 \times D + 1 \times \Omega = \Omega$. The amplitude of $\Delta\psi$ of the first (inphase) nutation term reads $A_i = -17\rlap{.}''206$; the one of $\Delta\epsilon$ reads $B_i = 9\rlap{.}''205$. The first line of Table 6.1 thus leads to

$$\Delta\psi = -17\rlap{.}''206 \sin \Omega \, ; \qquad \Delta\epsilon = 9\rlap{.}''205 \cos \Omega .$$

Accordingly the argument of the second partial nutation reads $\arg_2 = 2F - 2D + 2\Omega$, etc. To get the total values for the nutation in longitude and obliquity, we have to sum over all partial nutations.

Besides the inphase terms, sine terms in $\Delta\psi$ and cosine terms in $\Delta\epsilon$, we generally also face out-of-phase terms. Also in every more precise nutation series, the amplitudes will slowly vary with time. Finally, also the influences of the major planets have to be taken into account for precise nutation series. The dominant planetary term has a period of 311921.26 days and amplitudes $A_i = -0.3084$ mas, $A_i'' = 0.5123$ mas, $B_i = 0.1647$ mas, and $B_i'' = 0.2735$ mas. To get an idea about the orders of magnitudes of out-of-phase terms, time-dependent amplitudes, and planetary influences, we refer to Table 9.3.

Exercise 6.9: Compute the nutation angles $\Delta\epsilon$ and $\Delta\psi$ with *AstroRef* using the functions `Nutation1980` (nutation theory IAU 1980) as well as `Nutation` (nutation theory IAU 2000) for 1 January, 12^h TT, of the years 2000 until 2010.

Exercise 6.10: From Table 6.1, select all partial nutations with amplitudes larger than $0\rlap{.}''1$ to generate a plot of the nutation ellipse in the following sense: keep the mean pole at the origin of the coordinate system, i.e., only consider the nutational motion. In this coordinate system, plot the coordinates of the true pole ($X_P = \Delta\psi \sin \epsilon, Y_P = \Delta\epsilon$) for the time span of the longest partial nutation period.

6.7 Apparent Places

For practical applications so-called apparent places of astronomical bodies are of great interest. Starting from a catalog place (see the Introduction, Chapter 1 or more precise Section 1.1.3), the corresponding apparent place involves corrections for proper motion, annual aberration, and parallax, as well as precession and nutation. In the past Apparent Places of Fundamental Stars (APFS) have been published by *Astronomisches Recheninstitut* (ARI), Heidelberg.

With *AstroRef*, apparent places of stars or major planets may be calculated using `AppPlaceStar` or `AppPlaceSolarSystem`, respectively. Furthermore, a very efficient algorithm for the apparent place of the Sun is available with `AppPlaceSunFast`.

Exercise 6.11: With the *AstroRef* function `AppPlaceSunFast`, compute the apparent place of the Sun for 1 January, 12^h TT, for the following years: 1600, 1800, 1900, and 2000. Compare the results with those from DE405 by using the IMCCE Ephemeris website (http://www.imcce.fr/en/ephemerides/generateur_ephemerides.php).

Exercise 6.12: Use the *AstroRef* function `AppPlaceSolarSystem` to compute apparent places (α, δ) for the Moon and all planets for 1 March 1980, 0^h UTC.

6.8 High-Precision Astrometry

6.8.1 Gravitational Light Deflection

We will now study the propagation of light rays in our canonical barycentric metric in more detail. We have already noted that light rays are given by null geodesics, i.e., curves of the form $x_\gamma^\mu(\lambda)$, that satisfy the geodesic equation (3.12) (suppressing the index γ mostly in the following)

$$\frac{d^2 x^\alpha}{d\lambda^2} + \Gamma^\alpha_{\mu\nu} \frac{dx^\mu}{d\lambda} \frac{dx^\nu}{d\lambda} = 0,$$

especially

$$\frac{d^2 t}{d\lambda^2} = -\frac{1}{c} \Gamma^0_{\mu\nu} \frac{dx^\mu}{d\lambda} \frac{dx^\nu}{d\lambda}.$$

Now,

$$\frac{dx^i}{d\lambda} = \frac{dx^i}{dt} \frac{dt}{d\lambda}$$

and

$$\frac{d^2 x^i}{d\lambda^2} = \frac{d^2 x^i}{dt^2} \left(\frac{dt}{d\lambda}\right)^2 + \frac{dx^i}{dt} \left(\frac{d^2 t}{d\lambda^2}\right)$$

6.8 High-Precision Astrometry

so that

$$\frac{d^2 x^i}{dt^2} = \left(\frac{d\lambda}{dt}\right)^2 \left[\frac{d^2 x^i}{d\lambda^2} - \left(\frac{d^2 t}{d\lambda^2}\right) \frac{dx^i}{dt}\right] \tag{6.72}$$

$$= -\Gamma^i_{\mu\nu} \frac{dx^\mu}{dt} \frac{dx^\nu}{dt} + \frac{1}{c} \Gamma^0_{\mu\nu} \frac{dx^\mu}{dt} \frac{dx^\nu}{dt} \frac{dx^i}{dt}$$

or,

$$\frac{d^2 x^i_\gamma}{dt^2} = \left(\Gamma^0_{\mu\nu} \frac{1}{c} \frac{dx^i_\gamma}{dt} - \Gamma^i_{\mu\nu}\right) \frac{dx^\mu_\gamma}{dt} \frac{dx^\nu_\gamma}{dt}. \tag{6.73}$$

The null condition for a light ray reads

$$0 = ds^2 = g_{\mu\nu} dx^\mu_\gamma dx^\nu_\gamma = g_{\mu\nu} \frac{dx^\mu_\gamma}{dt} \frac{dx^\nu_\gamma}{dt}. \tag{6.74}$$

A light ray emitted at coordinate time t_e at a point $\mathbf{x}_e$ in an initial direction given by the Euclidean unit vector $\mathbf{n}$ ($\dot{\mathbf{x}}_\gamma(-\infty) = c\mathbf{n}$) can be written in the form

$$\mathbf{x}_\gamma(t) = \mathbf{x}_e + \mathbf{n} c (t - t_e) + \mathbf{x}_P \equiv \mathbf{x}_N + \mathbf{x}_P, \tag{6.75}$$

where we have assumed that a light particle in the Newtonian approximation moves along a straight line $\mathbf{x}_N$ with the vacuum speed of light and $\mathbf{x}_P$ describes a post-Newtonian correction to it. With the Christoffel symbols from (3.15), we can assess the orders of magnitude of the various terms in (6.73). With

$$\frac{dx^0_\gamma}{dt} = c, \qquad \frac{dx^i_\gamma}{dt} = c n^i + \cdots$$

we find

$$\frac{1}{c} \Gamma^0_{00} \frac{dx^i}{dt} \frac{dx^0}{dt} \frac{dx^0}{dt} = \mathcal{O}(c^{-1}),$$

$$\frac{1}{c} \Gamma^0_{0j} \frac{dx^i}{dt} \frac{dx^0}{dt} \frac{dx^j}{dt} = -\frac{1}{c^2} w_{,j} \frac{dx^j}{dt} \frac{dx^i}{dt} + \mathcal{O}(c^{-2}),$$

$$\frac{1}{c} \Gamma^0_{jk} \frac{dx^i}{dt} \frac{dx^j}{dt} \frac{dx^k}{dt} = \mathcal{O}(c^{-1})$$

$$\Gamma^i_{00} \frac{dx^0}{dt} \frac{dx^0}{dt} = -w_{,i} + \mathcal{O}(c^{-2}),$$

$$\Gamma^i_{0j} \frac{dx^0}{dt} \frac{dx^j}{dt} = \mathcal{O}(c^{-1}),$$

$$\Gamma^i_{jk} \frac{dx^j}{dt} \frac{dx^k}{dt} = \frac{1}{c^2} \left(\delta_{ij} w_{,k} + \delta_{ik} w_{,j} - \delta_{jk} w_{,i}\right) \frac{dx^j}{dt} \frac{dx^k}{dt} + \mathcal{O}(c^{-2})$$

$$= \frac{1}{c^2} \left(2 w_{,k} \frac{dx^k}{dt} \frac{dx^i}{dt} - w_{,i} \left|\frac{d\mathbf{x}}{dt}\right|^2\right) + \mathcal{O}(c^{-2}). \tag{6.76}$$

From this we can write the post-Newtonian geodesic equation (6.73) in the form

$$\frac{d^2 x^i}{dt^2} = w_{,i}\left(1 + \frac{1}{c^2}\left|\frac{d\mathbf{x}}{dt}\right|^2\right) - \frac{4}{c^2}\frac{dx^i}{dt}\left(\frac{d\mathbf{x}}{dt}\cdot\nabla w\right). \tag{6.77}$$

In a similar way (6.74) yields to post-Newtonian order

$$0 = \left(-1 + \frac{2w}{c^2}\right)c^2 + \delta_{ij}\left(1 + \frac{2w}{c^2}\right)\frac{dx^i}{dt}\frac{dx^j}{dt}, \tag{6.78}$$

that according to

$$\frac{dx^i}{dt} = c\,n^i + \frac{dx_P^i}{dt} + \cdots$$

can be written in the form

$$\mathbf{n}\cdot\frac{1}{c}\frac{d\mathbf{x}_P}{dt} = -\frac{2w}{c^2}. \tag{6.79}$$

Considering this in the geodesic equation to post-Newtonian order, we obtain

$$\frac{d^2 \mathbf{x}_P}{dt^2} = 2\left[\nabla w - 2\mathbf{n}(\mathbf{n}\cdot\nabla w)\right]. \tag{6.80}$$

Defining

$$\mathbf{x}_{P\parallel}(t) \equiv \mathbf{n}\cdot\mathbf{x}_P(t),$$

$$\mathbf{x}_{P\perp}(t) \equiv (1 - \mathbf{n}\cdot\mathbf{n})\,\mathbf{x}_P(t),$$

we get

$$\frac{1}{c}\frac{d\mathbf{x}_{P\parallel}}{dt} = -\frac{2w}{c^2} \tag{6.81}$$

and

$$\frac{d^2 \mathbf{x}_{P\perp}}{dt^2} = \frac{d^2 \mathbf{x}_P}{dt^2} - \mathbf{n}\left(\mathbf{n}\cdot\frac{d^2 \mathbf{x}_P}{dt^2}\right)$$

$$= 2\nabla w - 4\mathbf{n}(\mathbf{n}\cdot\nabla w) - 2\mathbf{n}(\mathbf{n}\cdot\nabla w) + 4\mathbf{n}(\mathbf{n}\cdot\nabla w)$$

or

$$\frac{d^2 \mathbf{x}_{P\perp}}{dt^2} = 2\left[\nabla w - \mathbf{n}(\mathbf{n}\cdot\nabla w)\right]. \tag{6.82}$$

We will now consider the light deflection in the gravitational field of a mass monopole located at the origin of our coordinate system. Since we need the scalar gravitational potential only to Newtonian order, we get

$$w \simeq \frac{GM}{r}$$

6.8 High-Precision Astrometry

with
$$r \simeq r_N = |\mathbf{x}_e + \mathbf{n}\, c(t - t_e)|.$$

Since
$$\nabla w = -\frac{GM}{r^3}\mathbf{x}$$

we find
$$\frac{d^2 \mathbf{x}_{P\perp}}{dt^2} = -\frac{2\,GM}{r^3}[\mathbf{x} - \mathbf{n}(\mathbf{n}\cdot\mathbf{x})] = -\frac{2\,GM}{r^3}\mathbf{d}, \qquad (6.83)$$

where
$$\mathbf{d} = \mathbf{n} \times (\mathbf{x} \times \mathbf{n}) = \mathbf{n} \times (\mathbf{x}_e \times \mathbf{n}) \qquad (6.84)$$

describes a vector that points from the gravitating mass to the point of closest approach of the unperturbed light ray. Since $r^2 = \mathbf{x}^2 = \mathbf{x}_e^2 + 2c\,\mathbf{n}\cdot\mathbf{x}_e(t - t_e) + c^2(t - t_e)^2$ we get

$$\frac{1}{c}\dot{\mathbf{x}}_{P\perp} = -\frac{2\,GM}{c}\mathbf{d} \int_{t_e}^{t} dt\, \left[\mathbf{x}_e^2 + 2c\,\mathbf{n}\cdot\mathbf{x}_e(t - t_e) + c^2(t - t_e)^2\right]^{-3/2},$$

i.e., we are looking at an integral of the form

$$\int dx\, \left[Ax^2 + Bx + C\right]^{-3/2} = \frac{2(2Ax + B)}{\Delta\sqrt{Ax^2 + Bx + C}}$$

with $\Delta \equiv 4AC - B^2$. In our case $\Delta = 4c^2\left[\mathbf{x}_e^2 - (\mathbf{n}\cdot\mathbf{x}_e)^2\right]$. Since

$$d^2 = [\mathbf{n} \times (\mathbf{x}_e \times \mathbf{n})]^2 = [\mathbf{x}_e - \mathbf{n}(\mathbf{n}\cdot\mathbf{x}_e)]^2 = \mathbf{x}_e^2 - (\mathbf{n}\cdot\mathbf{x}_e)^2$$

one obtains $\Delta = 4c^2 d^2$ and $2Ax + B = 2c^2(t - t_e) + 2c\,\mathbf{n}\cdot\mathbf{x}_e = 2c\,\mathbf{x}_N\cdot\mathbf{n}$. From this one derives

$$\frac{1}{c}\dot{\mathbf{x}}_{P\perp} = -\frac{2\,GM}{c^2}\frac{\mathbf{d}}{d^2}\left(\frac{\mathbf{x}_N(t)\cdot\mathbf{n}}{r_N(t)} - \frac{\mathbf{x}_e\cdot\mathbf{n}}{r_e}\right),$$

that, together with (6.81), leads to the relation ($r_e = |\mathbf{x}_e|$)

$$\frac{1}{c}\dot{\mathbf{x}} = \left(1 - \frac{2m}{r}\right)\mathbf{n} - \frac{2m}{d^2}\mathbf{d}\left(\frac{\mathbf{x}_N(t)\cdot\mathbf{n}}{r} - \frac{\mathbf{x}_e\cdot\mathbf{n}}{r_e}\right) \qquad (6.85)$$

with
$$m = \frac{GM}{c^2}.$$

In the following, we consider the situation where a remote light source at $\mathbf{x}_Q$ is observed from the Earth ($\mathbf{x}_E$; $|\mathbf{x}_E| \ll |\mathbf{x}_Q|$), defining the Euclidean unit vectors (Fig. 6.16):

$$\mathbf{p} = -\mathbf{n}, \qquad \mathbf{e} = \mathbf{x}_E/|\mathbf{x}_E|, \qquad \mathbf{q} = \mathbf{x}_Q/|\mathbf{x}_Q|. \qquad (6.86)$$

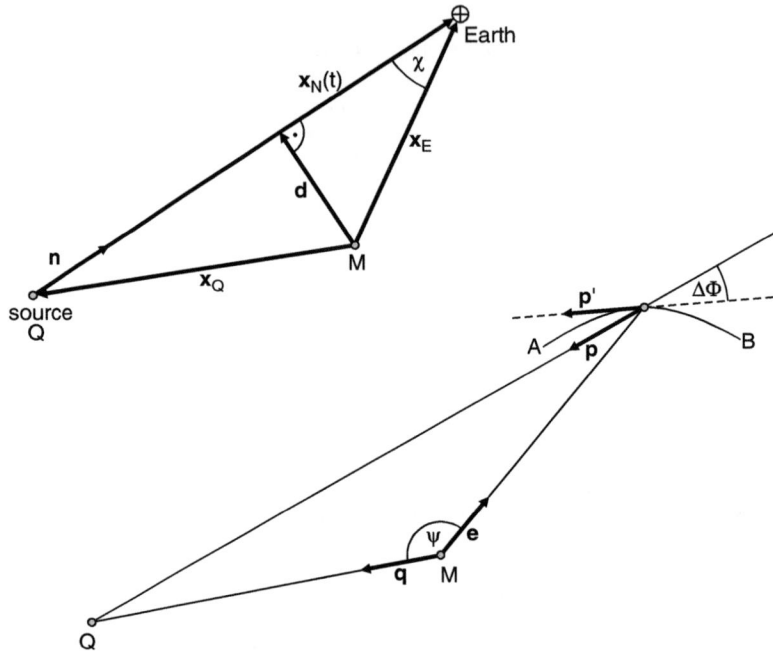

Fig. 6.16 Upper panel: geometry in the problem of light deflection in the gravitational field of a spherical mass M. Lower panel: the true light ray extends from A to B and is observed in a direction $\mathbf{p}'$

Here, $\mathbf{p}$ denotes a unit vector that points from the Earth to the unperturbed light ray. In that case we have $\mathbf{x}_N = \mathbf{x}_E$ and $\mathbf{x}_e = \mathbf{x}_Q$ with $\mathbf{x}_e \cdot \mathbf{n}/r_e = -1$ and $\mathbf{q} = \mathbf{p}$. The second term in (6.85), proportional to $\mathbf{d}$, describes the gravitational light deflection that leads to a post-Newtonian correction of the direction vector $\mathbf{p}$:

$$\delta \mathbf{p} = \frac{2m}{d^2} \mathbf{d}(\cos \chi + 1) \tag{6.87}$$

with $\cos \chi = -\mathbf{e} \cdot \mathbf{p}$ (Fig. 6.16). Since

$$\mathbf{d} = \mathbf{p} \times (\mathbf{x}_E \times \mathbf{p}) = r_E (\mathbf{e} - (\mathbf{p} \cdot \mathbf{e})\mathbf{p})$$

and

$$d^2 = r_E^2 (1 + \mathbf{e} \cdot \mathbf{p})(1 - \mathbf{e} \cdot \mathbf{p})$$

the equation for gravitational light deflection in the case of a very remote star can be written as

$$\mathbf{p}' = \mathbf{p} + \left(\frac{2\,GM}{c^2 r_E}\right) \frac{\mathbf{e} - (\mathbf{p} \cdot \mathbf{e})\mathbf{p}}{(1 + \mathbf{p} \cdot \mathbf{e})}. \tag{6.88}$$

Fig. 6.17 The five basic vectors in the formalism of high-precision astrometry; from Klioner (2003a)

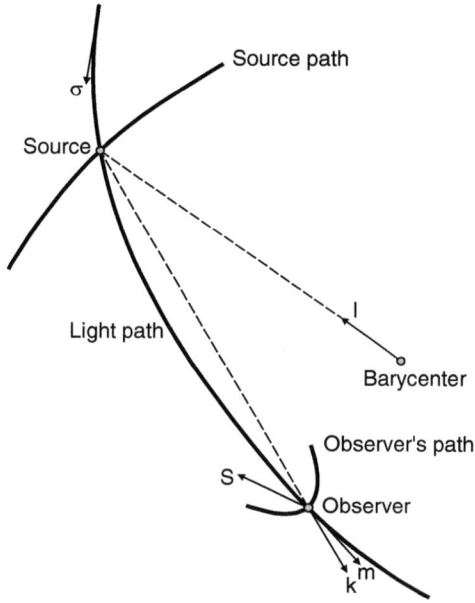

For the light deflection in the gravitational field of the Sun, the factor $(2\,GM_S/c^2 r_E) \simeq 2 \cdot 10^{-8}$. At the limb of the Sun, the deflection angle amounts to $1\rlap{.}''745$.

The effect of gravitational light deflection caused by the Sun on an observed position may be determined with the *AstroRef* function `LightDeflectionSun`.

6.8.2 The Klioner Formalism

A theory of astrometry at microarcsecond precision has been formulated by Klioner (2003a; see also Klioner & Kopeikin 1992). His formalism is based upon 5 basic Euclidean unit vectors: **s**, **m**, **σ**, **k**, and **l** (Fig. 6.17).

- **s** is the observed direction in the kinematically nonrotating local reference system of the observer.
- **m** is the BCRS unit vector tangential to the light ray at the moment of observation.
- **σ** is the BCRS unit vector tangential to the light ray at $t = -\infty$.
- **k** is the BCRS unit vector from the source to the observer.
- **l** is the BCRS unit vector from the barycenter to the source.

This implies that for a light source at finite distance, the emitted light ray is mathematically extended to past (timelike) infinity, such that the vector **σ** is well defined. High-precision astrometry can then be described by means of four consecutive transformations: **s** → **m** → **σ** → **k** → **l**.

6.8.2.1 Relativistic Aberration

The transformation from **s** to **m** accounts for relativistic aberration; it reads

$$\mathbf{s} = \left[\mathbf{s}' + \tilde{\gamma}\tilde{\boldsymbol{\beta}} + (\tilde{\gamma} - 1)(\tilde{\boldsymbol{\beta}} \cdot \mathbf{s}')\tilde{\boldsymbol{\beta}}/\tilde{\boldsymbol{\beta}}^2\right] \cdot \frac{1}{\tilde{\gamma}(1 + \tilde{\boldsymbol{\beta}} \cdot \mathbf{s}')} \quad (6.89)$$

where $\mathbf{s}' = -\mathbf{m}$, $\tilde{\gamma} = (1 - \tilde{\boldsymbol{\beta}}^2/c^2)^{-1/2}$, $\tilde{\boldsymbol{\beta}} = (\mathbf{v}_{\text{obs}}/c)\left(1 + 2w(\mathbf{x}_{\text{obs}})/c^2\right)$, $\mathbf{x}_{\text{obs}}$, and $\mathbf{v}_{\text{obs}}$ are the BCRS position and velocity of the observer. Without the gravitational w-term, this expression agrees with the aberration formula (6.37) from special relativity. Note that both vectors, **s** and **s**', are Euclidean unit vectors.

6.8.2.2 Gravitational Light Deflection

A second transformation $\mathbf{m} \to \boldsymbol{\sigma}$ accounts for the gravitational light deflection for remote sources. Considering only one mass monopole with

$$w \simeq \frac{GM}{r}$$

and one remote star that is observed, the BCRS metric leads to the following post-Newtonian result for the gravitational light deflection:

$$\mathbf{m} = \boldsymbol{\sigma} - \frac{2\,GM}{c^2}\frac{\mathbf{d}}{d^2}\left(1 + \frac{\boldsymbol{\sigma} \cdot \mathbf{x}_e}{|\mathbf{x}_e|}\right) \quad (6.90)$$

with

$$\mathbf{d} = \boldsymbol{\sigma} \times (\mathbf{x}_e \times \boldsymbol{\sigma}).$$

A third transformation, from $\boldsymbol{\sigma}$ to **k**,

$$\boldsymbol{\sigma} = \mathbf{k} + \frac{2m}{d^2}\mathbf{d}(|\mathbf{x}_{\text{obs}}| - |\mathbf{x}_e| + |\mathbf{x}_{\text{obs}} - \mathbf{x}_e|) \quad (6.91)$$

accounts for the gravitational light deflection for light sources located inside the solar system (Klioner 2003b). Here, $\mathbf{x}_{\text{obs}}$ is the BCRS vector of the observer.

6.8.2.3 Parallax

The fourth transformation, from **k** to **l**, describes the parallax. Let $(t_e, \mathbf{x}_e)$ and $(t_{\text{obs}}, \mathbf{x}_{\text{obs}})$ be the BCRS coordinates of the events of emission and observation, $\mathbf{R} = \mathbf{x}_{\text{obs}}(t_{\text{obs}}) - \mathbf{x}_e(t_e)$; then **k** and **l** are defined as

$$\mathbf{k} = \mathbf{R}/|\mathbf{R}|, \qquad \mathbf{l} = \mathbf{x}_e(t_e)/|\mathbf{x}_e(t_e)|.$$

6.8 High-Precision Astrometry

The relation between **k** and **l** is

$$\mathbf{k} = \eta(-\mathbf{l} + \mathbf{\Pi}) \tag{6.92}$$

with

$$\eta = \frac{|\mathbf{x}_e|}{|\mathbf{R}|} = |-\mathbf{l} + \mathbf{\Pi}|^{-1}.$$

Here,

$$\mathbf{\Pi}(t_{\text{obs}}) = \pi(t_{\text{obs}}) \frac{\mathbf{x}_{\text{obs}}(t_{\text{obs}})}{\text{AU}},$$

where AU is the astronomical unit and the parallax of the source, $\pi(t_{\text{obs}})$, is defined as

$$\pi(t_{\text{obs}}) = \frac{1 \text{ AU}}{|\mathbf{x}_e(t_e)|},$$

so that

$$\mathbf{\Pi} = \frac{\mathbf{x}_{\text{obs}}(t_{\text{obs}})}{|\mathbf{x}_e(t_e)|}.$$

Exercise 6.13: Show that to second order in $|\mathbf{\Pi}|$ the expression for **k** can be written in the form

$$\mathbf{k} = -\mathbf{l}\left(1 - \frac{1}{2}|\pi|^2\right) + \pi(1 + \mathbf{l} \cdot \mathbf{\Pi}) + \mathcal{O}(|\pi|^3)$$

with

$$\pi = \mathbf{l} \times (\mathbf{\Pi} \times \mathbf{l}).$$

Show first that

$$\eta = (1 - 2\mathbf{\Pi} \cdot \mathbf{l} + \mathbf{\Pi}^2)^{-1/2} \simeq 1 + \mathbf{\Pi} \cdot \mathbf{l} - \frac{1}{2}\mathbf{\Pi}^2 + \frac{3}{2}(\mathbf{\Pi} \cdot \mathbf{l})^2.$$

6.8.2.4 Proper Motion and Radial Velocity

To describe proper motion and the radial velocity of the light source, one might employ a simple model for its space motion in BCRS coordinates (e.g., Klioner 2000, Dravins et al. 1999):

$$\mathbf{x}_e(t_e) = \mathbf{x}_e(t_e^0) + \mathbf{v}\Delta t_e, \tag{6.93}$$

where $\Delta t_e = t_e - t_e^0$ and **v** is the BCRS velocity of the source at t_e^0. Here, t_e^0 corresponds to some initial epoch of observation t_{obs}^0. If t_e denotes the emission time of a certain photon, we have to sufficient approximation

$$c(t_{\text{obs}} - t_e) = |\mathbf{x}_{\text{obs}}(t_{\text{obs}}) - \mathbf{x}_e(t_e)|. \tag{6.94}$$

For some fictitious observer at the barycenter, B, one has

$$c(t_B - t_e) = |\mathbf{x}_e(t_e)|. \tag{6.95}$$

From the last two equations we can derive a relation between t_{obs} and t_B:

$$t_B = t_{\text{obs}} + \frac{|\mathbf{x}_e(t_e)| - |\mathbf{x}_e(t_e) - \mathbf{x}_{\text{obs}}(t_{\text{obs}})|}{c}$$

$$\simeq t_{\text{obs}} + \frac{1}{c}\mathbf{l} \cdot \mathbf{x}_{\text{obs}}(t_{\text{obs}}). \tag{6.96}$$

Let t_B^0 be the reference epoch for some astrometric catalog; then the corresponding emission time, t_e^0, is given by

$$c\left(t_B^0 - t_e^0\right) = \left|\mathbf{x}_e\left(t_e^0\right)\right|. \tag{6.97}$$

In the XV representation the source position is written in the form (6.93). The problem is to get a suitable representation for Δt_e. From (6.94)–(6.97), we get

$$\Delta t_e = t_e - t_e^0 = t_{\text{obs}} - t_B^0 + \frac{1}{c}\mathbf{l} \cdot \mathbf{x}_{\text{obs}}(t_{\text{obs}}) + \frac{1}{c}\left(\left|\mathbf{x}_e\left(t_e^0\right)\right| - |\mathbf{x}_e(t_e)|\right)$$

and the last term is (to first order in v) given by $-c^{-1}(\mathbf{l}_0 \cdot \mathbf{v})\Delta t_e$, where $\mathbf{l}_0$ refers to time t_e^0, so that finally

$$\Delta t_e \simeq \frac{\tau}{1 + \mathbf{l}_0 \cdot \mathbf{v}/c} \tag{6.98}$$

with

$$\tau = t_{\text{obs}} - t_B^0 + \frac{1}{c}\mathbf{l} \cdot \mathbf{x}_{\text{obs}}(t_{\text{obs}}). \tag{6.99}$$

The source's BCRS spatial coordinates are then given by

$$\mathbf{x}_e(t_e) = \mathbf{x}_e(t_e^0) + \mathbf{v}_{\text{app}}\,\tau, \tag{6.100}$$

where the apparent source velocity, $\mathbf{v}_{\text{app}}$, is given by

$$\mathbf{v}_{\text{app}} \equiv \frac{\mathbf{v}}{1 + \mathbf{l}_0 \cdot \mathbf{v}/c}. \tag{6.101}$$

In the XV representation the six quantities $\mathbf{x}_e(t_e^0)$ and $\mathbf{v}$ are used to characterize the source position. By inserting (6.93) into the definition for $\mathbf{l}$ and $\pi(t_{\text{obs}})$, we obtain

$$\mathbf{l} = \mathbf{l}_0 + \dot{\mathbf{l}}_0\,\Delta t_e, \qquad \pi(t_{\text{obs}}) = \pi_0 + \dot{\pi}_0\,\Delta t_e. \tag{6.102}$$

Expressions for $\mathbf{l}_0$ and π_0 are given below; expressions for $\dot{\mathbf{l}}_0$ and $\dot{\pi}_0$ can be found in Klioner (2003b).

In the PPM representation (parallax and proper motion) one employs the following quantities referring to some reference epoch t_B^0:

6.8 High-Precision Astrometry

- α_0 (right ascension)
- δ_0 (declination)
- π_0 (parallax)
- $\mu_{\alpha 0}$ (apparent proper motion in α)
- $\mu_{\delta 0}$ (apparent proper motion in δ)
- μ_{r0} (apparent radial velocity times parallax over 1 AU)

Then,

$$\mathbf{l}_0 = \frac{\mathbf{x}_e(t_e^0)}{|\mathbf{x}_e(t_e^0)|} = \begin{pmatrix} \cos\delta_0 \cos\alpha_0 \\ \cos\delta_0 \sin\alpha_0 \\ \sin\delta_0 \end{pmatrix}. \tag{6.103}$$

Defining

$$\boldsymbol{\mu} \equiv \mathbf{v}_{\text{app}} \frac{\pi_0}{\text{AU}} \tag{6.104}$$

with

$$\pi_0 \equiv \frac{\text{AU}}{|\mathbf{x}_e(t_e^0)|}$$

we can decompose the space motion vector $\boldsymbol{\mu}$ by using the following set of orthonormal vectors:

$$\mathbf{e}_{(r)}^0 = \mathbf{l}_0; \quad \mathbf{e}_{(\alpha)}^0 = \begin{pmatrix} -\sin\alpha_0 \\ +\cos\alpha_0 \\ 0 \end{pmatrix}; \quad \mathbf{e}_{(\delta)}^0 = \begin{pmatrix} -\sin\delta_0 \cos\alpha_0 \\ -\sin\delta_0 \sin\alpha_0 \\ \cos\delta_0 \end{pmatrix} \tag{6.105}$$

in the form

$$\boldsymbol{\mu} = \mu_{r0} \mathbf{e}_{(r)}^0 + \mu_{\alpha 0} \mathbf{e}_{(\alpha)}^0 + \mu_{\delta 0} \mathbf{e}_{(\delta)}^0 \tag{6.106}$$

with

$$\mu_{r0} = \boldsymbol{\mu} \cdot \mathbf{e}_{(r)}^0; \quad \mu_{\alpha 0} = \boldsymbol{\mu} \cdot \mathbf{e}_{(\alpha)}^0; \quad \mu_{\delta 0} = \boldsymbol{\mu} \cdot \mathbf{e}_{(\delta)}^0.$$

The last equations show how the six quantities from the PPM representation can be obtained from $\mathbf{x}_e(t_e^0)$ and $\mathbf{v}_0$. If the PPM quantities are given, $\mathbf{l}_0$ is obtained from (6.103) and $\mathbf{x}_e(t_e^0) = \mathbf{l}_0(\text{AU}/\pi_0)$; $\boldsymbol{\mu}$ is obtained from (6.106) giving the apparent space motion vector $\mathbf{v}_{\text{app}}$. Finally,

$$\mathbf{v}_0 \simeq \frac{\mathbf{v}_{\text{app}}}{1 - \mathbf{l}_0 \cdot \mathbf{v}_{\text{app}}/c}. \tag{6.107}$$

Chapter 7
Celestial Reference System

7.1 Concepts

7.1.1 Barycentric Celestial Reference System

According to IAU 2000 Resolution B1.3 (e.g., Soffel et al. 2003), the Barycentric Celestial Reference System (BCRS) is defined by the form of the BCRS metric tensor that agrees with our canonical barycentric one (3.2). As we have noted earlier, this form of the metric tensor implies that all kinds of material or energy outside the solar system have been ignored. This leads to the asymptotic condition

$$\lim_{r \to \infty} g_{\mu\nu} = \mathrm{diag}(-1, +1, +1, +1).$$

7.1.1.1 The BCRS and the Cosmic Expansion

There have been several attempts to include the cosmic expansion, described by some cosmic scale factor $a(t)$, into this metric (e.g., Cooperstock et al. 1998; Soffel and Klioner 2004; Klioner and Soffel 2004). One tries to match the cosmological Friedmann metric (e.g., Weinberg 1972) to the BCRS metric. In Klioner and Soffel (2004), it is shown that the cosmic expansion leads to an additional term in g_{00} of the form

$$\frac{1}{c^2}\frac{\ddot{a}(t)}{a(t)},$$

i.e., to an additional central cosmic tidal force with perturbing function

$$R = \frac{1}{2}\frac{\ddot{a}}{a}|\mathbf{x}|^2. \tag{7.1}$$

The quantity $\ddot{a}/a$ can be calculated as

$$\frac{\ddot{a}}{a} = -qH(t)^2 \qquad (7.2)$$

where $H(t)$ and $q(t)$ are the Hubble constant and the deceleration parameter, respectively. Using present values for $q \sim -0.6$ and $H \sim 71$ km/s/Mpc, one gets

$$\frac{\ddot{a}}{a} = 3.2 \times 10^{-23} \, \text{s}^{-2}.$$

This is the value for $\ddot{a}/a$ for the present epoch. The time dependence of this quantity is different for different cosmological models. In a flat de Sitter universe, $\ddot{a}/a$ is time independent, $\ddot{a}/a = \Lambda c^2/3$, where Λ is the cosmological constant. In general, however, $\ddot{a}/a$ is a time-dependent quantity. At the distance of Pluto with $r = 40$ AU, the maximal value of the disturbing cosmological acceleration amounts to 2×10^{-23} m/s^2. Klioner and Soffel (2004), using the perturbing function (7.1), have studied the consequences of the cosmic expansion upon planetary orbits: inclination, ascending node, and semi-latus rectum ($p = a(1 - e^2)$) are constant for a radial perturbation; secular perturbations of the argument of perihelion ω and the mean anomaly M for Pluto are about 10^{-5} μas per century and three orders of magnitude smaller for Mercury. Amplitudes for periodic perturbations in semimajor axis, $\Delta a/a$, and eccentricity, Δe, are about 10^{-17} for Pluto and 10^{-23} for Mercury. Therefore, perturbations of the solar system dynamics caused by the expansion of the universe are completely negligible over time, much less than the age of the universe (13.6 billion years).

7.1.2 Geocentric Celestial Reference System

IAU 2000 Resolution B1.3 further defines a Geocentric Celestial Reference System (GCRS), whose spatial coordinates **X** are considered to be kinematically nonrotating with respect to the BCRS spatial coordinates. By replacing the δ_{ai}-term in the relation between x^μ and X^α by a general (slowly time dependent) rotation matrix R_{ai}, we get a more general coordinate transformation with

$$X^\alpha = R_{ai}\left[r_E^i + \frac{1}{c^2}(\ldots)\right] + \mathcal{O}(c^{-4}), \qquad (7.3)$$

where $r_E = x - x_E$. In the case $R_{ai} = \delta_{ai}$, the geocentric coordinates **X** are said to be kinematically nonrotating (with respect to the BCRS), and the orientation of the spatial BCRS coordinates implies a corresponding orientation of the spatial GCRS coordinates.

IAU 2000 Resolution B1.3 recommends writing the metric tensor of the GCRS in the same form as the BCRS metric tensor but with different metric potentials

7.1 Concepts

$W(T, \mathbf{X})$ and $W^a(T, \mathbf{X})$:

$$G_{00} = -1 + \frac{2W}{c^2} - \frac{2W^2}{c^4} + \mathcal{O}(c^{-5}),$$

$$G_{0a} = -\frac{4}{c^3} W^a + \mathcal{O}(c^{-5}), \qquad (7.4)$$

$$G_{ab} = \delta_{ab}\left(1 + \frac{2}{c^2} W\right) + \mathcal{O}(c^{-4})$$

And also, the field equations are the same as the BCRS ones with all variables related with corresponding GCRS variables. Of special interest is the formal linearity of the GCRS field equations that implies a unique splitting of the metric potentials W and W^a in the form

$$W(T, \mathbf{X}) = W_E(T, \mathbf{X}) + W_{\text{ext}}(T, \mathbf{X}),$$
$$W^a(T, \mathbf{X}) = W_E^a(T, \mathbf{X}) + W_{\text{ext}}^a(T, \mathbf{X}). \qquad (7.5)$$

Here, W_E and W_E^a are the metric potentials of the Earth. For practical applications outside the Earth, the scalar potential can be expanded in terms of spherical harmonics in the form

$$W_E(T, \mathbf{X}) = \frac{GM_E}{R}\left[1 + \sum_{l=2}^{\infty}\sum_{m=0}^{l}\left(\frac{R_E}{R}\right)^l P_{lm}(\cos\theta)(C_{lm}\cos m\phi + S_{lm}\sin m\phi)\right],$$

where (R, θ, ϕ) are the spherical coordinates of the GCRS position vector $\mathbf{X}$, M_E is the post-Newtonian mass of the Earth and (C_{lm}, S_{lm}) are post-Newtonian potential coefficients. For the case of a spherical, nonrotating Earth, the metric potentials read

$$W_E = \frac{GM_E}{R}, \qquad W_E^a = 0.$$

A transformation of these GCRS potentials into the BCRS leads to the potentials w_E and w_E^i if the masses of all other bodies are neglected.

We now come to the GCRS coordinate acceleration of a satellite, $d^2 Z_S^a / dT^2$, in the spherically symmetric gravitational field of the Earth. The nonvanishing Christoffel symbols can be derived from (3.15) with $W_E = GM_E/R$:

$$\Gamma^0_{0a} = \frac{1}{c^2}\frac{GM_E}{R^3} X^a$$

$$\Gamma^a_{00} = \frac{1}{c^2}\frac{GM_E}{R^3} X^a - \frac{4}{c^4}\frac{G^2 M_E^2}{R^4} X^a \qquad (7.6)$$

$$\Gamma^a_{bc} = -\frac{1}{c^2}\frac{GM_E}{R^3}\left[\delta_{ab}X^c + \delta_{ac}X^b - \delta_{bc}X^a\right].$$

All other Christoffel symbols vanish in that case. Inserted into (3.16), interpreted as analogous equation in the GCRS, one obtains

$$\frac{d^2 Z_S^a}{dT^2} = -c^2 \Gamma_{00}^a - \Gamma_{bc}^a V_S^b V_S^c + 2\Gamma_{0b}^0 V_S^b V_S^a$$
$$= -\frac{GM_E}{R_S^3} Z_S^i + \frac{GM_E}{c^2 R_S^3} \left\{ \left[4\frac{GM_E}{R_S} - \mathbf{V}_S^2 \right] Z_S^a + 4(\mathbf{Z}_S \cdot \mathbf{V}_S) V_S^a \right\}. \quad (7.7)$$

This is the post-Newtonian GCRS acceleration of a satellite in the Schwarzschild field of the Earth.

To a good approximation,

$$W_E^a(T, \mathbf{X}) = -\frac{G}{2} \frac{(\mathbf{X} \times \mathbf{S}_E)^a}{R^3}, \quad (7.8)$$

where $\mathbf{S}_E$ denotes the Earth's total angular momentum (spin); corresponding gravitomagnetic type effects are usually called Lense-Thirring effects. The Lense-Thirring acceleration of a satellite in the GCRS is given by (e.g., Damour et al. 1994):

$$\frac{d^2 Z_S^a}{dT^2} = \frac{4}{c^2}(W_{a,b}^E - W_{b,a}^E) V_S^b = -\frac{4}{c^2} \mathbf{V}_S \times (\nabla \times \mathbf{W}_E)^a. \quad (7.9)$$

It is useful to expand the external potentials further:

$$W_{\text{ext}} = W_{\text{tid}} + W_{\text{iner}}, \quad W_{\text{ext}}^a = W_{\text{tid}}^a + W_{\text{iner}}^a, \quad (7.10)$$

where the tidal parts depend quadratically upon X^a and the inertial parts linearly. Of special importance is this term:

$$W_{\text{iner}}^a = -\frac{1}{4} c^2 \epsilon_{abc} \Omega_{\text{iner}}^b X^c \quad (7.11)$$

It describes a Coriolis force that appears because the GCRS does not represent a local inertial system. This Coriolis force is dominated by *geodetic precession*, with

$$\Omega_{\text{iner}} \simeq \Omega_{\text{GP}} = \frac{3}{2} \frac{GM_S}{c^2} \frac{\mathbf{x}_E \times \mathbf{v}_E}{r_E^3}, \quad (7.12)$$

where M_S denotes the solar mass. This geodetic precession results from the orbital motion of the Earth about the barycenter; it amounts to about $2''$ per century. Due to the eccentricity of the Earth's orbit, Ω_{iner} also leads to an annual nutation term.

Since the GCRS was defined to be kinematically nonrotating, the geodesic precession enters the precession–nutation matrix. If the Earth were spherically symmetric, the external torques resulting from the gravitational action of other

7.1 Concepts

bodies would vanish, and only the relativistic geodesic motion would contribute to precession–nutation.

The form of the metric tensors in the BCRS and the GCRS impose strong constraints on the coordinate transformations between the two systems. The GCRS coordinates as functions of $t =$ TCB and x^i can be written in the form

$$\begin{aligned}T = {}& t - \frac{1}{c^2}[A(t) + \mathbf{v}_E \cdot \mathbf{r}_E] \\ & + \frac{1}{c^4}\left[B(t) + B^i(t)r_E^i + B^{ij}(t)r_E^i r_E^j + C(t,\mathbf{x})\right] + \mathcal{O}(c^{-5}), \\ X^a = {}& \delta_{ai}\left[r_E^i + \frac{1}{c^2}\left(\frac{1}{2}v_E^i \mathbf{v}_E \cdot \mathbf{r}_E + w_{ext}(\mathbf{x}_E)r_E^i + r_E \mathbf{a}_E \cdot \mathbf{r}_E - \frac{1}{2}a_E^i r_E^2\right)\right] \\ & + \mathcal{O}(c^{-4}),\end{aligned}$$ (7.13)

with ($\mathbf{r}_E = \mathbf{x} - \mathbf{x}_E$)

$$\begin{aligned}\frac{d}{dt}A(t) &= \frac{1}{2}v_E^2 + w_{ext}(\mathbf{x}_E), \\ \frac{d}{dt}B(t) &= -\frac{1}{8}v_E^4 - \frac{3}{2}v_E^2 w_{ext}(\mathbf{x}_E) + 4v_E^i w_{ext}^i(\mathbf{x}_E) + \frac{1}{2}w_{ext}^2(\mathbf{x}_E), \\ B^i(t) &= -\frac{1}{2}v_E^2 v_E^i + 4w_{ext}^i(\mathbf{x}_E) - 3v_E^i w_{ext}(\mathbf{x}_E), \\ B^{ij} &= -v_E^i \delta_{ai}Q^a + 2\frac{\partial}{\partial x^j}w_{ext}^i(\mathbf{x}_E) - v_E^i \frac{\partial}{\partial x^j}w_{ext}(\mathbf{x}_E) + \frac{1}{2}\delta^{ij}\dot{w}_{ext}(\mathbf{x}_E), \\ C(t,\mathbf{x}) &= -\frac{1}{10}r_E^2(\dot{\mathbf{a}}_E \cdot \mathbf{r}_E).\end{aligned}$$ (7.14)

Here, $\mathbf{x}_E(t)$, $\mathbf{v}_E(t)$, and $\mathbf{a}_E(t)$ denote the barycentric coordinate position, velocity, and acceleration of the geocenter; the dots denote the total derivative with respect to $t =$ TCB; and the external potentials are given by

$$w_{ext} = \sum_{A \neq E} w_A, \qquad w_{ext}^i = \sum_{A \neq E} w_A^i.$$ (7.15)

This 4-dimensional coordinate transformation from BCRS coordinates to GCRS coordinates presents a generalized Lorentz transformation. Without the gravitational potentials and the acceleration terms, it reduces to the Lorentz transformation (2.23) expanded to the corresponding orders. Neglecting the $1/c^4$ terms in the $T - t$ relation, (7.13) one obtains

$$T = t - \frac{1}{c^2}\left(\int_{t_0}^t \left(\frac{v_E^2}{2} + w_{ext}(\mathbf{x}_E)\right)dt + \mathbf{v}_E \cdot \mathbf{r}_E\right) + \mathcal{O}(c^{-4}),$$ (7.16)

that reduces to the old TCB − TCG relation.

Without the external gravitational potential, according to (2.30), we have

$$T = t\left(1 + \frac{1}{2}\beta^2\right) - \frac{\mathbf{v}\cdot\mathbf{x}}{c^2} + \mathcal{O}(c^{-4}).$$

Replacing the barycentric vector $\mathbf{x}$ by the Newtonian expression $\mathbf{X} + \mathbf{v}t$, we obtain

$$T = t\left(1 - \frac{1}{2}\beta^2\right) - \frac{\mathbf{v}\cdot\mathbf{X}}{c^2} + \mathcal{O}(c^{-4})$$

in accordance with (7.16).

7.2 Observational Methods

Besides ground-based astrometry mentioned earlier and astrometric space missions described below, it is mainly the method of very long baseline interferometry (VLBI) that is employed for the realization of celestial reference systems.

7.2.1 Very Long Baseline Interferometry

Radio interferometry involving very long baselines (VLBI) was made possible by a number of innovative products such as atomic clocks with high stability on short timescales, high-performance correlators, etc. Figure 7.1 shows a functional diagram of a single-baseline VLBI system, where the distance between individual radio telescopes can amount to several thousand kilometers. For geodetic VLBI measurements, the radio signals observed typically result from quasars, very distant and bright cosmic radio sources. Most quasars show significant redshifts in their spectra implying large escape velocities from the Earth. If

$$z = \frac{\Delta\lambda}{\lambda}$$

is the measured redshift of a quasar, then it shows a recession velocity of about ($v \ll c$)

$$v \simeq z \cdot c.$$

Using the Hubble law

$$v = H_0 d,$$

where H_0 is the present value of the Hubble constant ($H_0 \simeq 70 \, \text{km/s/Mpc}$), one finds that the distance d to quasars is of the order of several billion light-years.

Radio signals from quasars are received by at least two radio telescopes. The primary observable used in geodetic VLBI is the group delay, i.e., the time difference between the arrival time of a plane wave front at the two stations. At present several

7.2 Observational Methods

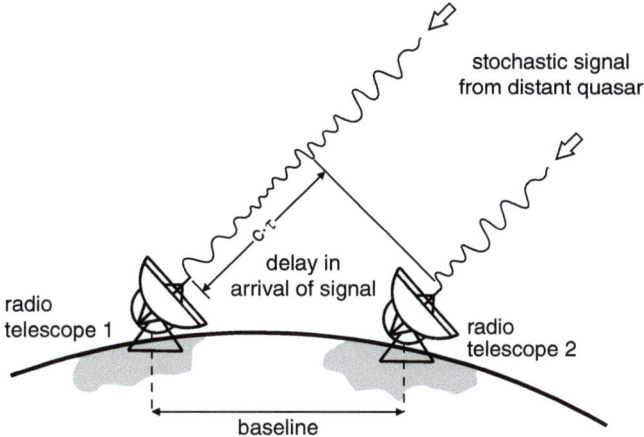

Fig. 7.1 Scheme of a single-baseline VLBI system; from www.fs.wettzell.de

Fig. 7.2 VLBI 20-m radio telescope of the German geodetic fundamental station in Wettzell. In the background, the two radio antennas from the twin telescope can be seen. Courtesy of Alexander Neidhardt

dozens of VLBI stations are in operation. In 1999 the *International VLBI Service for Geodesy and Astrometry* (IVS; a network of VLBI stations, analysis centers, data centers, etc.) was founded. Figure 7.2 shows the VLBI radio antenna of the German geodetic fundamental station in Wettzell. The global distribution of IVS stations is shown in Fig. 7.3.

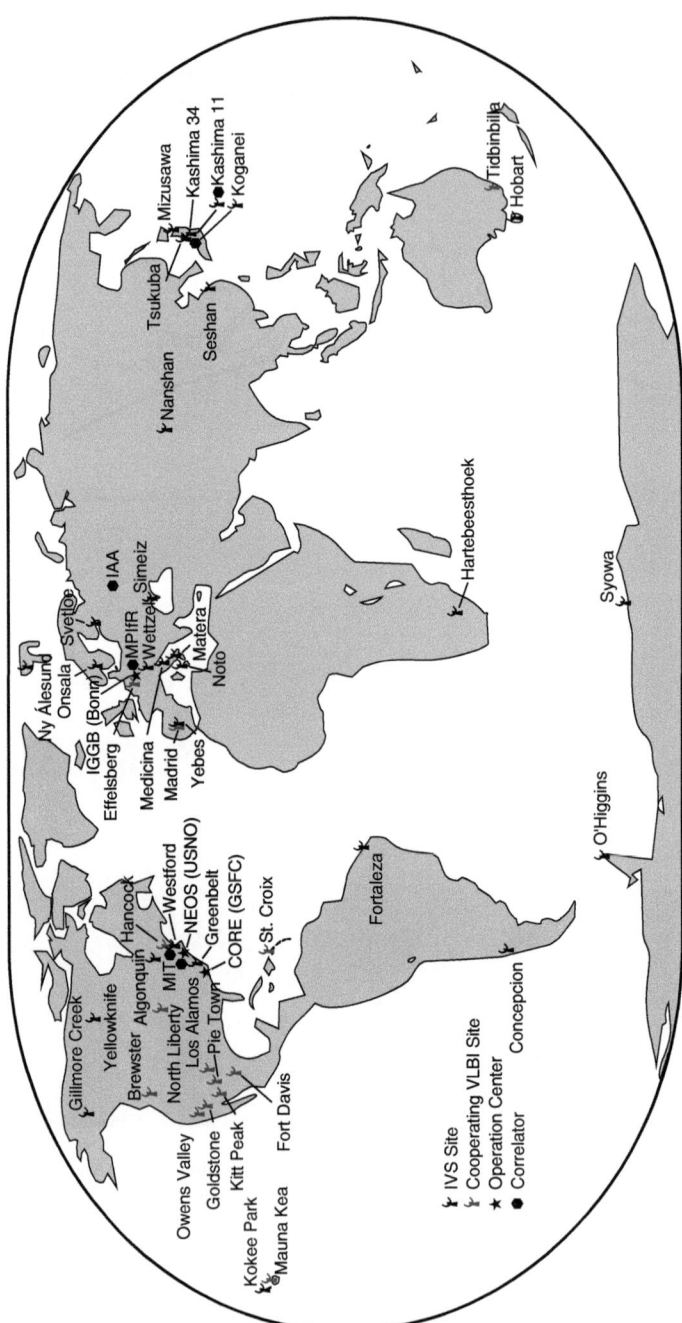

Fig. 7.3 Global distribution of IVS stations

7.2 Observational Methods

Typical geodetic VLBI frequencies are in the X-band at 8.4 GHz and S-band at 2.3 GHz. In the future, VLBI observations at higher frequencies may also be used. Since VLBI antennas have no direct cable connection, atomic clocks are employed for the production of time tags serving as phase reference. Signals are transformed into the MHz region and stored together with the time tags. Information from the various stations is then collected in a so-called correlator where the cross-correlation functions R_{12} are determined. Let $S_1(t)$ and $S_2(t)$ be the signals arriving at antennas 1 and 2 at time t; the cross-correlation function is given by

$$R_{12}(\tau) = \lim_{T \to \infty} \frac{1}{T} \int_{-T/2}^{+T/2} S_1(t) S_2(t-\tau) \, dt. \tag{7.17}$$

Because of the stochasticity of the signals, $R_{12}(\tau)$ is only different from zero if τ is in the vicinity of the geometric time delay, i.e., the time the signal takes to reach the second antenna after it has reached the first one. The precision of such measurements is given by

$$\sigma_r = \frac{1}{2\pi \, (\text{SNR}) \, B_s}, \tag{7.18}$$

where B_s is the so-called synthesized bandwidth of the system, obtained by recording on several 2-MHz frequency channels distributed around the center frequency of, e.g., 8.4 GHz, and thus extending the bandwidth significantly.

The signal-to-noise ratio, SNR, depends upon the received flux density (typically in Jansky, $1 \, \text{Jy} = 10^{-26} \, \text{W m}^{-2} \, \text{Hz}^{-1}$), the system temperature, the involved bandwidth, and the integration time of the signals in the correlator. Presently, with cooled receivers, precisions of the group delay are of the order of a few picoseconds, corresponding to an accuracy for baselines of a few millimeters.

The main source of errors in VLBI measurements results from the time delay in the troposphere that is usually described by means of atmospheric mapping functions. The mapping function, $m(a)$, is defined as the ratio of the propagation delay at geometric elevation a to the corresponding delay in zenith direction. *True* mapping functions, which serve as standard of comparison for a model, have been obtained by ray tracing through the atmosphere using the state given by vertical profiles of the pressure, temperature, and relative humidity obtained from radiosonde profiles (Niell 2000). Two widely used mapping functions are the *mapping temperature test*, MTT (Herring 1992) and the *Niell mapping functions*, NMF (Niell 1996). Recently, Böhm and Schuh (2004) have derived improved mapping functions called *Vienna mapping functions*, VMF, by means of ray tracing using numerical weather models.

VLBI measurements mainly serve for the determination of intercontinental baselines and their temporal variations of Earth orientation parameters and the realization of the International Celestial Reference System (ICRS). Due to its direct connection to the quasi-inertial celestial reference frame of extragalactic radio sources, VLBI is the only space geodetic technique that allows to measure long-term UT1 and precession/nutation.

Several software packages have been developed over the years for VLBI processing and/or analysis. The *Calc/Solve* analysis package has been under development and in use for over 30 years with most of the development work being done by the VLBI group at Goddard Space Flight Center (GSFC). The OCCAM software was originally developed at the University of Bonn, Germany, and then continuously improved by others, e.g., Titov in Australia (Titov 2004; Titov et al. 2004). The software *SteelBreeze* was developed as a tool for geodetic VLBI data analysis at the Main Astronomical Observatory of the National Academy of Sciences of Ukraine by Sergei Bolotin. Finally *QUASAR* (Gubanov et al. 2004; Kurdubov 2007) is the VLBI analysis software package developed by the Institute of Applied Astronomy of the Russian Academy of Sciences. Since 2009 a new VLBI software called Vienna VLBI Software (VieVS) has been developed at Vienna University of Technology (Böhm et al. 2011) which is based on Matlab^{TM} and offers a lot of additional features such as Monte Carlo simulations, scheduling, and also a global solution analyzing in one step all VLBI data that have been observed so far.

7.2.2 Astrometric Space Missions

7.2.2.1 Hipparcos

Besides the geodetic space techniques, dedicated astrometric space missions like Hipparcos play an important role for the realization of astronomical reference frames (e.g., Kovalevsky 1995; Kovalevsky and Seidelmann 2004). For the history of astrometry see, e.g., websites maintained by Erik Høg: http://www.astro.ku.dk/~erik.

The acronym Hipparcos stands for *High-Precision Parallax Collecting Satellite*. The satellite itself was launched on 8 August 1989 into some eccentric transfer orbit. The transition into a geostationary orbit, however, failed, so the orbit had a perigee height of 500 km, an apogee height of 36, 500 km, and an orbital period of $10^h 40^m$. Communication with the satellite was realized with three ground stations: Odenwald (Germany), Perth (Australia), and Goldstone (USA). Near perigee, the satellite went through the radiation belt so that measurements were possible only during 65–70% of the orbital period. Radiation damage finally destroyed parts of the satellite; observations finished in March 1993.

Hipparcos was an instrument of global astrometry (see Fig. 7.4). A mirror was cut in the middle, and both parts were joined together such that two light rays incident at an angle of $\Delta\varphi_0 = 58°31''25$ were mapped into the focal plane. Light rays arriving through two baffles first hit the beam combiner, the mirror just mentioned, before they arrive at the focal plane. The optics is indicated in Fig. 7.5. In the focal plane, several grids can be found: a central grid with a total of 2,688 slits and, at two sides of the main grid, so-called star mappers that mainly served for the purpose of attitude control (Fig. 7.6). Operation of the Hipparcos mission required the use of an input catalog, containing positions, magnitudes, proper motions, and parallaxes

7.2 Observational Methods

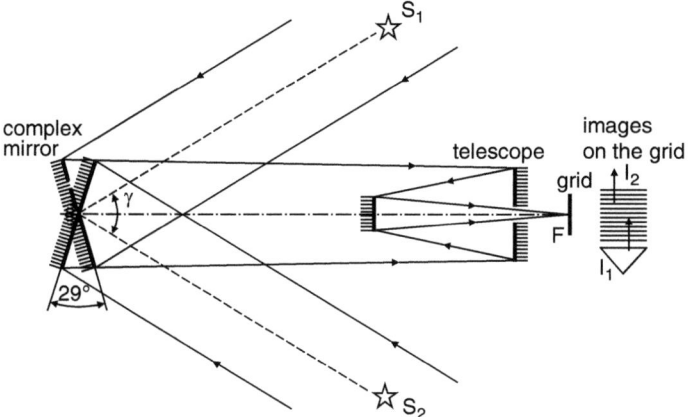

Fig. 7.4 The basic principle of Hipparcos: light from two stars S_1 and S_2 from two different fields of view produces images I_1 and I_2 in the focal plane. The grid produces a certain light intensity pattern that is time dependent due to the rotation of the satellite about its spin axis; from Kovalevsky (1995)

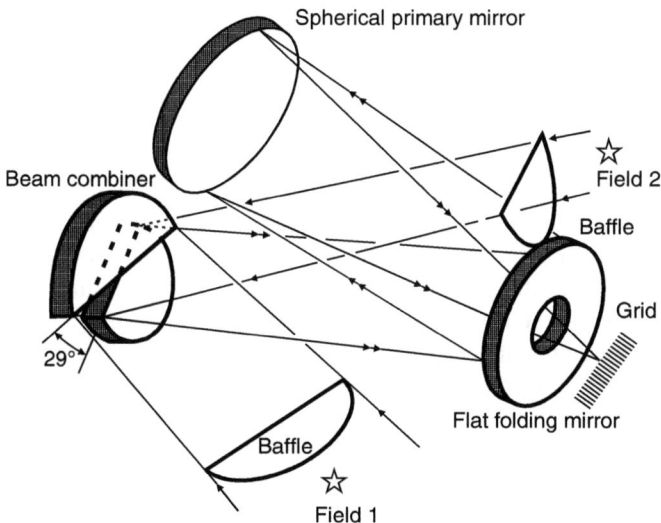

Fig. 7.5 Configuration of the Hipparcos optics; from Kovalevsky (1995)

of about 118,000 stars, 48 minor planets, 3 moons (Europa, Titan, Iapetus), and 1 quasar (3C 273).

The satellite rotated about its vertical axis and thus scanned the whole sky such that all interesting stars were observed as often as possible. A constraint was that the angle between a field of view and the Sun had to be larger than 45° to avoid stray light. The sidereal rotational period of the satellite was 2^h08^m so that a stellar

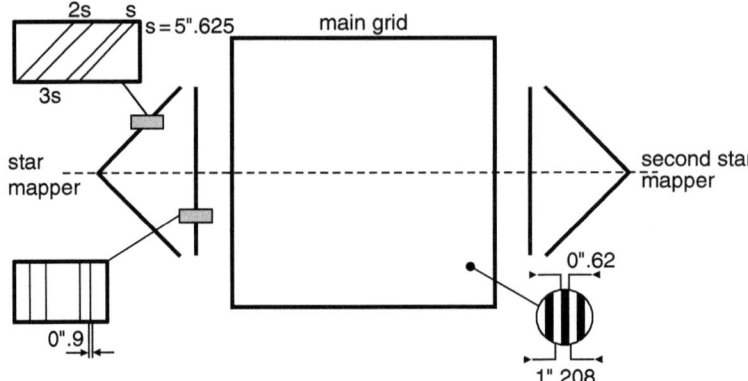

Fig. 7.6 Various grids in the focal plane of Hipparcos; from Kovalevsky (1995)

image crossed the main grid within 19 s. The satellite's spin axis rotated within 57 days around the direction to the Sun. Since the satellite was spinning around an axis nearly perpendicular to the two fields of view, many stars were observed repeatedly in a course of a few hours which was important for the construction of an all-sky astrometric catalog (during the whole mission, every star was observed 30–150 times). The positions of these stars along a great circle were linked very accurately which proved at the same time a calibration of the basic angle and of all the slits in the focal plane.

Measurements involving the main grid led to the Hipparcos catalog that is considered as the optical realization of the ICRS. The Hipparcos catalog contains positions and proper motions of 118,218 stars with accuracies of a few milliarcseconds and a few milliarcseconds per year respectively. The limiting magnitude was $V = 12.4$ mag; the catalog is complete up to $V = 7.3$ mag. The information from one of the star mappers was used to first create the Tycho-1 catalog that contains positions and proper motions of about one million stars with accuracies of about 25 mas and 25 mas/year, respectively. Meanwhile a Tycho-2 catalog with positions, proper motions, and two-color photometry for about 2.5 million stars was released (Høg et al. 2000). For stars brighter than $V = 9$ mag, the astrometric error is 7 mas. The overall error for all stars is 60 mas.

7.2.2.2 The Astrometric Project Gaia

After the shutdown of the Hipparcos satellite in 1993, ESA suggested a new improved astrometric satellite mission named *Gaia* (Fig. 7.7). Similarly to Hipparcos, Gaia consists of two telescopes providing two observing directions with a fixed angle of 106.5° (the *basic angle*) between them. Each of the two telescopes consists of three curved, rectangular mirrors; a beam combiner; and two flat rectangular mirrors (Fig. 7.8). The two telescopes focus their light onto the focal

7.2 Observational Methods

Fig. 7.7 Illustration of the Gaia satellite; from ESA, C. Carreau. Credit: ESA

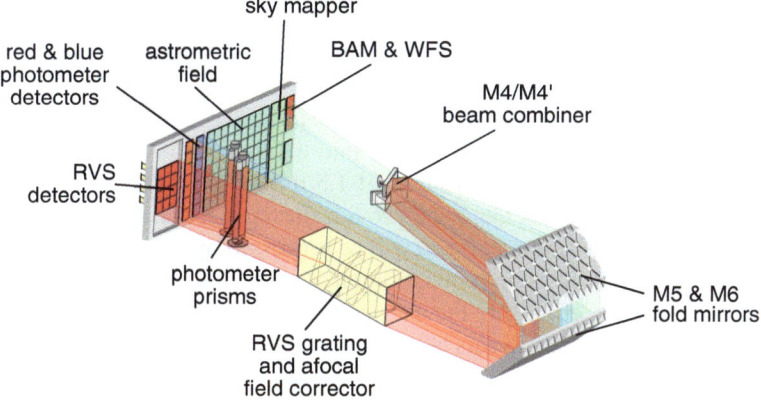

Fig. 7.8 Part of the optics of the space mission Gaia with the focal plane assembly at left. Courtesy of EADS Astrium

plane that houses the largest digital CCD array that was ever built, an ensemble of 106 individual CCDs (Fig. 7.9). Out of these 106 CCDs, 102 are dedicated to star detections, and they are grouped into four fields: star mapper (SM) CCDs, astrometric field (AF) CCDs, photometric field (blue: BP and red: RP) CCDs, and spectroscopic field CCDs. The strips SM1 and SM2 are used for initial star acquisition. The strips AF1 to AF9 constitute the astrometric field for precise position determinations. The strips BP and RP allow spectral measurements in the range 330–680 nm and 640–1,000 nm. Finally, the strips RVS1 to RVS3 allow fine spectroscopy in the range 847–874 nm.

The spacecraft rotates continuously around an axis perpendicular to the two telescopes' line of sight so that each celestial object will be observed about 70 times

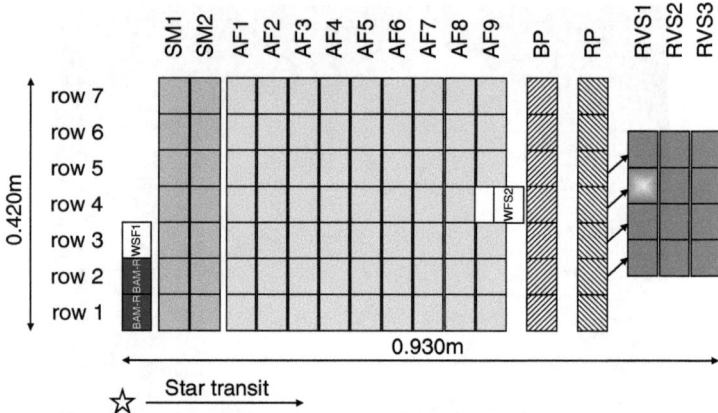

Fig. 7.9 CCDs in the focal plane of Gaia. Each CCD is about 45 mm × 60 mm in size and has about 9 megapixels. Courtesy of EADS Astrium

during the mission that is expected to last for 5 years. Gaia will compile a catalog of approximately one billion stars to magnitude 20 with an accuracy of 10 μas at 10 mag, 20 μas at 15 mag, and 200 μas at 20 mag.

Presently (2012), the launch of the Gaia satellite is scheduled for mid 2013, where a Soyuz rocket should transport it into a Lissajous orbit around the Sun–Earth L_2 Lagrangian point at a distance of 1.5 million kilometers from the Earth. The overall data volume that Gaia will provide during its 5-year mission is of order 200 TB; this enormous amount of data will be processed by the Data Processing and Analysis Consortium (DPAC), a collaboration of about 400 scientists and engineers.

About 4 months after launch, the astrometric observations are expected to start. First results are expected 2–3 years later; due to the time-consuming data processing, final results should be available only in 2021/2022.

7.3 Classical Celestial Reference System

The classical celestial reference system with equinox-based spatial coordinates was basically realized by means of stellar catalogs. The first stellar catalog that was based upon telescope observations presumably was compiled by John Flamsteed, a British astronomer. It came into print in 1728 after Flamsteed's death. After that, stellar catalogs were issued for different purposes. A list of stellar catalogs has been compiled by the *Centre de Données astronomique de Strasbourg* (CDS, http://cdsweb.u-strasbg.fr/cats/cats.html). Some important stellar catalogs are listed in Table 7.1. Prior to the introduction of the ICRS, the fundamental catalog FK5 was the realization of the classical celestial reference system.

7.3 Classical Celestial Reference System

Table 7.1 A compilation of some important classical astrometric catalogs

Name of the catalog	Date	Approximate number of stars	Precision of positions (mas)	Precision of proper motions (mas/year)	Maximal stellar magnitude	Average star density per square degree
FK4	1964	1,500	50	3	6	1/27
SAO	1966	260,000	1,500	–	9	6.3
AGK3	1975	183,000	210	10	9	8.8
FK5	1988	1,500	50	2	6	1/27
FK5 Sup	1991	3,000	80	3	9	1/14
IRS	1990	36,000	220	5	9	0.9
ACRS	1992	320,000	220	6	8	10.5
PPM	1991	380,000	290	6	9	11
GSC 1.0	1988	20,000,000	1,500	–	16	500
GSC 1.1	1992	20,000,000	700	–	16	500
GSC 1.2	1996	20,000,000	400	–	16	500
GSC 3.2	2006	945,593,000	200	–	20	23,000
AC2000	1996	4,600,000	300	–	13	115
Hipparcos	1997	118,000	1	1	12	3
Tycho	1997	1,058,000	25	–	12	25
Tycho-2	2000	2,500,000	50	2	11	61
2MASS	2006	440,000,000	80	–	15	11,000
USNO-A1.0	1997	488,000,000	250	–	24	12,000
USNO-B1.0	2003	1,049,000,000	200	–	24	25,000
UCAC3	2009	100,766,000	20	5	16	2,400

7.3.1 Fundamental Catalogs FK3, FK4, and FK5

Over the last century the classical celestial reference system was realized by a series of so-called *fundamental catalog* (of stars) (e.g., Walter and Sovers 2000). The 5th catalog of that series, the FK5 (Fricke et al. 1988), was the IAU-adopted system until it was replaced by the ICRS and its optical realization, the Hipparcos catalog. The series of FK catalogs were compiled from transit circle catalogs based on absolute observations of stars, i.e., without the need to relate to an a priori known celestial coordinate system. To succeed, the data model of such absolute observations includes Earth rotation parameters, nutation, and precession, as well as the dynamics of solar system objects. There were no "fixed fiducial points," like in the ICRF which is based on distant quasars.

In 1938 the International Astronomical Union (IAU) recommended the adoption of the fundamental catalog FK3, containing 1950.0 coordinates of 1,535 fundamental stars.

The FK4, adopted in 1964, contains the B1950.0 coordinates of the FK3 stars and 1987 additional ones. Most of the stars are brighter than 7.0 mag. The CDS version of FK4 contains seven data files, six for different equinoxes (1950, 1955, 1860, 1965, 1970, 1975) and one for the additional stars.

The FK5 (Fricke et al. 1988), printed in 1988, contains improved values for positions and proper motions of the classical 1,535 as well as additional 3,117 fundamental stars. The machine-readable version of FK5 contains values for epoch and equinox J2000.0 and B1950.0. The FK5 was compiled from a total of about 300 individual catalogs. With respect to presently achievable accuracies (e.g, with Hipparcos), the FK5 shows relatively large systematic errors, which were estimated at the time to be in the range of 30–100 mas. The true systematic errors were later found to be even larger than that when compared to Hipparcos results.

However, even the Hipparcos catalog is not without problems. Due to the short time span of the observations of 3.58 years, the parallax and proper motions of many stars could not be well determined and separated from orbital motions in case of multiple stars. Significant discrepancies between Hipparcos proper motions and those derived from about 100 years epoch span (Tycho-2 catalog) were found.

The latest version, FK6 (Wielen et al. 1999), combines all available information for about 4,000 "astrometrically excellent stars."

7.3.2 Astrographic Catalog and Derived Catalogs

Around 1900 the entire sky was observed with photographic plates for the first time (Astrographic Catalog, AC). It was not until the 1980s that all those measures of over four million stars became available in machine-readable format. For the astrometric re-reduction of these data, the *Astrographic Catalog of Reference Stars* (ACRS) was constructed at the US Naval Observatory. Almost the same, about 300, 000 stars were compiled differently in the *Positions and Proper Motions* (PPM) project at Heidelberg.

The PPM catalog presents an extension of the FK5 with respect to a larger number of stars. The catalog PPM north contains J2000.0 positions and proper motions of 181, 731 stars north of $\delta = -2.5°$. The mean error is of order $0''.27$ in position and $0''.43$/cen in proper motion. PPM south contains positions and proper motions of 197, 179 stars south of $-2.5°$ declination. In addition to that, a supplement catalog contains values of additional 90, 000 stars. The idea of the PPM catalog was a realization of the IAU 1976 celestial reference system with a sufficiently dense and precise catalog.

The final reduction of the AC data was based on the Hipparcos reference system and is called AC2000.2 (Urban et al. 2001).

7.3.3 Hipparcos and Tycho Catalogs

The ESA space mission *Hipparcos* produced a highly accurate stellar catalog (the Hipparcos catalog). Although based on stars, dedicated additional observing by various techniques provided a link to the extragalactic system (Kovalevsky et al. 1997), with the highest weight coming from VLBI observations of a dozen radio stars. In 2000, the IAU adopted the Hipparcos catalog as the optical realization of the ICRF, replacing the FK series of catalogs. Later, an IAU resolution was passed to define the Hipparcos Celestial Reference System (HCRS) as a subset of the "good" Hipparcos stars. Results from a new reduction of the Hipparcos data were published by van Leeuwen (2007), lowering the systematic error floor and obtaining smaller astrometric errors, particularly for the bright stars.

Besides the Hipparcos catalog, Tycho and Tycho-2 catalogs have been introduced above. Typical proper motion errors of the Tycho-2 catalog stars are about 2 mas/year and position errors at mean epoch (near 1991) are about few mas to 100 mas, depending upon magnitude.

7.3.4 2MASS

Around the year 2000, the *2-Micron All-Sky Survey* (2MASS) was performed at near-infrared wavelengths with two identical telescopes (north and south). The resulting catalog of over 440 million stars (Skrutskie et al. 2006) is very reliable and provides, besides accurate J-, H-, and Ks-band photometry, also accurate positions at its epoch (errors about 80 mas). However, it does not give proper motions. The reference frame of the 2MASS catalog is based on Tycho-2 stars.

7.3.5 *Schmidt Plate Survey Catalogs*

By about 1990, all usable Schmidt plates from the various all-sky surveys of the 1950s–1970s (Palomar, ESO, Australia) were digitized at various places. Scans performed at the *Space Telescope Science Institute* lead to the series of *Guide Star Catalogs* (GSC), mainly tailored to support the *Hubble Space Telescope* mission. The latest release is GSC 2.3 (Spagna et al. 2006). Plate scans performed at Edinburgh lead to the *SuperCOSMOS* data set with pixel data served on the web (Hambly et al. 2001, 2004). At the US Naval Observatory Flagstaff Station (NOFS), the *Precision Measuring Machine* (PMM) was used to digitize those plates to construct the USNO-A and USNO-B catalogs (Monet et al. 2003) of about one billion stars. Positional precision of those catalogs is about 250 mas, with systematic errors on the same level. In order to better control remaining astrometric systematic errors in the Schmidt plate data, various efforts recently were undertaken. Utilizing the 2MASS catalog for calibrations lead to such catalogs as PPMXL (Röser et al. 2010) and XPM (Fedorov et al. 2010) by reanalyzing existing data like the USNO-B. The reference system of all those catalogs can be traced back to the Tycho-2 system.

7.3.6 UCAC

In 1998 the US Naval Observatory (USNO) began a new, all-sky, astrometric survey, the *USNO CCD Astrograph Catalog* (UCAC) project. Observations were completed in 2004. This was the first of its kind to use a CCD detector instead of photographic plates. A series of catalogs (UCAC1 to UCAC4) were published (Zacharias et al. 2000, 2004, 2010, 2012) containing accurate positions of up to over 100 million stars. Proper motions for many of those stars were derived by utilizing early epoch catalogs, some of which are unpublished and were constructed from new plate measurements undertaken for the UCAC project.

The accuracy of the astrometric data strongly depends on the magnitude of the stars. For the stars in the about 9–14 mag range (bandpass between V and R), the positional errors per coordinate are about 20 mas at mean UCAC epoch (about 2000), with errors increasing to 100 mas at the 16th magnitude limit. Proper motion errors range from about 2–10 mas/year, depending on their observing history and quality of early epoch positions. The Lick Observatory Northern Proper Motion (NPM) and Yale/San Juan Southern Proper Motion (SPM) data form the basis for the proper motions of the vast majority of faint stars in UCAC. Tycho-2 stars were directly used as reference stars for UCAC catalog reductions.

7.4 International Celestial Reference System

In the past, fundamental astronomical celestial reference systems have been realized through specific stellar catalogs with associated precession–nutation models and conventions on proper motion, aberration, and parallax. Introduced in 1998, the ICRS is defined directly by a catalog of distant stationary sources. This catalog, the International Celestial Reference Frame (ICRF), contains positions of selected extragalactic radio sources: quasars, BL Lac objects, and active galactic nuclei. Relevant radio sources are categorized as defining sources, candidate sources, and others. The ICRF1 catalog contains the positions of 608 radio sources with 212 defining sources and 294 candidate sources and a further 102 sources set aside for the purpose of densification. Defining sources have to meet a number of criteria: the formal position error must be less than 1 mas, it should have been observed at least 20 times during a time span of more than 20 years and should not show significant intrinsic structure and proper motion. It is important to note that most quasars show structural elements of a few milliarcseconds that strongly depend upon radio frequency. In the future, VLBI observations in K-band (18–26.5 GHz), K_a-band (26.5–40 GHz), and Q-band (30–50 GHz) will play an important role in addition to the classical S- (2–4 GHz) and X- (8–12 GHz) bands, since there the structure problems are less serious.

Figure 7.10 shows the isophotal lines (lines of equal radio intensity) of the quasar $1156 + 295$ in the S-band at 2.324 GHz. Units for right ascension (on the x-axis)

Fig. 7.10 Isophotes of the quasar 1156+295 at 2.324 GHz (S-Band). Spatial structures of milliarcsecond size are clearly visible (USNO)

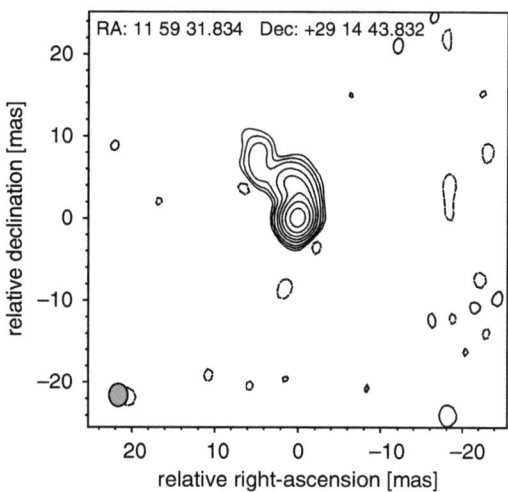

and declination (on the y-axis) are mas. Spatial structures of milliarcsecond size are clearly visible. With respect to the ICRF, the intrinsic structure of a quasar is characterized by a frequency-dependent structure index (e.g., Charlot 1990; Fey and Charlot 1997). A structure index of 1 implies excellent astrometric quality and a well-defined hot spot. A quasar with a structure index of 2 has a clear structure with good astrometric quality. Defining sources have a structure index of 1 or 2.

The ICRF1 had an estimated noise floor of 250 µas and an estimated axis stability of about 20 µas. This presented roughly an order of magnitude improvement over the FK5 (Fricke et al. 1988). Even so, it had its limitations and deficiencies. The distribution of defining sources was very nonuniform, with most being in the northern hemisphere.

Using nearly 30 years of VLBI observations, the ICRF2 was realized in 2009 (Fey et al. 2009). It contains precise positions of 3,414 compact astronomical radio sources, more than five times the number of original ICRF1 sources. The noise floor of ICRF2 is of order 40 µas, and the axis stability is about 10 µas. A total number of approximately 6.5 million S-/X-band group delay measurements were used for the ICRF2. The ICRF2 has a total of 295 defining sources shown in Fig. 7.11. The whole ICRF2 catalog is contained in Fey et al. (2009).

7.4.1 Orientation of the ICRS

The *origin* of the ICRS is simply the solar system barycenter. However, the *orientation* of the ICRS spatial axes is a more complicated matter. Whereas pre-ICRS celestial reference systems depended on the orientation of the Earth in space, namely, through the mean equator and equinox of a specified epoch, one of the new paradigms for astronomical reference systems is that this connection be broken.

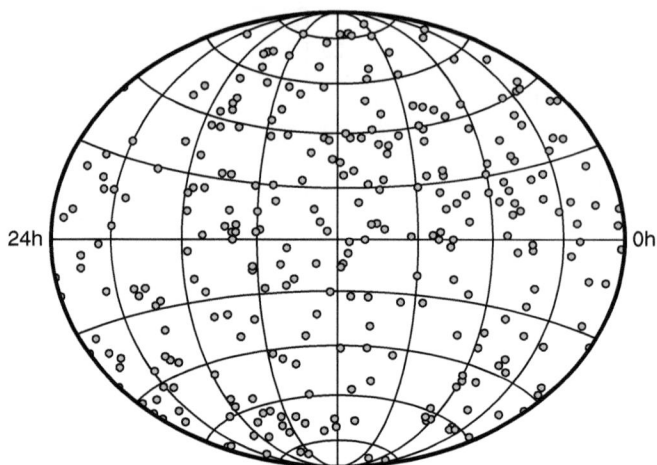

Fig. 7.11 Distribution of ICRF2 defining sources

Thus, the orientation of the ICRS is now entirely conventional, and in principle any convention could have been chosen, wholly unrelated to the Earth's axis or the ecliptic. However, for continuity, an orientation was chosen that is close to mean pole and equinox J2000.0, so that for many applications, the distinction between ICRS and mean J2000.0 coordinates can be neglected. Nevertheless, because different realizations of the ICRS exist, in the form of different ICRF catalogs, the relative orientations of the various coordinate systems with respect to each other have to be determined experimentally and indicated explicitly.

For the relation of the ICRS with the classical equinox-based astronomical reference systems, the ICRS positions of the J2000.0 mean celestial pole and equinox are needed. This information constitutes the so-called *frame bias*. VLBI observations have placed the J2000.0 mean pole about 18 mas from the ICRS pole, while LLR observations have shown that J2000.0 and ICRS right ascensions differ by about 15 mas (McCarthy and Petit 2003). The overall difference in orientation is about 24 mas.

7.4.2 Realization of the ICRF in the Optical

The Hipparcos reference system, realized by the corresponding optical catalog, was oriented according to the ICRF within ±0.6 mas at epoch 1991.25. The rotational velocity between the two systems is less than ±0.25 mas/year. In this sense the Hipparcos system presents a realization of the ICRS in the optical. The essential parts of the Hipparcos and Tycho catalogs are contained in a CD-ROM package called *Celestia 2000* (http://astro.estec.esa.nl/hipparcos/celestia/celestia-pr.html).

Chapter 8
Terrestrial Reference System

8.1 Concepts

A terrestrial reference system (TRS) is a spatial reference system co-rotating with the Earth in its diurnal motion in space. In such a system, positions of points anchored on the Earth solid surface have coordinates which undergo only small variations with time, due to geophysical effects (tectonic, tidal deformations, loading effects, etc.). A terrestrial reference frame (TRF) is a set of physical points with precisely determined coordinates in a specific coordinate system attached to a TRS. Such a TRF is said to be a realization of the TRS (http://itrf.ensg.ign.fr/trs_trf.php).

8.1.1 Local Terrestrial Systems

Historically first local coordinates were realized by means of standard geodetic methods (point positioning) and instruments like theodolites (Figs. 8.1 and 8.2).

Point positioning means coordinate determination of points in the real world with respect to a coordinate system. Such points may be, e.g., land marks or satellites. The positions are computed from measurements which relate those points to others for which the positions within the coordinate system are known. The known points may be a direct consequence from the realization of the coordinate system, or they may be generated by triangulation.

Modern surveying networks are built up hierarchically. Today, most of measurements are carried out by making use of some Global Navigation Satellite System (GNSS). The millimeter accuracy requirements needed for surveying may only be achieved by differential GNSS methods which determine coordinate differences instead of absolute positions. For the higher order surveying networks, static GNSS methods are used, while for ordinary measurements real-time kinematic GNSS methods are employed.

Fig. 8.1 Historical theodolite, made by Utzschneider & Liebherr, Munich, 1820 (from: University of Helsinki, observatory museum). Courtesy of Sami Maisala

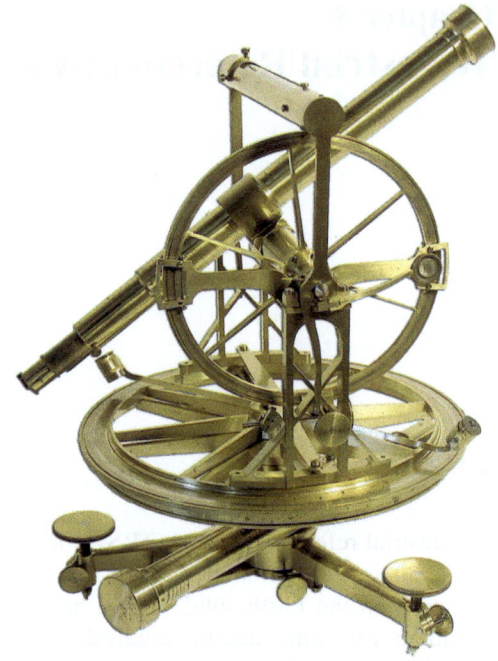

Fig. 8.2 Topcon 5″ Digital Theodolite, model DT-205L. Courtesy of Topcon

However, there are also special measurements like, e.g., high-precision engineering measurements for which the accuracies achievable with current GNSS are not sufficient. Such measurements are done using precise geodetic instruments which determine angles and distances (tachymeters) or other very special methods.

8.1.2 Global Terrestrial Systems

Global terrestrial reference systems cover the entire Earth. In recent times the most important representative of such a global system is the International Terrestrial Reference System (ITRS) discussed below. The ITRS and ITRF solutions, starting with the ITRF92, are maintained by the International Earth Rotation and Reference Systems Service (IERS).

At highest precision also for the ITRS concept, relativity has to be taken into account. Here, the starting point is the GCRS with its coordinates $(T, \mathbf{X})$. The ITRS coordinates are then constructed by a series of ordinary rotations, as described in Chapter 9.

8.2 Observational Methods

8.2.1 Classical Methods

The classical methods of astronomical geodesy are based upon optical observations of stars or the Sun often accompanied by time measurements. Here we will discuss only a few selected classical methods.

8.2.1.1 Longitude Determination with the Passage Instrument

The astronomical coordinates of the second kind define celestial longitude, Λ, and latitude, Φ, for an observer on the Earth. The celestial latitude is given by $\Phi = \delta_{\text{To}}$, where the index To refers to the observer's topocenter. Celestial longitude is reckoned from the Greenwich celestial meridian in eastward direction.

For two locations 1 and 2 on the Earth, the difference in longitude (taken positive in eastward direction) is given by

$$\Delta \Lambda = \Lambda_2 - \Lambda_1 = \text{LAST}_2 - \text{LAST}_1, \quad (8.1)$$

i.e., $\Delta \Lambda$ is determined by the difference of local apparent sidereal time. The longitude of some place is given by

$$\Lambda = \text{LAST} - \text{GAST}, \quad (8.2)$$

where GAST is Greenwich apparent sidereal time. In this way longitude is counted positive to the east from the Greenwich meridian and negative to the west. If accuracy is such that effects from polar motion have to be considered, the reference meridian has to be defined carefully (see below). The problem therefore is to determine local apparent sidereal time, LAST. This can, e.g., be achieved by observing the passage of a star through the local astronomical meridian by means of a passage instrument that is mounted so that only stars in the meridian can be observed. With such an instrument a vertical-hair passage can be observed with an accuracy of about 5 ms. We had

$$h = \text{LAST} - \alpha$$

and the hour angle vanishes in the meridian, $h = 0$, so that

$$\text{LAST} = \alpha.$$

Hence, if the right ascension α of a star is known, it is equal to the local apparent sidereal time if the star is found in the meridian. From this we get

$$\Lambda = \alpha - \text{GAST}. \tag{8.3}$$

8.2.1.2 Simultaneous Determination of Longitude and Latitude by Means of the Lines of Position

In the lines of position method zenith distances together with times are measured for a certain number of stars. We had (5.3)

$$\cos z = \sin \Phi \sin \delta + \cos \Phi \cos \delta \cos h.$$

The hour angle h is given by the local apparent sidereal time, LAST, of the observer:

$$h = \text{LAST} - \alpha$$

and

$$\text{LAST} = \text{GAST} + \Lambda.$$

We had already discussed Greenwich apparent sidereal time above, and we will assume the stellar coordinates (α, δ) to be known. The methods aim at a determination of (astronomical) longitude and latitude (Λ, Φ) of the observer.

Let us first consider only one single star. Then at time t_1 somewhere on Earth at location L_1, the star will appear in zenith direction. The set of positions where the star is observed with zenith distance z_1 at t_1 then defines a circle C_1 around L_1. If one observes a second star under a zenith distance z_2 at time t_2, then we get a second circle C_2 that usually will intersect C_1 in two points. Under idealized conditions the

8.2 Observational Methods

measurement of (z, t) for a third star will lead to a third circle that will intersect the first two ones in exactly one point. This common intersection point of the three circles will then define the position (Λ, Φ) of the observer.

Nowadays one starts with approximate values (Λ_0, Φ_0) from which one derives a calculated value $z_0 = z_0(\Lambda_0, \Phi_0)$. We then write

$$\Lambda = \Lambda_0 + \delta\Lambda; \quad \Phi = \Phi_0 + \delta\Phi; \quad z = z_0 + \delta z.$$

Here z is a measured value of zenith distance and z_0 the calculated approximate one. For the iteration process for a single star, one might start with the first-order relation

$$dz = -dh \sin A \cos \Phi - d\delta \cos p - d\Phi \cos A \tag{8.4}$$

with $d\delta = 0$ and $dz = \delta z$. Assuming the time error dh to be given essentially by $\delta\Lambda$, i.e., $dh \simeq \delta\Lambda$, we get for one single star the first-order relation

$$\delta\Phi \cos A + \delta\Lambda \cos \Phi \sin A = -\delta z.$$

Considering also a so-called zero point error, δi, we get for a certain number of stars:

$$\delta\Phi \cos A_i + \delta\Lambda \cos \Phi \sin A_i + \delta i = -\delta z_i. \tag{8.5}$$

Besides $\delta\Phi$ and $\delta\Lambda$ the zero point error δi presents a third unknown quantity that describes a constant error of zenith distance. For theodolite measurements, e.g., the zero point error may arise from the vertical collimation error of the instrument. If three stars are observed, the system of equations can be solved uniquely. If more than three stars are observed, the problem can be viewed as an equalization problem with

$$\delta\Phi \cos A_i + \delta\Lambda \cos \Phi \sin A_i + \delta i = -\delta z_i + V_i. \tag{8.6}$$

With the condition $\sum_i (V_i V_i) = \text{minimum}$, the solve for parameters can be deduced.

The name of the method (lines of position method) becomes obvious from the following graphic method to derive (Λ, Φ): from the point $P_0 = (\Lambda_0, \Phi_0)$, we plot the azimuth lines A_i of the observed stars and along these lines the corresponding δz_i values ($\delta z < 0$ towards the star, $\delta z > 0$ away from the star). Perpendicular to the azimuth lines, in a distance δz of P_0, we plot the corresponding line of position (tangents to the circles mentioned above). Let us assume the y-axis of this plot to point towards north and the x-axis towards east as in Fig. 8.3. Without any errors the lines of position would intersect in exactly a point P providing the correction $\Delta\Lambda \cos \Phi_0$ on the x-axis and the correction $\Delta\Phi$ on the y-axis (Fig. 8.4). If only the instrumental zero point error δi was present, the lines of position would be tangents to a circle with the radius δi around P. Due to additional measurement errors, however, the lines of position will generally not be perfect tangents to the circle around P, but the center of the inscribed circle of the figure defined by the lines of position will provide some good approximation to the sought-after corrections.

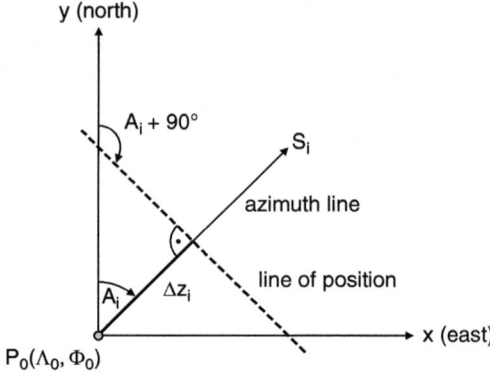

Fig. 8.3 Construction of the line of position

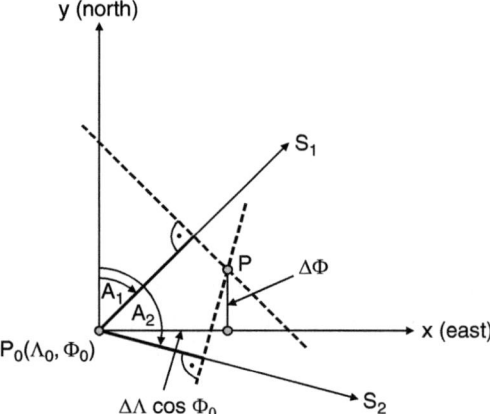

Fig. 8.4 Corrections to the approximate position in the line of position method

8.2.1.3 Azimuth Determination of a Target

The azimuth of a target is that horizontal angle between the meridian plane and the vertical plane running through the zenith of the observer and the aiming point. If the horizontal angle θ between the aiming point and some astronomical body (star, Sun) is measured, we can easily get the azimuth of the target, A_t, according to

$$A_t = A_a - \theta, \tag{8.7}$$

if the azimuth of the astronomical A_a body is known (Fig. 8.5).

If the observer's position (Λ, Φ) and the astronomical coordinates (α, δ) of the astronomical body at measured time (e.g., CET) are known, the azimuth A_a can be obtained from

$$\tan A_a = \frac{\sin h_a}{\sin \Phi \cos h_a - \cos \Phi \tan \delta}, \tag{8.8}$$

8.2 Observational Methods

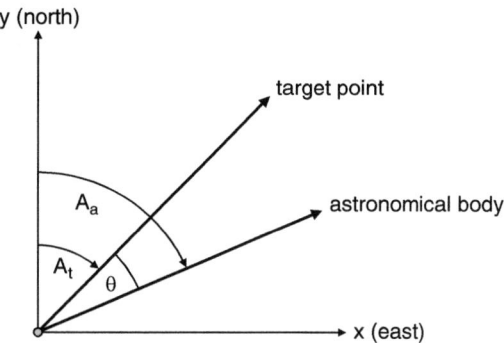

Fig. 8.5 Azimuth determination of a target

where $h_a = \text{GAST} + \Lambda - \alpha$ and GAST has to be derived from the measured time. Especially useful for such measurements at the northern hemisphere is Polaris (αUMi) since its azimuthal motion presently is less than $0\rlap{.}''31$ per second of time ($\Phi = 50°, \delta = 89°15'$). Generally azimuth determinations with the Sun are less accurate. Ephemerides provide coordinates for the heliocenter that cannot be observed directly. Usually the Sun is observed at its limb so that one also needs the apparent solar radius at the time of observation. Another problem is the unavoidable effect of solar radiation upon the observing device leading to instrumental problems.

8.2.2 Satellite Laser Ranging

Laser distance measurements to satellites, the so-called *satellite laser ranging* (SLR), determine the light travel time of laser pulses from an observing station to the satellite and back. It is thus a two-way method. The emission of a laser pulse starts a time interval counter. The signal that is reflected from the satellite is received again from the ground station, amplified, and stops the counter. From the round-trip travel time, Δt, the distance d between ground station and satellite can be derived:

$$d = \frac{\Delta t}{2} c. \tag{8.9}$$

Besides a precise time measurement, SLR requires large laser power, strong focusing of the laser light, and a very sensitive reception device. A large pulse rate is also favorable. Due to the satellite's motion and the Earth's rotation in space, the emission and reception device has to be tracked continuously which requires knowledge about the satellite orbit.

For SLR to work, the satellite has to be equipped with laser reflectors (Fig. 8.6) that reflect an incident laser beam exactly back into the incoming direction. The most well-known satellite of this kind is the *Laser Geodynamic Satellite* LAGEOS (Fig. 8.7). LAGEOS is a completely passive satellite with a circumference of 60 cm,

Fig. 8.6 Laser retroreflectors or triple prisms

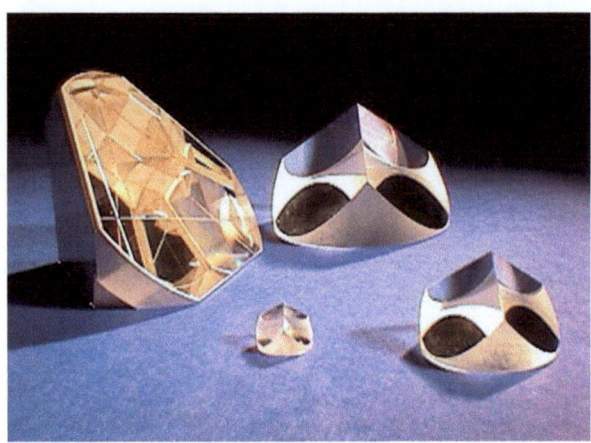

Fig. 8.7 The Laser Geodynamical Satellite LAGEOS

an outer shell of aluminum, and a heavy core made of beryllium and copper. Its spherical surface is completely covered with 426 laser reflectors. The orbit of LAGEOS is nearly circular ($e \sim 0.004$), and its semimajor axis is about twice the Earth's radius. Another well-known laser dynamical satellite is *Satellite de Taille Adaptée avec Réflecteurs Laser pour les Etudes de la Terre* (STARLETTE). For SLR measurements, one typically works with neodymium lasers with powers of about a gigawatt and pulse lengths of about 100 ps or 3 cm in space at a rate of a few pulses per second together with a computer-controlled guiding system. Geodetic work with artificial satellites started in the USA in 1961/1962. In 1965 SLR measurements with a precision of a few centimeters became possible. Presently, the LAGEOS orbit can be determined at the centimeter level. This high accuracy is achieved not only by the high pulse rate and sharp filtering in time (1 μs), in space ($5''$), and in frequency (0.1 nm) but also by a precise control of the telescope on the basis of high-precision satellite ephemerides.

Fig. 8.8 The SLR station of GFZ German Research Centre for Geosciences Potsdam. Image: GFZ Deutsches GeoForschungszentrum

Meanwhile SLR stations can be found in many places on Earth (Fig. 8.8). SLR mainly serves for a precise determination of station positions, baselines, and their temporal variations (crustal motion); it is thus an important tool for the realization of the ITRS.

8.2.3 Global Positioning System

The Navigation System with Time And Ranging Global Positioning System (NAVSTAR GPS) belongs to the group of GNSS. Like any GNSS, the GPS provides position and time information to the user. Although GPS was originally designed as a military navigation system by the US Armed Forces, it nowadays also serves civil purposes including geodetic measurements.

Between 24 and 32 GPS satellites with semimajor axes of about 26, 600 km emit precisely timed microwave signal patterns which are correlated with similar patterns generated by a GPS receiver (Fig. 8.9). This way the time offset between satellite and receiver is determined which is then converted to a so-called *pseudorange*. The reason for this term is that besides the offset induced by the pure signal travel time, there is generally a systematic offset between the satellite clock and the receiver clock which acts on the distance solution. But, since all satellite clocks operate in a common synchronized system, there is only one systematic clock offset between the receiver and the satellite system. Therefore, during a position solution, four unknowns have to be determined, i.e., three geocentric station coordinates and the receiver clock offset. Hence, if the signals of at least four satellites may be received at the same time, a solution can be calculated. Modern GPS receivers have the ability to receive simultaneously the signals of all satellites in sight providing the opportunity of least squares fits for the unknowns. This leads to much more accurate results. Furthermore, the improved calculation power of modern receivers enables real-time solutions which opened new fields of application.

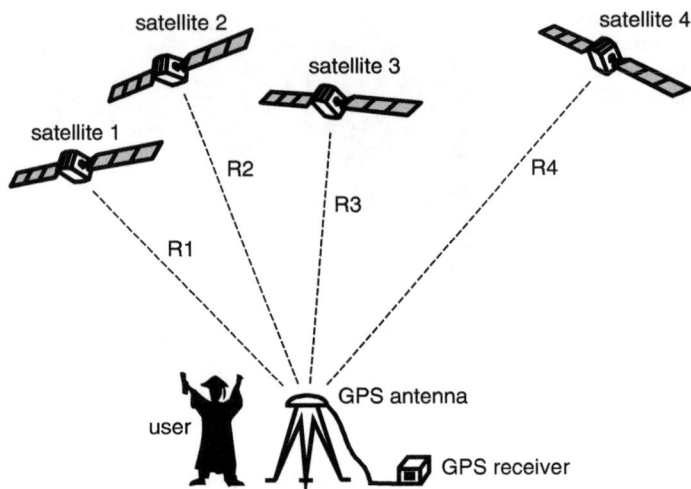

Fig. 8.9 Basic principle of position determination and navigation with GPS

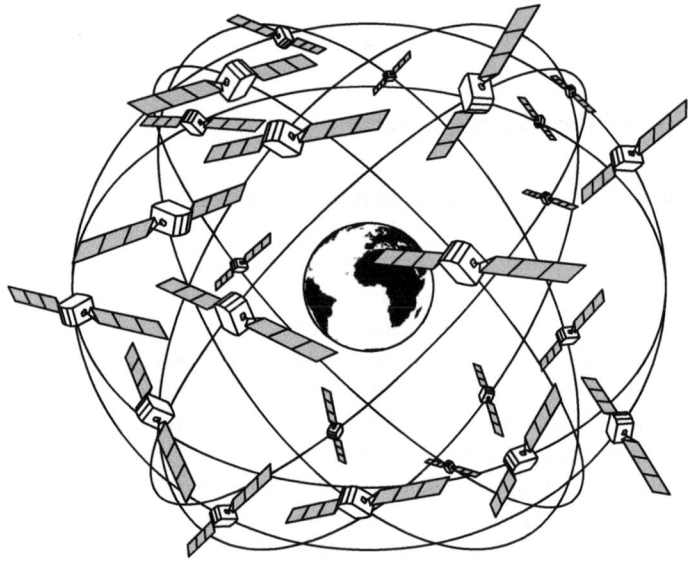

Fig. 8.10 Constellation of satellites in the GPS. The (up to 32) space vehicles are in six orbital planes

8.2.3.1 Space Segment

The 24–32 GPS satellites are denoted as *space segment*. The satellites are equally spread on six nearly circular orbital planes with semimajor axes of about 26, 600 km and inclinations of 55° (Fig. 8.10). The arrangement of the orbital planes permits the observation of at least six satellites at any time almost everywhere on the Earth. However, positioning problems may arise for the polar zones because of the limited inclination.

8.2.3.2 Control Segment

Tasks of the GPS *control segment* are:

- Control of the satellite system
- Determination of the GPS system time
- Forecast of the satellite orbital parameters and satellite time (broadcast ephemerides)
- Upload of orbit and time information into the data storage of each satellite

All ground stations carrying out these tasks belong to the control segment.

US Air Force stations and shared stations of the National Geospatial-Intelligence Agency (NGA) are responsible for monitoring the orbits of the satellites. The tracking data are sent to the master control station at Schriever Air Force Base in Colorado Springs. After processing, the navigational data are sent back to the satellites. Furthermore, the atomic clocks aboard the satellites are synchronized to within a few nanoseconds.

Additionally, for civil purposes everywhere in the world, commercial, scientific, or public GPS information services have emerged. The International Association of Geodesy (IAG), e.g., launched the International GNSS Service (IGS) in 1990. This service provides data needed for using GPS and other GNSS at the geodetic accuracy level.

8.2.3.3 User Segment

The user's GPS receiver is called the *user segment*. GPS receivers come in a variety of formats, from devices integrated into cars, phones, and watches to dedicated devices (e.g., from Trimble®, Garmin™, and Leica Geosystems®).

A GPS receiver consists of an antenna, a stable quartz clock, and a processing unit. The receiver must be able to monitor several satellites at the same time, each of them tracked in a separate channel. Sometimes, more than one channel is used for one satellite. Modern devices typically have between 20 and several dozen channels, sometimes more than 200 channels in multiconstellation receivers.

Furthermore, geodetic receivers may include additional receivers for terrestrial broadcasted differential correction data for high-precision measurements.

8.2.3.4 GPS Signals, Code, and Carrier

All GPS satellites emit carrier, code, and data signals which are used for the observations (Tab. 8.1). At least the two carrier signals L1 and L2 with frequencies

Table 8.1 Important properties of GPS code and carrier phases

	Code	Carrier
Chip length/wavelength	P-code: 29.3 m	L1: 19.05 cm
	C/A-code: 293.0 m	L2: 24.45 cm
Noise	P-code: 0.6–1.0 m	1–3 mm
– Classical receiver	C/A-code: 5.0–10.0 m	
– Modern development	dm	1–3 mm
Propagation effects	Ionospheric	Ionospheric
	delay	acceleration
Ambiguity	Unambiguous	Ambiguous

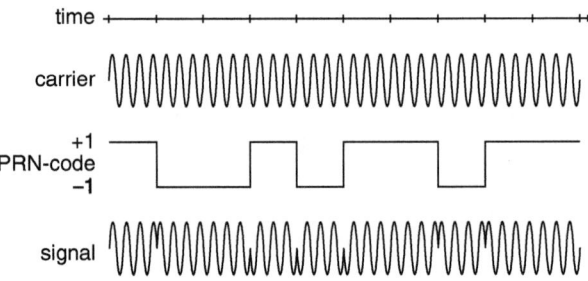

Fig. 8.11 Carrier and code signal in GPS

of 1.57542 GHz and 1.2276 GHz, respectively, are broadcasted simultaneously. A method called *code division multiple access* (CDMA) allows the receiver to distinguish between the signals of different satellites. The CDMA method encodes the message data with a pseudorandom noise (PRN) sequence that is unique for any satellite. The PRN code consists of a sequence of $+1$ and -1 with pseudorandom character (Fig. 8.11). Since the receiver knows about all actually possible PRN sequences, it is able to find the corresponding one and finally to decode the message data. There are two different kinds of PRN sequences that are used: the coarse/acquisition code (C/A-code) at 1.023 million chips per second and the precise code (P-code) at 10.23 million chips per second. Here, the term "chips" relates to the bits which do not carry information but are arranged in a pseudorandom manner. The carrier frequency L1 is modulated by C/A- and P-code. L2, however, is only modulated by P-code. Recent GPS satellites additionally broadcast a third frequency called L5 and additional user signals. Finally, it has to be noted that since GPS is a military system, there are some possibilities to artificially degrade the accuracy for nonmilitary users.

The pseudoranges are found by cross correlation of the C/A- and P-codes generated by the satellite and the receiver. This technique provides solutions in real time. However, the accuracy is limited to decimeter to meter range. Therefore,

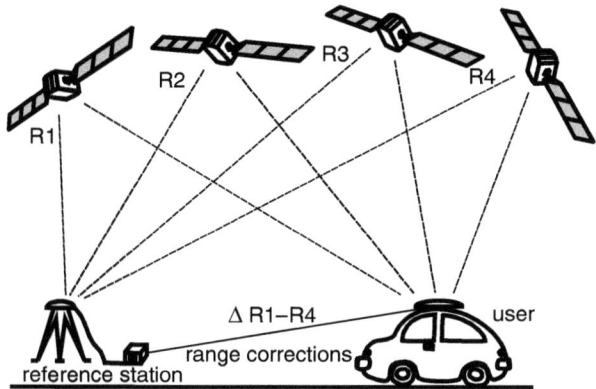

Fig. 8.12 DGPS with range corrections; from Seeber et al. (1996)

geodetic GPS receivers also include phase measurements of the carrier waves leading to accuracies of a few millimeters. As a prerequisite for such measurements, the carrier signal has to be reconstructed. This can be achieved without loss of signal quality by subtracting the code from the received signal. But also special codeless techniques have been developed and are used regularly.

8.2.3.5 Differential GPS

On the geodetic level GPS signals are severely influenced by atmospheric conditions. To increase the accuracy of measurements, differential GPS (DGPS) has been developed. The improvement results from the fact that two close-by GPS receivers will experience almost the same position offsets induced by atmospheric influences, and therefore, the position difference between the two stations will only be very little affected.

A typical DGPS setup is depicted in Fig. 8.12. One receiver is placed on a point with known terrestrial coordinates, therefore working as a reference station. The terrestrial position of the second receiver is unknown. Now the coordinate offset is determined as the difference of the two GPS positions. Finally the offset is simply added to the known terrestrial coordinate. With real-time kinematic GPS, the necessary corrections are sent from the reference station to the second receiver by a radio link. If DGPS measurements employ carrier phases, accuracies of a few millimeters can be achieved.

8.2.4 GLONASS

GLONASS (Globalnaja Nawigazionnaja Sputnikowaja Sistema; Russian for GNSS) is the Russian GNSS that shows a lot of similarities to GPS. In 2011, GLONASS consisted of 24 operational and 6 backup satellites. The system is run by the Russian space forces as system operator for the Russian government. Due to its broad field of application, it has significant merits also for the civil user. GLONASS operates with two different navigation signals: the navigation signal SP for ordinary precision and the high-precision signal HP. Every GLONASS user can permanently use the SP services for position and time determination. The achievable accuracies are 57–70 m horizontally, 70 m vertically, for velocity components 15 cm/s, and of order 1 µs for time. In differential mode accuracies can be increased significantly. The SP signal in L1 operates in several channels with frequencies $1,602$ MHz + $n \times 0.5625$ MHz, where n is the channel number ($n = -7, \ldots, 6$). Thus, every satellite broadcasts with its own individual frequency. Only antipodal satellites share the same frequency. GLONASS satellites are equipped with cesium clocks and synchronized with highly stable H masers ($\sigma \simeq 5 \times 10^{-14}$) on the ground, generating the GLONASS system time T_{GL}. The system time is directly related with UTC by

$$T_{GL} = UTC + 3^h.$$

8.2.5 DORIS

The measuring system *Doppler Orbitography by Radiopositioning Integrated on Satellite* (DORIS) is a 2-frequency Doppler system that comes into operation on several space platforms. For DORIS the broadcast stations are on ground the receivers on the space platforms. DORIS started operation in 1990. DORIS stations can be found on all large tectonic plates of the Earth. Presently with DORIS, one attains positional accuracies of about 2 cm. DORIS is coordinated at Institut Géographique National (IGN) in Champs-sur-Marne, France (http://lareg.ensg.ign.fr).

8.2.6 GALILEO

To achieve European independence from the military-controlled navigation systems GPS and GLONASS, EU and ESA have started to assemble the European GNSS named *GALILEO* (Fig. 8.13). States like China or India are participating. Initial date of the system becoming operational was 2012, but later it was moved for a number of times and is unlikely to be ready before 2018.

The system design is based upon 30 satellites (27 plus 3 spares) orbiting the Earth at a height of $23,616$ km. The first test satellite *Giove A* started operation in December 2005, the second one, *Giove B*, in April 2008. Together with a terrestrial control Segment, a global coverage should be realized. Planned is a free service (open service) of general interest for positioning, navigation, and time synchronization. Further services with costs will be realized like the commercial service for increased positional accuracy; a safety-of-life service for air and railway traffic; a service for police, coast guard, or the secret service; and a tracing and emergency medical service.

Fig. 8.13 The planned European satellite navigation system GALILEO. Image: ESA

8.2.7 Gyroscopes

A gyroscope with known or negligibly small external torques is a local rotational sensor that due to the Sagnac effect delivers information about the rotational motion of the gyroscope's platform with respect to (local) inertial axes.

8.2.7.1 Passive Sagnac Interferometers

There are several ways of constructing Sagnac interferometers, which eventually may serve for the application in geodetic astronomy. Passive Sagnac interferometers based upon monochromatic light waves were the first to demonstrate their suitability for the highly resolved measurement of rotations. The historic experiments of Sagnac (1913) and Michelson and Gale (1925) are famous examples. It is noteworthy that the Michelson–Gale experiment, for example, just resolved Earth rotation, while the requirements for the sensor resolution in geodetic astronomy are nine orders of magnitude higher. Worse than that, the sensor drift needs to be controlled to a few parts per billion over several months at the same time. Even fiber optic gyros (FOGs), which are modern versions of the passive Sagnac interferometer concept, do not perform well enough to provide a viable solution.

The operation principle of a FOG is fairly simple, while the actual sensor design itself is highly complex in order to obtain high sensor stability and resolution (Lefevre 1993). Figure 8.14 illustrates the basic concept. A light beam with a narrow spectral band width is generated by a light source. Ideally, a monochromatic laser beam would be required, but due to substantial interference as a result of scattered light, this is not possible. The light beam is then passed on to an equal intensity beam splitter. The resultant two light beams are guided around a monomode fiber coil in opposite direction. After passing through the fiber, both beams are superimposed again by the same beam splitter and steered onto a photodetector. If the entire apparatus is at complete rest, each of the beams travels the same distance, and

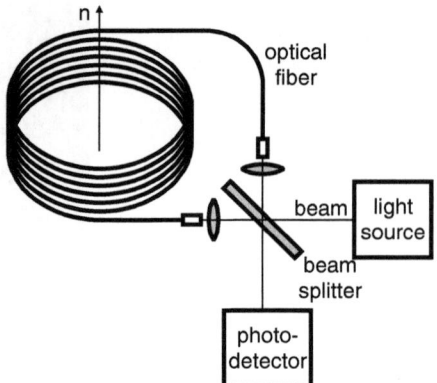

Fig. 8.14 Operation principle of a fiber optic gyroscope

there will be no phase difference between them. However, if the FOG is rotating about the normal vector **n** of the fiber coil, the two beams no longer travel the same distance, and a small phase shift between the light beams is observed. Because the signals travel at the speed of light, the obtained phase shift remains to be very small. Therefore, a modulation technique, pulsed operation, and $\pi/2$—phase shifting for one sense of propagation are employed to achieve a maximum of instrumental sensitivity. Furthermore, the sensor is operated in a closed loop configuration in order to ensure a wide dynamic range. Details on the general sensor design of fiber optical gyroscope are given in Lefevre (1993).

A full description of the Sagnac effect is based on general relativity (Soffel 1989; Höling 1990), but in this case a classical interpretation yields the same result (Milonni and Eberly 1988). The observed phase difference is

$$\delta\phi = \frac{8\pi A}{\lambda c}\mathbf{n}\cdot\mathbf{\Omega}, \qquad (8.10)$$

where A is the area circumscribed by the light beams, λ the effective optical wavelength of the two beams, **n** the normal vector upon A, and $\mathbf{\Omega}$ the rate of rotation of the interferometer. The inner product $\mathbf{n}\cdot\mathbf{\Omega}$ is related with the orientation of the sensor relative to the vector of rotation. Equation (8.10) relates the obtained phase difference to the rate of rotation of the entire apparatus and can be interpreted as the gyroscope equation (Stedman 1997). Because glass fibers with a length of several hundred meters are used, the scale factor can be made very large by winding the fiber to a coil, and the sensitivity for rotational excitations is therefore much larger than that for a single loop. For that case the scale factor in the gyroscope equation can be written as $4\pi LR/(\lambda c)$, where R is the radius of the coil and L the length of the fiber. In this way Earth rotation can be continuously observed to about an accuracy of 10% even on a relatively modest FOG of about the size of a small cell phone.

The realization of very large FOGs is not without problems. For a large-scale installation at the geodetic observatory in Wettzell, approximately 2 km of fiber has been wound around a large Zerodur disc with 4.25-m diameter inside an underground temperature-stabilized laboratory. Although this created a very large-scale factor, Earth's rotation could not be resolved below the level of one part per million (ppm). At the same time, the sensor stability suffered considerably because of the increased length of the fiber, despite an overall temperature variability of less than 4 mK/day.

8.2.7.2 Active Sagnac Interferometers

Ring lasers are essentially close in design to either the Sagnac or the Michelson–Gale construction. The major difference is the fact that the light amplifying mechanism is placed inside the optical cavity of the respective interferometer. Together with the lasing condition for a "ring" cavity, where an integer number of waves has to fit into the resonator to obtain coherent amplification, this translates the phase difference of the passive interferometer into a frequency difference for the actual ring laser (Macek and Davies 1963). As a proof of concept, the applicability of ring lasers was already shown as early as 1963, 3 years after the first demonstration of an optical maser (see Fig. 8.15). In the 1970s, small ring lasers quickly became the preferred rotation sensor for inertial navigation but remained at a resolution level 5 orders of magnitude short of the requirements of space geodesy. On top of that, the sensor drift was too high by about the same amount. Nevertheless, the concept of constructing gas lasers based on the neon transition at $\lambda = 632.8$ nm with helium aiding the pumping process remained the most successful instrumental approach. In order to gain sensitivity and stability, several changes in concept were necessary.

First of all, the cavity, although still manufactured monolithically from the low thermal expansion material Zerodur, was upscaled dramatically from approximately $0.017\,m^2$ to first $1\,m^2$ (Schreiber et al. 2003) and then to $16\,m^2$ (the G ring in Wettzell). The latter ring produced the best results (Schreiber et al. 2009), because it represents the best compromise to date between mechanical stability of the ring laser body on one side and a large-scale factor on the other side. Constructions enclosing an area as large as $834\,m^2$ (Hurst et al. 2009) while still representing a working rotation sensor proofed to be mechanically too unstable. The phase difference of (8.10) then becomes the ring laser equation for the Sagnac frequency shift, Δf_{Sagnac} (P: perimeter):

$$\Delta f_{\text{Sagnac}} = \frac{4A}{\lambda P}\mathbf{n}\cdot\mathbf{\Omega}. \tag{8.11}$$

By means of superimposing both beams behind a mirror, the corresponding interference pattern is produced by the two counter traveling beams. Any change of the rate of rotation and the projection of the normal vector of the gyroscope area onto the axis of rotation induces a corresponding shift of the Sagnac frequency which can be measured with high precision. All this requires the scale factor of the ring laser platform to be more stable than the observable, which is a considerable requirement.

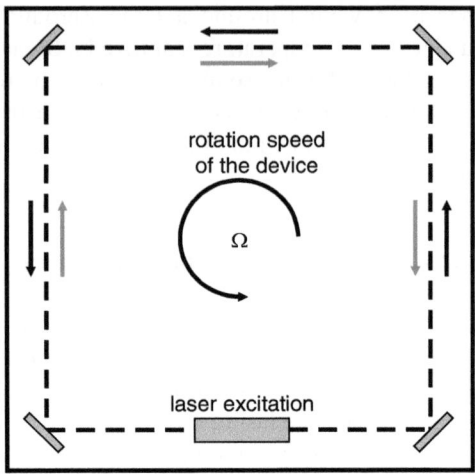

Fig. 8.15 Schematic diagram of a square ring laser gyroscope; from http://www.fs.wettzell.de

Envisaged accuracies of such a ring laser are in the range of 10^{-9} of the Earth's rotation rate (6 nrad per day). Such an accuracy implies enormous technical requirements. The stability of geometrical quantities like circumference and area has to be guaranteed at a level of better than 1 nm for the entire length of the 16-m cavity in case of the G ring. The reflecting mirrors have to be of extraordinary quality, with a loss in the range of 5 ppm, since the two counter traveling beams interact with each other and produce some coupling effects. This effect is known as *lock-in*, and variations in the amount of coupling depend critically upon the constancy of the distances of the mirrors that have to be mounted on some bedplate with extreme thermal stability. At present, the G-ring laser at the Geodetic Observatory Wettzell is the most stable and most sensitive ring laser of this kind (other even larger instruments can be found in the Cashmere Cavern, near Christchurch, New Zealand). A Zerodur disc (baseplate) with a diameter of 4.25 m, a thickness of 25 cm, and weight of 9 t serves as a stable baseplate for the G ring. The whole sensor is located in a temperature-stabilized pyramid-shaped underground laboratory and is operated under remote control (Fig. 8.16). A pressure-stabilizing vessel encloses the construction and maintains the integral length of the resonator constant in a feedback loop arrangement. When operating properly, G ring can resolve Earth rotation to 5 parts in 10^9, which brings the gyroscope to the point that it can resolve the Chandler and the annual wobble of the Earth (Schreiber et al. 2011).

8.3 International Terrestrial Reference System

The ITRS was defined by the IUGG Resolution No. 2 adopted in Vienna, 1991. The ITRS is established by a net of observing stations involved in VLBI, SLR, GPS, and DORIS observations. The origin of the ITRS should agree with the

8.3 International Terrestrial Reference System

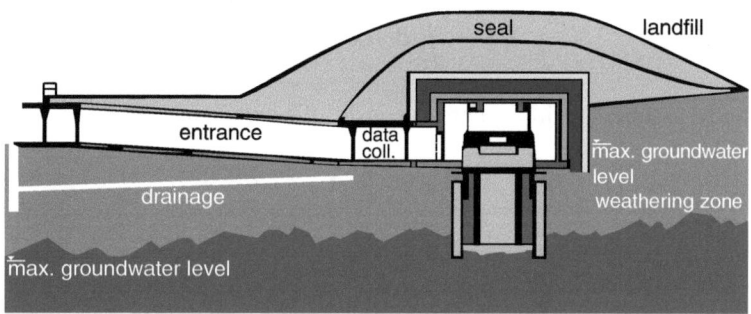

Fig. 8.16 A transversal section through the laboratory where the G-ring laser gyroscope is found at the German geodetic observatory in Wettzell

Earth's center of mass including the oceans and atmosphere. The time variable of the ITRS should be terrestrial coordinate time (TCG; see below). The spatial ITRS coordinates are realized by values for the station positions. The terrestrial pole (the ITRS z-axis) should lie near (within 5 mas) the mean rotation pole for the years 1900–1905 to ensure historical continuity. This historical mean pole is often called the conventional international origin (its acronym CIO is not used in this book because it could be mismatched with the similar acronym for the celestial intermediate origin from Sect. 9.4.1). Originally, the conventional international origin was realized by defining (terrestrial) longitude and latitude to five stations of the International Latitude Service (ILS). The ITRS pole is also called the IERS reference pole. By convention, the x-axis of the ITRS should lie near (within 5 mas) the *Bureau International de l'Heure* (BIH) 1984.0 reference meridian (McCarthy 1996), which lies about 100 m east of the Greenwich prime meridian. The x-axis of the ITRS lies on the IERS reference meridian. Both the IERS reference pole and meridian are implicitly defined by the set of station coordinates.

Several realizations of the ITRS have been worked out so far: ITRF-92, 93, 94, 96, 2000, 2005, and 2008. At some fixed epoch, different realizations of the ITRS are related by a Helmert transformation described by a set of 7 parameters: (T_1, T_2, T_3) for translation, (R_1, R_2, R_3) for rotation, and one parameter (D) for scale changes. The relation between two realizations is given in the form

$$\begin{pmatrix} X' \\ Y' \\ Z' \end{pmatrix} = \begin{pmatrix} X \\ Y \\ Z \end{pmatrix} + \begin{pmatrix} T_1 \\ T_2 \\ T_3 \end{pmatrix} + \begin{pmatrix} D & -R_3 & R_2 \\ R_3 & D & -R_1 \\ -R_2 & R_1 & D \end{pmatrix} \begin{pmatrix} X \\ Y \\ Z \end{pmatrix}, \qquad (8.12)$$

where the various parameters can be taken from IERS publications. The temporal change of ITRS axes with time is related with a so-called *no-net-rotation* (NNR) condition for the lithosphere of the Earth. The NNR condition is also called Tisserand condition, implying the total angular momentum $\mathbf{L}_\mathcal{L}$ of the lithosphere ($\mathcal{L}$) to vanish in the ITRS, i.e.,

$$\mathbf{L}_{\mathcal{L}} = \int_{\mathcal{L}} dm (\mathbf{x} \times \mathbf{v}) = 0. \tag{8.13}$$

We will now assume the lithosphere to consist of a certain number of plates, labeled by a number n, so that

$$\mathbf{L}_{\mathcal{L}} = \sum_{n} \mathbf{L}_n.$$

Assuming the geocentric distance R = const. for the lithosphere and the surface mass density to be constant, we get for $\mathbf{v}_n = \boldsymbol{\omega}_n \times \mathbf{x}$ (rigid rotation of plate n) the following conditions for each plate (dropping the index n; I_{ij} are the components of the moments of inertia tensor; Λ_G and Φ_G are geographic longitude and latitude):

$$L_x = I_{xx}\omega_x + I_{xy}\omega_y + I_{xz}\omega_z$$
$$= \int_{\mathcal{L}} \cos\Phi_G \, d\Phi_G d\Lambda_G \, [(\sin^2\Phi_G + \cos^2\Phi_G \sin^2\Lambda_G)\,\omega_x$$
$$- (\cos^2\Phi_G \sin\Lambda_G \cos\Lambda_G)\,\omega_y - (\sin\Phi_G \cos\Phi_G \sin\Lambda_G)\,\omega_z] = 0$$

$$L_y = I_{yx}\omega_x + I_{yy}\omega_y + I_{yz}\omega_z$$
$$= \int_{\mathcal{L}} \cos\Phi_G \, d\Phi_G d\Lambda_G \, [(-\cos^2\Phi_G \sin\Lambda_G \cos\Lambda_G)\,\omega_x$$
$$+ (\cos^2\Phi_G \cos^2\Lambda_G + \sin^2\Phi_G)\,\omega_y - (\sin\Phi_G \cos\Phi_G \sin\Lambda_G)\,\omega_z] = 0$$

$$L_z = I_{zx}\omega_x + I_{zy}\omega_y + I_{zz}\omega_z$$
$$= \int_{\mathcal{L}} \cos\Phi_G \, d\Phi_G d\Lambda_G \, [(-\sin\Phi_G \cos\Phi_G \cos\Lambda_G)\,\omega_x$$
$$- (\sin\Phi_G \cos\Phi_G \cos\Lambda_G)\,\omega_y - (\cos^2\Phi_G)\,\omega_z] = 0 \tag{8.14}$$

Table 8.2 displays the angular velocity vectors of the various tectonic plates of the NNR-NUVEL1A plate model (DeMets et al. 1994).

The NUVEL1 model for plate motion has clear deficiencies since all data refer to the edges of continental plates related with a long-time average over seafloor spreadings deduced from paleomagnetic data. Figure 8.17 shows differences between VLBI drift velocities and NUVEL1A plate-model velocities. The *Deutsches Geodätisches Forschungsinstitut* (DGFI) in Munich since 1988 has computed *actual plate kinematics and crustal deformation models* (APKIM) in approximately annual rhythm (e.g., Drewes 1998, 2006; Drewes and Angermann 2001). These models are based on recent GPS, SLR, VLBI, and DORIS data. Figure 8.18 shows velocity differences between the NUVEL1A and the APKIM2005 plate models. Likely in the near future, the NUVEL1A model will be replaced by a more realistic plate tectonic model such as APKIM. Besides

8.3 International Terrestrial Reference System

Table 8.2 Angular velocity vectors of the various plates of the NNR-NUVEL1A plate model

Plate abbr.	Rot. vector (geogr.)			Rot. vector (cart.)			Plate name
	Φ_G	Λ_G	Ω	ω_x	ω_y	ω_z	
AFRC	50.569	−73.978	0.2909	0.891	−3.099	3.922	Africa
ANTA	62.986	244.264	0.2383	−0.821	−1.701	3.706	Antarctica
ARAB	45.233	−4.464	0.5455	6.685	−0.521	6.760	Arabia
AUST	33.852	33.175	0.6461	7.839	5.124	6.282	Australia
CARB	25.014	266.989	0.2143	−0.178	−3.385	1.581	Caribbea
COCO	24.487	244.242	1.5103	−10.425	−21.605	10.925	Cocos
EURA	50.631	247.725	0.2337	−0.981	−2.395	3.153	Eurasia
INDI	45.505	0.345	0.5453	6.670	0.040	6.790	India
NOAM	−2.438	−85.895	0.2069	0.258	−3.599	−0.153	N. America
NAZC	47.804	259.870	0.7432	−1.532	−8.577	9.609	Nazca
PCFC	−63.045	107.325	0.6408	−1.510	4.840	−9.970	Pacific
SOAM	−25.325	235.570	0.1164	−1.038	−1.515	−0.870	S. America
JUFU	−30.054	58.870	0.6658	5.200	8.610	−5.820	Juan de Fuca
PHIL	−38.011	−35.360	0.8997	10.090	−7.160	−9.670	Philippine
RIVR	20.428	253.128	1.9781	−9.390	−30.960	12.050	Rivera
SCOT	−25.273	261.234	0.1705	−0.410	−2.660	−1.270	Scotia

The values for Φ_G and Λ_G are given in degrees, for Ω in degrees per million years, and for ω_x, ω_y and ω_z in milliradians per million years

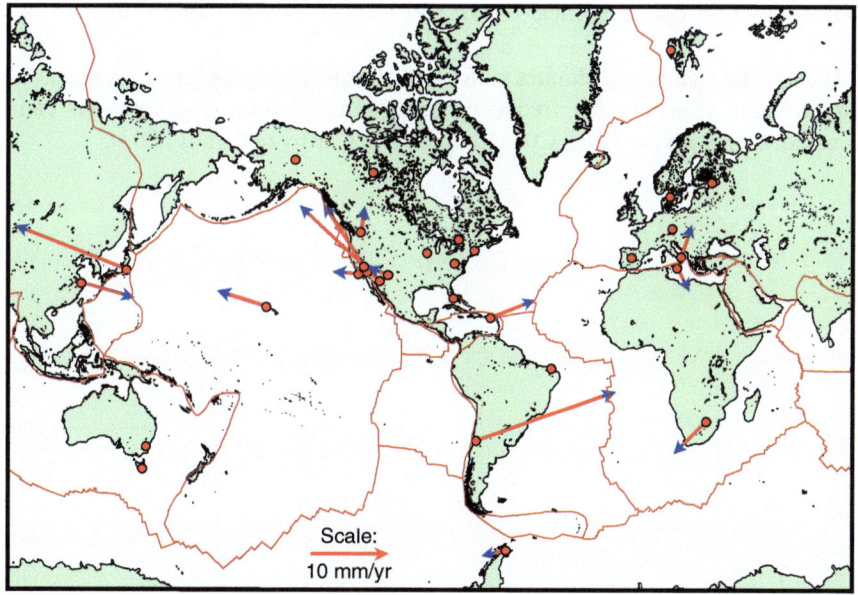

Fig. 8.17 Differences between VLBI drift velocities and NUVEL1A plate-model velocities (solution KB 2007dn, version 1); from Goddard Space Flight Center GSFC, NASA, http://lupus.gsfc.nasa.gov/dataresults_nuvel_2007dn.htm

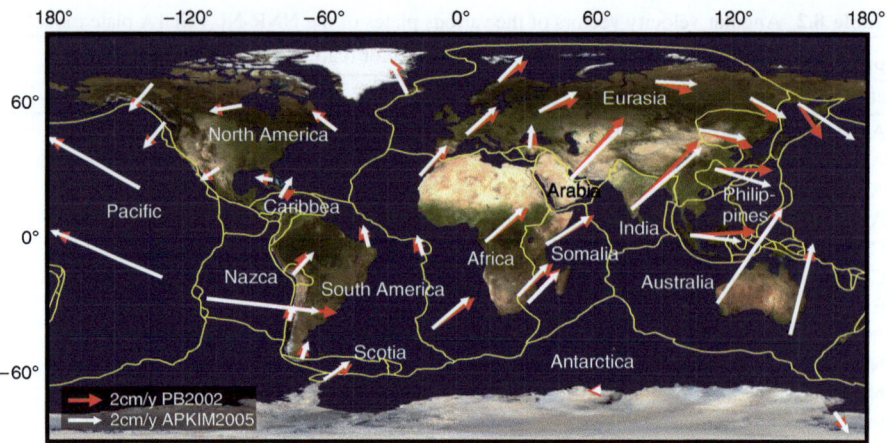

Fig. 8.18 Velocity differences between the NUVEL1A and the APKIM2005 plate models

NUVEL1A and the APKIM series, a series of further tectonic plate motion models have appeared in the literature: HS2-NUVEL1A (Gripp and Gordon 1990), HS3-NUVEL1A (Gripp and Gordon 2002), REVEL2000 (Sella et al. 2002), CGPS of 2004 (Prawirodirdjo and Bock 2004), and GSRM v1.2 of 2004 (Kreemer et al. 2003).

Clearly the station coordinates suffer high-frequency variations, e.g., due to tidal displacements. Such higher frequency parts of the station displacements can be accessed, e.g., with the IERS C03 (Chap. 7) (McCarthy and Petit 2003).

Chapter 9
From the GCRS to the ITRS

9.1 Polar Motion

Polar motion means motion of the Earth's rotation axis with respect to the Earth's surface. The possibility for such a motion of the Earth's rotation axis was first suggested by Leonhard Euler in 1765. It should reflect itself in elevation variations of the pole with a period of about 10 months. The observational verification by means of latitude variations, however, remained unsuccessful for a long time. It was only after 1884 when Karl Friedrich Küstner detected the latitude variations. To confirm this detection, the International Latitude Service, ILS, was founded in 1899, and within the next few years, the latitude variations were confirmed without doubts. However, it was found that the motion was not in accordance with Euler's theory of a rigid Earth with a period of about 304 days (the Euler period). Later Seth Carlo Chandler found that the polar motion contains two dominant contributions:

1. A motion of the rotation pole about the principle axis of inertia with a period of about 430 days (the Chandler period).
2. An analogous motion with a period of 1 year.

The first motion was interpreted by Simon Newcomb as resulting from the rotation of an elastic gyroscope, so what the Euler period is for a rigid Earth, the Chandler period is for an elastic Earth. The annual motion results from meteorological and geophysical perturbations. Figure 9.1 shows the polar motion for the years 2007–2011. One sees the spiral motion between two circles of 3 and 9 m as beating between the annual and the Chandler period. In Fig. 9.2, different causes for polar motion are depicted. It should be noted here that nowadays the concept of the Earth's rotation axis has been replaced by the conventional intermediate pole (discussed below) and that polar motion is described with respect to that pole.

Let $\mathbf{X}_I$ be a vector in the quasi-inertial system (the GCRS), $\mathbf{X}_A$ the same vector in the true astronomical system at epoch T. It is given by the true astronomical coordinates (α, δ) at time T. We have

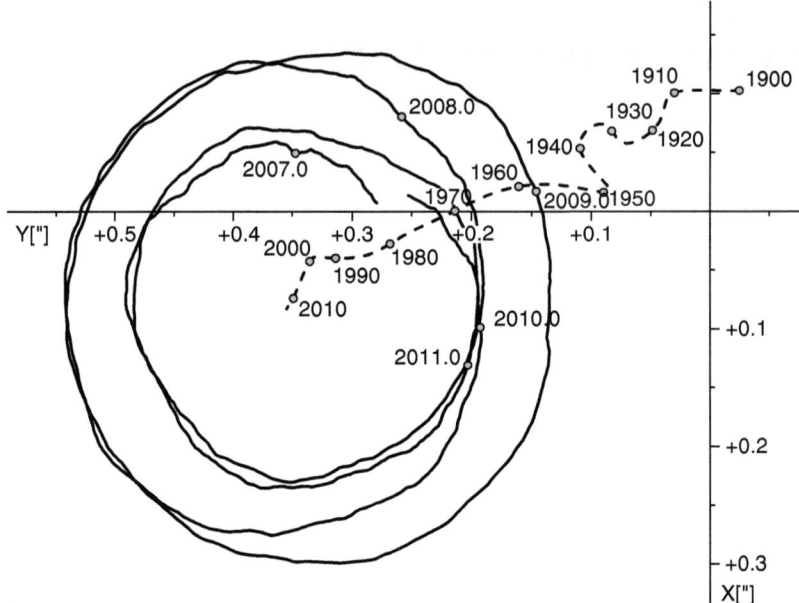

Fig. 9.1 *Solid line*: polar motion, 2007–2011 (Standard EOP data files, IERS). *Dashed line*: mean pole displacement, 1900–2010 (EOP C01, IERS)

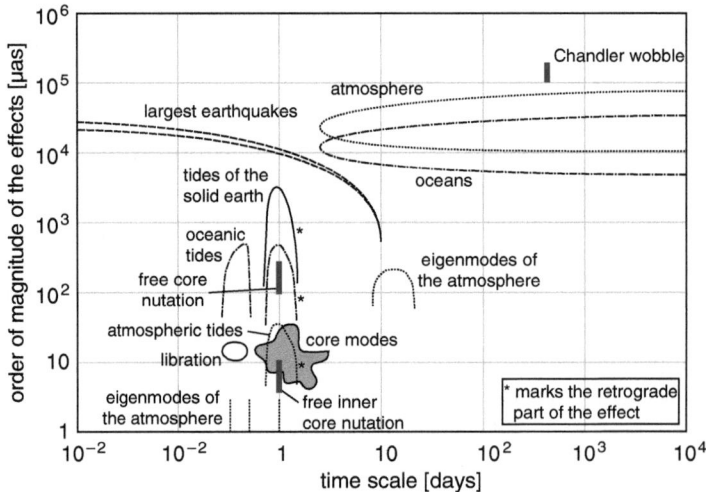

Fig. 9.2 Different causes for polar motion; from NASA's Goddard Space Flight Center, modified

9.1 Polar Motion

Fig. 9.3 The location of the NCP in the terrestrial system

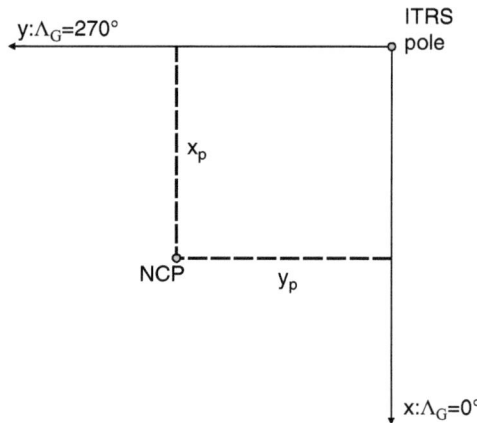

$$\mathbf{X}_A = \mathcal{N}\mathcal{P}\,\mathbf{X}_I \tag{9.1}$$

and
$$\mathbf{X}_T = \mathcal{S}\,\mathbf{X}_A, \tag{9.2}$$

if $\mathbf{X}_T$ denotes the corresponding geographic or terrestrial vector. If the x-axis is rotated from the true equinox to the Greenwich meridian, we obtain a vector $\mathbf{X}_{Gr}$:

$$\mathbf{X}_{Gr} = \mathcal{R}_3(\text{GAST})\,\mathbf{X}_A. \tag{9.3}$$

The angle of rotation is GAST, i.e., Greenwich (apparent) sidereal time. With two further rotations we get to the vector in terrestrial (geographic) coordinates:

$$\mathbf{X}_T = \mathcal{R}_2(-x_p)\,\mathcal{R}_1(-y_p)\,\mathbf{X}_{Gr}$$

or
$$\mathbf{X}_T = \mathcal{R}_2(-x_p)\,\mathcal{R}_1(-y_p)\,\mathcal{R}_3(\text{GAST})\,\mathbf{X}_A, \tag{9.4}$$

i.e.,
$$\mathcal{S} = \mathcal{R}_2(-x_p)\,\mathcal{R}_1(-y_p)\,\mathcal{R}_3(\text{GAST}). \tag{9.5}$$

The two angles (x_p, y_p) are called pole coordinates. Signs are such that the terrestrial x-axis points towards the Greenwich meridian and the y-axis towards $\Lambda_G = 270°$. For that reason in Fig. 9.4 we see the positive segment $-y_p$. The two pole coordinates describe the position of the NCP in the ITRS (Fig. 9.3). Using the *AstroRef* package, the pole coordinates may be accessed by calling GetPoleCoords.

With a first rotation, $\mathcal{R}_1(-y_p)$, the old y-axis is transformed into the new y-axis, and the pole moves from NCP to P'. With a second transformation about the new y-axis with angle $-x_p$, we get into the ITRS, and the pole moves from P' to the ITRS pole (Fig. 9.4).

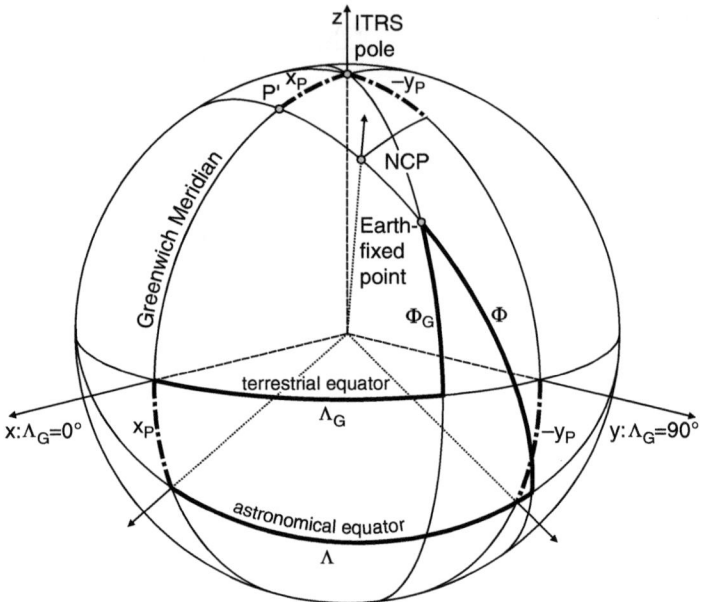

Fig. 9.4 Terrestrial coordinates

To summarize, the classical transformation from the GCRS ($\mathbf{X}_I$) to the ITRS ($\mathbf{X}_T$) can be written in the form

$$\mathbf{X}_T = \mathcal{W}^* \, \mathcal{R} \, \mathcal{Q} \, \mathbf{X}_I. \tag{9.6}$$

Here, $\mathcal{Q}$ is the precession–nutation matrix, $\mathcal{Q} = \mathcal{N}\mathcal{P}$ with

$$\mathcal{N} = \mathcal{R}_1(-\epsilon) \, \mathcal{R}_2(-\Delta\psi) \, \mathcal{R}_1(\epsilon_A)$$

and

$$\mathcal{P} = \mathcal{R}_3(-z_A) \, \mathcal{R}_2(+\theta_A) \, \mathcal{R}_3(-\zeta_A).$$

The matrix $\mathcal{R}$ is defined as

$$\mathcal{R} = \mathcal{R}_3(\text{GAST})$$

and the polar motion matrix, $\mathcal{W}^*$, is given by

$$\mathcal{W}^* = \mathcal{R}_2(-x_p) \, \mathcal{R}_1(-y_p).$$

Note that later the order of polar motion matrices was exchanged, i.e., $\mathcal{W}^*$ was replaced by $\mathcal{W}$ with

$$\mathcal{W} = \mathcal{R}_1(-y_p) \, \mathcal{R}_2(-x_p). \tag{9.7}$$

Since x_p and y_p are small quantities, the differences between $\mathcal{W}^*$ and $\mathcal{W}$ are of order 0.1 μas for current polar motion amplitudes and thus negligible.

9.1 Polar Motion

Note that often in the literature the inverse transformation is written as

$$\mathbf{X_I} = \mathbf{Q\,R\,W\,X_T}. \quad (9.8)$$

Then W, R, and Q are the matrices inverse to $\mathcal{W}, \mathcal{R}$, and $\mathcal{Q}$, i.e.,

$$\mathbf{W} = \mathcal{W}^{-1} = \mathcal{R}_2(+x_p)\,\mathcal{R}_1(+y_p) \quad \text{etc.}$$

In Fig. 9.4, we can well recognize the polar-motion-induced latitude variations $\Phi_G - \Phi$ of an Earth-fixed point. Also the values for azimuth and longitude are affected by polar motion. Let Φ, Λ, and A be the observed values for astronomical latitude, longitude, and azimuth, related with the $\mathbf{X}_{Gr}$ system. The equation

$$\mathbf{X_T} = \mathcal{R}_2(-x_p)\,\mathcal{R}_1(-y_p)\,\mathbf{X}_{Gr}$$

reads

$$\begin{pmatrix} \cos\Phi_G \cos\Lambda_G \\ \cos\Phi_G \sin\Lambda_G \\ \sin\Phi_G \end{pmatrix} = \mathcal{U} \begin{pmatrix} \cos\Phi \cos\Lambda \\ \cos\Phi \sin\Lambda \\ \sin\Phi \end{pmatrix}$$

with

$$\mathcal{U} \equiv \mathcal{R}_2(-x_p)\,\mathcal{R}_1(-y_p) \simeq \begin{bmatrix} 1 & 0 & x_p \\ 0 & 1 & -y_p \\ -x_p & y_p & 1 \end{bmatrix}, \quad (9.9)$$

where we considered only first-order terms ($\cos x_p \simeq \cos y_p \simeq 1$, $\sin x_p \simeq x_p$, etc.). Written explicitly, we have

$$\cos\Phi_G \cos\Lambda_G = \cos\Phi \cos\Lambda + x_p \sin\Phi$$
$$\cos\Phi_G \sin\Lambda_G = \cos\Phi \sin\Lambda - y_p \sin\Phi \quad (9.10)$$
$$\sin\Phi_G = -x_p \cos\Phi \cos\Lambda + y_p \cos\Phi \sin\Lambda + \sin\Phi.$$

The effect of polar motion on astronomical latitude therefore is given by

$$\sin\Phi_G = \sin\Phi + \cos\Phi(y_p \sin\Lambda - x_p \cos\Lambda). \quad (9.11)$$

A Taylor expansion $\sin\Phi_G = \sin\Phi + (\Delta\Phi)_{PM} \cos\Phi + \cdots$ then, to first order in the pole coordinates, leads to

$$(\Delta\Phi)_{PM} = \Phi_G - \Phi = y_p \sin\Lambda - x_p \cos\Lambda. \quad (9.12)$$

In an analogous way, we get to first order

$$(\Delta\Lambda)_{PM} = \Lambda_G - \Lambda = -(x_p \sin\Lambda + y_p \cos\Lambda) \tan\Phi. \quad (9.13)$$

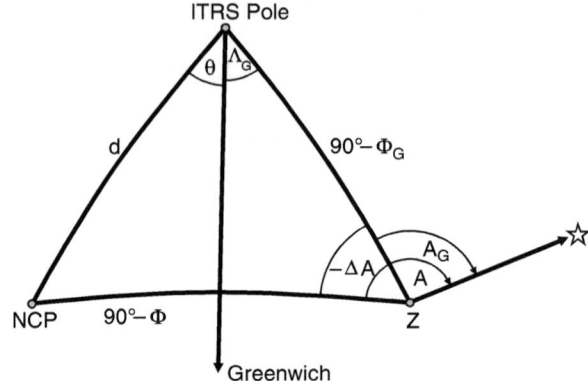

Fig. 9.5 Influence of polar motion upon the astronomical azimuth

We see that even for an observer located at the Greenwich prime meridian, the astronomical longitude varies with time. If effects from polar motion are considered, the reference meridian for astronomical longitude is a special kind of Greenwich meridian. It is defined as that hour circle that connects the celestial pole (the CIP; see below) with a point on the celestial equator defined by the intersection with the IERS reference meridian.

The effect of polar motion on astronomical azimuth can be deduced from Fig. 9.5. Here, Z denotes the observer's zenith. Let $(\Delta A)_{\text{PM}} = A_G - A$; from Fig. 9.5, we can read off

$$\frac{\sin(-\Delta A)}{\sin d} = \frac{\sin(\theta + \Lambda_G)}{\sin(90° - \Phi)}$$

or

$$\sin(\Delta A)_{\text{PM}} = -\frac{\sin d \sin(\theta + \Lambda_G)}{\cos \Phi}$$

or

$$(\Delta A)_{\text{PM}} \simeq -d(\sin \theta \cos \Lambda_G + \cos \theta \sin \Lambda_G) \sec \Phi.$$

Since $d \cos \theta = x_p; d \sin \theta = y_p$ we finally obtain

$$(\Delta A)_{\text{PM}} \simeq -(x_p \sin \Lambda_G + y_p \cos \Lambda_G) \sec \Phi. \qquad (9.14)$$

9.2 Universal Time, UT1, and Length-of-Day Variations

At present, every astronomical timescale is basically derived from TAI. However, in ordinary life the orientation of the Earth with respect to the Sun plays an important role implying that so-called *Earth timescales* are of great use. Such Earth timescales are simply angles (i.e., no physical times) describing the orientation of Earth in space.

9.2 Universal Time, UT1, and Length-of-Day Variations

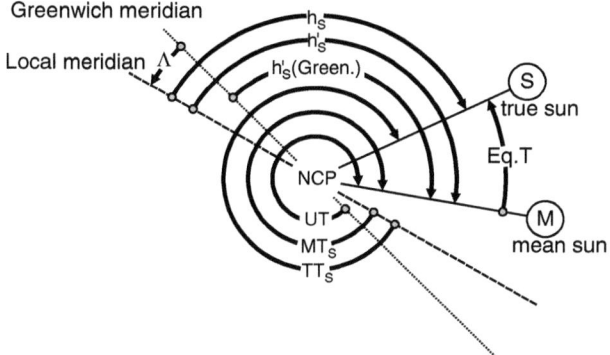

Fig. 9.6 Various Earth timescales

9.2.1 Universal Time

True solar time TT_S is defined by

$$TT_S = h_{Sun} + 12^h, \qquad (9.15)$$

where h_{Sun} is the hour angle of the true Sun. The 12^h is added so that $0^h\, TT_S$ coincides with midnight (and not with noon where the Sun passes through the meridian and $h_{Sun} = 0$). The obliquity of the ecliptic and the eccentricity of the Earth's orbit about the Sun lead to variations of the TT_S length of day as measured with atomic clocks. For that reason mean solar time MT_S is introduced with a fictitious *mean Sun* that moves in the equatorial plane with constant speed. If h_{mSun} denotes the hour angle of the mean Sun, one has

$$MT_S = h_{mSun} + 12^h. \qquad (9.16)$$

The difference between true and mean solar time is the *Equation of Time*, Eq.T.,

$$Eq.T. = TT_S - MT_S = h_{Sun} - h_{mSun}. \qquad (9.17)$$

Figure 9.6 shows UT, MT_S, and TT_S, Fig. 9.7 the equation of time (Eq.T.). If the declination of the Sun is plotted versus the equation of time, we end with a figure called analemma (Fig. 9.8). It is found on many sundials to convert true solar time to mean solar time. Mean solar time is also called local mean time (LMT), and GMT stands for Greenwich mean time:

$$\begin{aligned} LMT &= h_{mSun} + 12^h \\ GMT &= h_{mSun}(Gr) + 12^h. \end{aligned} \qquad (9.18)$$

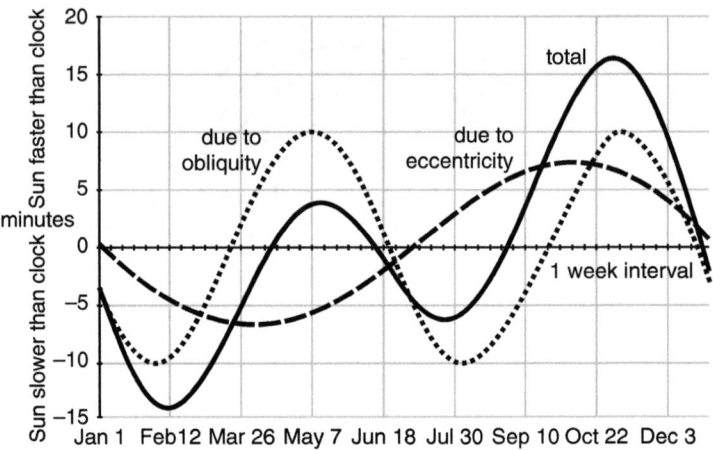

Fig. 9.7 *Solid line*: the equation of time for the year 2000; *dotted line*: contribution of the obliquity of the ecliptic; *dashed line*: contribution of the eccentricity of the Earth's orbit

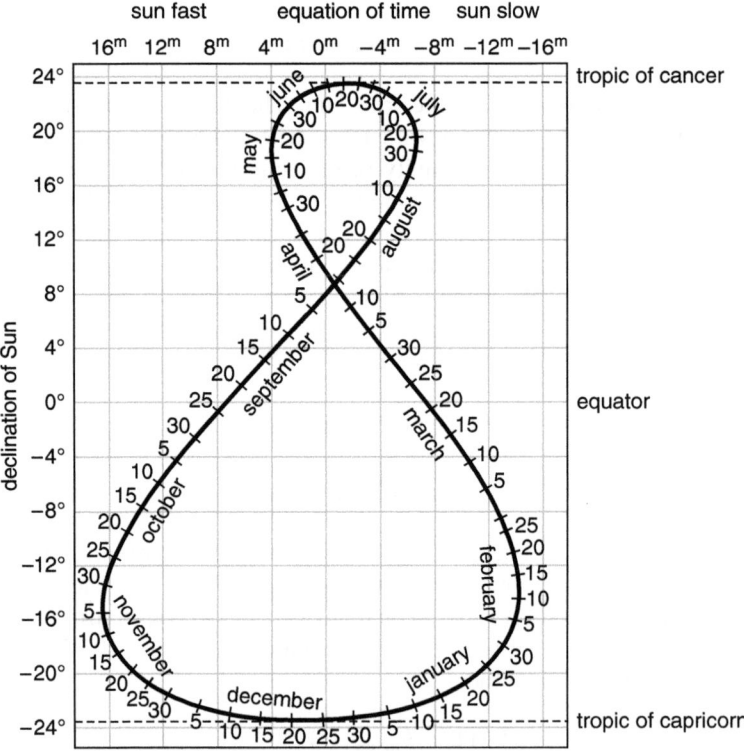

Fig. 9.8 Analemma for the year 2000; data from US Coast and Geodetic survey

9.2 Universal Time, UT1, and Length-of-Day Variations

Universal time UT1 is mean solar time referred to the Greenwich meridian, i.e.,

$$\text{UT1} = \text{GMT}. \tag{9.19}$$

UT1 has a direct relation to sidereal time. Let α_{mSun} be the right ascension of the mean Sun, then

$$\text{LMST} = h_{\text{mSun}} + \alpha_{\text{mSun}} \tag{9.20}$$

and

$$\text{GAST} = \text{UT1} + \alpha_{\text{mSun}} + \text{Eq.E.} - 12^h. \tag{9.21}$$

Universal time, UT1, can be written in different ways:

$$\begin{aligned}
\text{UT1} = \text{GMT} &= h_{\text{mSun}}(\text{Gr}) + 12^h \\
&= \text{GMST} + 12^h - \alpha_{\text{mSun}} \\
&= \text{LMST} + 12^h - \alpha_{\text{mSun}} - \Lambda \\
&= \text{GAST} + 12^h - \alpha_{\text{mSun}} - \text{Eq.E.} \\
&= \text{LAST} + 12^h - \alpha_{\text{mSun}} - \text{Eq.E.} - \Lambda,
\end{aligned} \tag{9.22}$$

where Λ is the actual astronomical longitude of the observer's location. If we take the geographic longitude, Λ_G, instead of Λ, we get the universal time UT0,

$$\text{UT0} = \text{LAST} + 12^h - \alpha_{\text{mSun}} - \Lambda_G = \text{UT1} - (\Delta\Lambda)_{\text{PM}} \tag{9.23}$$

or

$$\text{UT1} = \text{UT0} + (\Delta\Lambda)_{\text{PM}}. \tag{9.24}$$

The use of UT0 has the advantage that the reduction from local apparent sidereal time involves the constant value for the geographic longitude, whereas the astronomical value varies with time due to polar motion. According to Aoki et al. (1982),

$$\begin{aligned}
\alpha_{\text{mSun}} = {}& 67310^s\!.54841 + 8640184^s\!.812866\, T_U \\
& + 0^s\!.093104\, T_U^2 - 6^s\!.2 \times 10^{-6}\, T_U^3,
\end{aligned} \tag{9.25}$$

so that (GMST = UT1 + α_{mSun} − 12^h)

$$\begin{aligned}
\text{GMST of } 0^h \text{ UT1} = {}& 24110^s\!.54841 + 8640184^s\!.812866\, T_U \\
& + 0^s\!.093104\, T_U^2 - 6^s\!.2 \times 10^{-6}\, T_U^3,
\end{aligned} \tag{9.26}$$

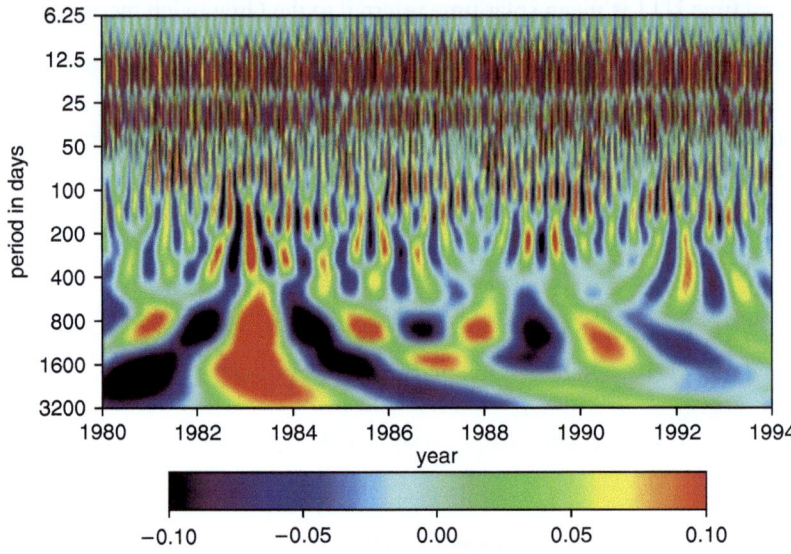

Fig. 9.9 Wavelet representation of LOD variations measured by VLBI; from Global Geophysical Fluid Center

where

$$T_U = d_U/36525 \tag{9.27}$$

and d_U represents the number of days of UT1 elapsed since JD 2 451 545.0 UT1 (1 January 2000, 12^h UT1), taking the values of $\pm 0.5, \pm 1.5, \pm 2.5$, etc. Note that (9.25) and (9.26) are based on the IAU 1976 precession and IAU 1980 nutation theory.

9.2.2 UT1 and Its Variations

Universal time is influenced by polar motion and length-of-day (LOD) variations (Fig. 9.9). Diverse effects such as solid Earth tides and changes in the atmospheric angular momentum (Fig. 9.10) that are responsible for variations in the length of day (LOD variations) are indicated in Fig. 9.11. For periods larger than about 1,000 days, the well- known ENSO (El Niño Southern Oscillation) plays an important role. For timescales of decades, the core mantle coupling and processes related with the magnetic solar cycle are of importance. Finally, due to tidal friction, the length of day steadily increases by about 2 ms per century.

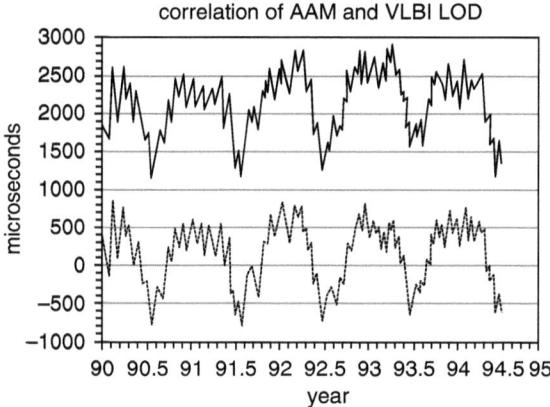

Fig. 9.10 Correlations of VLBI LOD variations (upper curve) with the atmospheric angular momentum (lower curve), AAM; from Global Geophysical Fluid Center

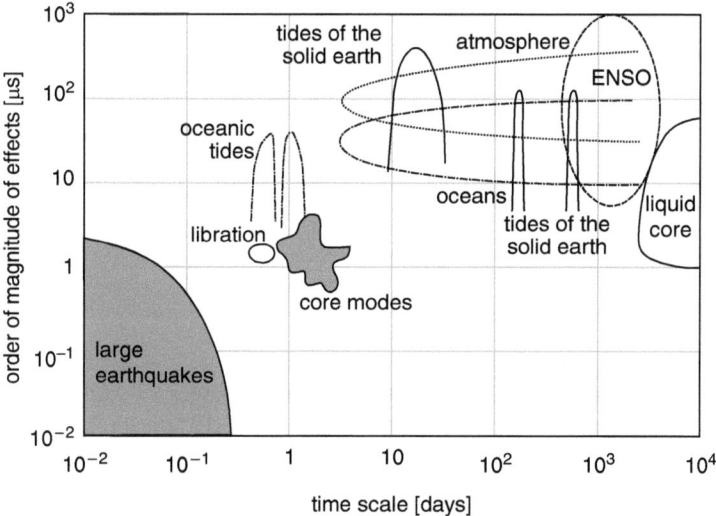

Fig. 9.11 Geophysical causes of LOD variations; from NASA's Goddard Space Flight Center, modified

9.3 Celestial Intermediate Pole

Of fundamental interest is the motion of the ITRS (with spatial coordinates X_T, Y_T, Z_T) in the GCRS (with spatial coordinates X, Y, Z) that in principle can be described with just three Euler angles ψ_E, θ_E, and ϕ_E (see Fig. 9.12). However, the introduction of intermediate poles whose motions can be accurately predicted

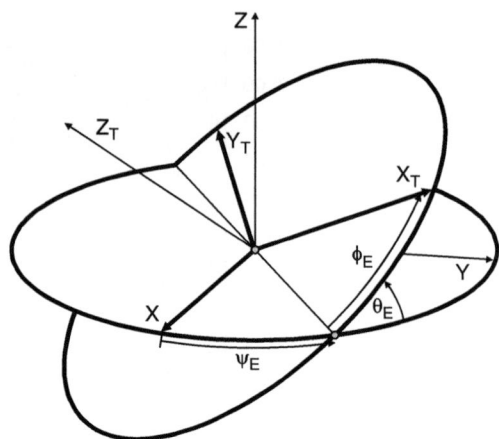

Fig. 9.12 The three Euler angles ψ_E, θ_E, and ϕ_E

has turned out to be very useful. The idea is to separate, more or less, the motion of the ITRS in the GCRS into two parts: a precession–nutation (P–N) part that can in principle be computed from gravitational torques exerted by the Moon, Sun, and planets and a geophysical part (polar motion) that results from the complex geophysics of the Earth's system and that can be predicted only poorly. Note, however, that the two intermediate poles, namely, the celestial intermediate pole (CIP) and the instantaneous rotation pole (IRP), described below, do not completely separate the external parts from the geophysical parts.

9.3.1 Instantaneous Rotation Pole

For the construction of the classical astronomical reference systems, the IRP was employed as intermediate pole. For the real Earth, the IRP presents a generalization of the corresponding concept for a rigid Earth where the rotation vector $\boldsymbol{\Omega}$ is uniquely determined by the velocity field

$$\mathbf{v} = \boldsymbol{\Omega} \times \mathbf{x}. \tag{9.28}$$

Clearly the Earth is not a rigid body, and the various observing stations are moving even in the ITRS. One way to define some IRP for the Earth is to consider the transformation from the GCRS with spatial coordinates $\mathbf{X}_I$ to the ITRS with coordinates $\mathbf{X}_T$:

$$\mathbf{X}_T = \mathcal{S}\mathcal{N}\mathcal{P}\mathbf{X}_I \equiv \mathcal{R}_{GI}\mathbf{X}_I,$$

which is given by some rotation matrix $\mathcal{R}_{GI}$. Now, for any rotation matrix, there is an associated rotation vector $\boldsymbol{\omega}$ that in our case is given by

9.3 Celestial Intermediate Pole

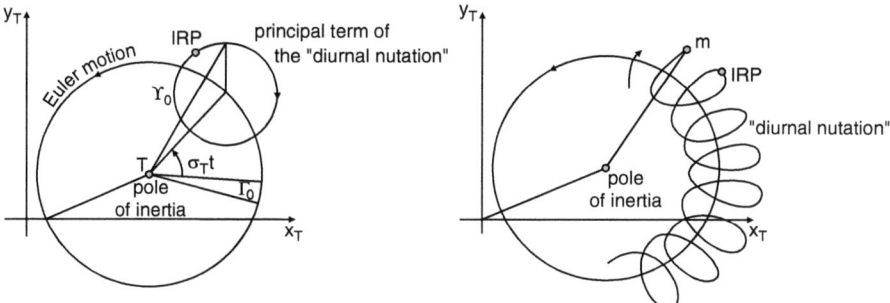

Fig. 9.13 Motion of the IRP in the ITRS

$$\frac{d\mathcal{R}_{GI}}{dT} \mathcal{R}_{GI}^T = \begin{bmatrix} 0 & -\omega_z & \omega_y \\ \omega_z & 0 & -\omega_x \\ -\omega_y & \omega_x & 0 \end{bmatrix}_{GI}. \tag{9.29}$$

The IRP might be identified with the orientation of this angular velocity ω_{GI}.

The IRP was the basic astronomical intermediate pole for precession–nutation from 1964 and had this role until 1984. Conceptually it divides the external astronomical parts into two parts:

- A precession–nutation part.
- An astronomically forced almost diurnal polar motion part.

In the literature we find different notations for the astronomically forced nearly diurnal polar motion terms. They are also called diurnal nutations or forced latitude variations (Fedorov 1963). In the ITRS this forced motion has a diurnal part with an amplitude of 8.7 mas and an almost diurnal part with amplitudes ≤ 7 mas. For a rigid Earth the motion of the IRP in the ITRS is depicted in Fig. 9.13.

In the GCRS the forced (almost) diurnal polar motion of the IRP results from the fact that the IRP undergoes a slightly different nutation as the ITRS pole. Such differences in the nutational motion are called Oppolzer motion (Oppolzer 1886; Jeffreys 1963) with corresponding Oppolzer terms in a nutation series. Such Oppolzer terms contain a constant term with an amplitude of 8.7 mas and long-period terms with periods of 18.6 years, 1 year, 182 days, 13.7 days, etc.

One can show that no geodetic space method is able to observe directly the IRP. To be more precise, these methods are unable to distinguish between certain nutations (and length-of-day variations) on the one hand and certain polar motion terms on the other, the reason being that these space methods are sensitive only to the total transformation from the GCRS to the ITRS. Above we wrote this transformation in the form

$$\mathbf{X}_T = \mathcal{R}_2(-x_p)\,\mathcal{R}_1(-y_p)\,\mathcal{R}_3(\theta)\,\mathcal{N}\,\mathcal{P}\,\mathbf{X}_I$$

with θ = GAST. We will now consider small variations in ϵ, ψ, θ, and x_p, y_p; to this end, we consider the matrix

$$\mathcal{C} \equiv \mathcal{U} \mathcal{R}_3(\theta + \delta\theta) \, \delta\mathcal{N} \tag{9.30}$$

with

$$\delta\mathcal{N} = \begin{bmatrix} 1 & -\cos\epsilon \, \delta\psi & -\sin\epsilon \, \delta\psi \\ \cos\epsilon \, \delta\psi & 1 & -\delta\epsilon \\ \sin\epsilon \, \delta\psi & \delta\epsilon & 1 \end{bmatrix},$$

$$\mathcal{R}_3(\theta + \delta\theta) = \begin{bmatrix} \cos\theta - \delta\theta \sin\theta & \sin\theta + \delta\theta \cos\theta & 0 \\ -\sin\theta - \delta\theta \cos\theta & \cos\theta - \delta\theta \sin\theta & 0 \\ 0 & 0 & 1 \end{bmatrix}$$

$$\tag{9.31}$$

and $(x = x_p + \delta x, y = y_p + \delta y)$

$$\mathcal{U} \equiv \mathcal{R}_2(-x) \, \mathcal{R}_1(-y) = \begin{bmatrix} 1 & 0 & x \\ 0 & 1 & -y \\ -x & y & 1 \end{bmatrix}. \tag{9.32}$$

To first order we obtain

$$\mathcal{C} = \begin{bmatrix} \cos\theta - z \sin\theta & \sin\theta + z \cos\theta & v \cos\theta - w \sin\theta + x \\ -\sin\theta - z \cos\theta & \cos\theta - z \sin\theta & -v \sin\theta - w \cos\theta - y \\ -v - x \cos\theta - y \sin\theta & w - x \sin\theta + y \cos\theta & 1 \end{bmatrix}$$

with

$$v = -\delta\psi \, \sin\epsilon, \qquad w = \delta\epsilon$$

and

$$x = x_p + \delta x, \qquad y = y_p + \delta y, \qquad z = \delta\theta - \delta\psi \, \cos\epsilon.$$

Nutation (plus LOD variations) and polar motion cannot be separated if

$$\mathcal{C}(\delta\psi, \delta\epsilon) = \mathcal{C}(\delta x, \delta y, z) \tag{9.33}$$

i.e., if

$$\begin{bmatrix} \cos\theta & \sin\theta & v \cos\theta - w \sin\theta \\ -\sin\theta & \cos\theta & -v \sin\theta - w \cos\theta \\ -v & w & 1 \end{bmatrix}$$

$$= \begin{bmatrix} \cos\theta - z \sin\theta & \sin\theta + z \cos\theta & \delta x \\ -\sin\theta - z \cos\theta & \cos\theta - z \sin\theta & -\delta y \\ -\delta x \cos\theta - \delta y \sin\theta & -\delta x \sin\theta + \delta y \cos\theta & 1 \end{bmatrix}.$$

9.3 Celestial Intermediate Pole

This condition leads to the following three relations:

$$\delta\theta = \delta\psi \cos\epsilon$$
$$\delta x = -\delta\psi \sin\epsilon \cos\theta - \delta\epsilon \sin\theta \qquad (9.34)$$
$$\delta y = -\delta\psi \sin\epsilon \sin\theta + \delta\epsilon \cos\theta.$$

From this we see that any geodetic space technique cannot distinguish between certain types of nutation and quasi-diurnal polar motion.

9.3.2 Definition of the CIP

For these reasons a conventional CIP was introduced by IAU 2000 Resolution B1.7 that conceptually replaces the IRP (see also McCarthy and Capitaine 2002, IERS C03/C10; Capitaine 1990).

The CIP acts as intermediate pole in the transformation between the GCRS and the ITRS. Its motion in the GCRS is described by the motion of the ITRS pole with periods larger than 2 days. The position is fixed by the IAU 2000A model for precession–nutation and the frame bias.

Frequencies in the GCRS and ITRS are depicted in Fig. 9.14. Periods longer than 2 days in the GCRS correspond to frequencies between -0.5 cpsd and $+0.5$ cpsd (cpsd: cycles per sidereal day) since $f_{\text{cpsd}} = 1/P_d$. In these units (cpsd), one has

$$f_{\text{ITRS}} = f_{\text{GCRS}} - 1.$$

For positive frequencies the motion is called *prograde*, i.e., in the sense of Earth's rotation; for negative frequencies, it is called *retrograde*. In the terrestrial system the motion of the CIP should have no periods between -48^h and -16^h. It is clear that this definition presents some kind of generalization of the IRP. As seen from space, i.e., in the GCRS, the external torque due to the gravitational action of Moon, Sun, and planets varies dominantly slowly with time, i.e., periods in the GCRS will dominantly be large. On the other hand, the Chandler wobble and many geophysical interactions are relatively slow in the ITRS which puts them in the retrograde diurnal band as seen from the GCRS.

The practical problem is how to realize the CIP definition in actual VLBI measurements. For usual campaigns with about three 24^h sessions per week, the separation into nutation and polar motion cannot be achieved in a clean way. More interesting are special continuous VLBI campaigns over 15 days within the International VLBI Service for Geodesy and Astrometry (IVS), the so-called *CONT* campaigns (e.g., Artz et al. 2010). Here, the definition of the CIP can be realized

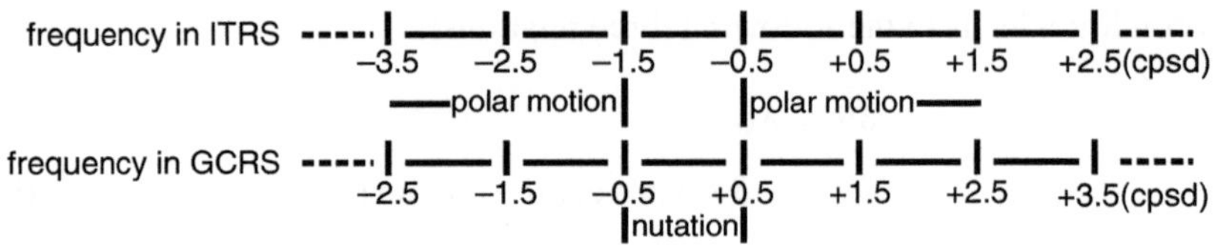

Fig. 9.14 Frequencies in the GCRS und ITRS. The frequency bands for nutation and polar motion are indicated (IERS C03/C10)

with the following strategy (Nilsson et al. 2010). The pole coordinates, x_p and y_p, as well as the celestial pole offsets, dX and dY, are written as a Fourier series:

$$x_p(t) = \sum_{k=0}^{N-1} [a_k \cos(\omega_k t) + b_k \sin(\omega_k t)],$$

$$y_p(t) = \sum_{k=0}^{N-1} [-a_k \cos(\omega_k t) + b_k \sin(\omega_k t)],$$

(9.35)

where $\omega_k = 2\pi(k - N/2)/T$, N is the number of estimation epochs and T is the length of a CONT campaign (e.g., 15 days). Similarly,

$$dX(t) = \sum_{k=0}^{N-1} [c_k \cos(\omega_k t) - d_k \sin(\omega_k t)],$$

$$dY(t) = \sum_{k=0}^{N-1} [c_k \cos(\omega_k t) + d_k \sin(\omega_k t)].$$

(9.36)

The representation of polar motion and nutation by Fourier series is then used in the least-squares adjustment with actual data sets, i.e., the quantities a_k, b_k, c_k, and d_k are estimated instead of x_p, y_p, dX, and dY. One can then force a_k and b_k to be zero for $-3\pi < \omega_k < -\pi$, and c_k and d_k to be zero for $-\pi < \omega_k < +\pi$, in the least-squares adjustment.

This definition of the CIP implies a certain order of astronomically forced and geophysical terms. Certain geophysical terms are now considered as nutation. Such geophysical nutation terms have periods larger than 2 days in the GCRS.

On the other hand, astronomically forced terms with periods of less than 2 days in the GCRS (formerly: daily and subdaily nutations) are now considered as polar motion (Table 9.1; Folgueira et al. 2001). Nutation and polar motion periods in the GCRS and ITRS are depicted in Fig. 9.15.

9.3 Celestial Intermediate Pole

Table 9.1 Classification of astronomically forced Earth orientation terms by their periods in the GCRS and ITRS

Geo-potential	Tidal-potential	Sum of all amplitudes (μas)	GCRS motion	ITRS motion
$U_{l,0}$ for $l = 2, 3, \ldots$	$u_{l,1}$	Nutation $> 10^7$ + precession	Long-periodic	Retrograde diurnal
$U_{3,1}$ $U_{4,1}$	$u_{3,0}$ $u_{4,0}$	91.3 periodic $1.0 +$ drift 5.1 μas/year	Prograde diurnal	Long-periodic
$U_{2,2}$ $U_{3,2}$	$u_{2,1}$ $u_{3,1}$	51.5 0.2	Prograde semidiurnal	Prograde diurnal
$U_{3,3}$	$u_{3,2}$	0.1	Prograde terdiurnal	Prograde semidiurnal
$U_{3,1}$	$u_{3,2}$	0.8	Retrograde diurnal	Retrograde semidiurnal
$U_{3,2}$	$u_{3,3}$	0.1	Retrograde semidiurnal	Retrograde terdiurnal

The diurnal and subdiurnal terms in the GCRS are now considered to be polar motion terms. In the first column $U_{l,j}$ stands for the Stokes coefficients $C_{l,j}$, $S_{l,j}$ (degree l, order j) and $u_{l,j}$ in the next column stands for the tidal potential (degree l, order j). The third column shows the sum of absolute values for all amplitudes larger than 0.01 μas

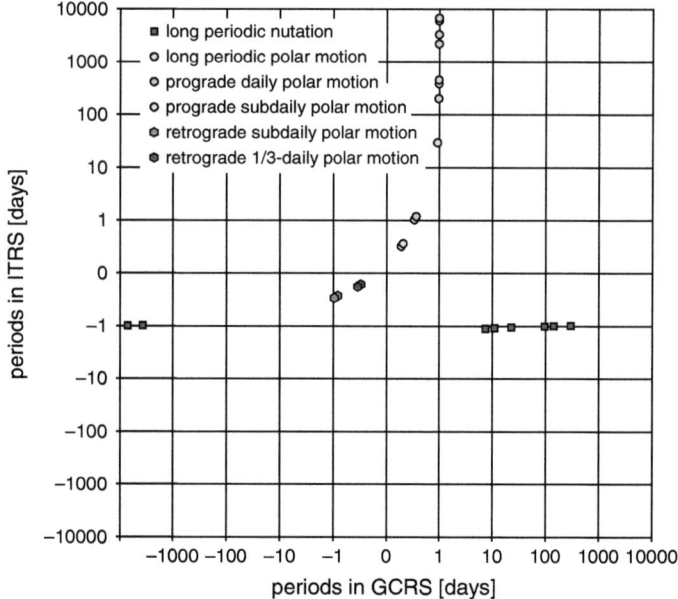

Fig. 9.15 Nutation and polar motion periods in the GCRS and ITRS

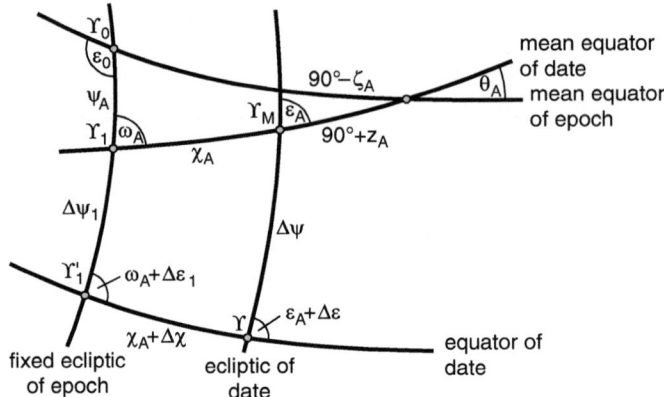

Fig. 9.16 Various precession–nutation angles

9.3.3 Motion of the CIP in the GCRS; CP Offsets

To a good approximation, the motion of the CIP in the GCRS is described by the IAU 2000A precession–nutation model.

VLBI observations show that deficiencies of this model are of order 0.5 mas. The main limitation is the omission of corrections for the free-core nutation, which can be extrapolated only a relatively short time into the future. Corrections to the IAU model are published regularly by the IERS as *celestial pole offsets*. Meanwhile one often considers the Cartesian components (X, Y) of the CIP unit vector in the GCRS instead of the P–N angles. The time-dependent CIP offsets are then given by corrections (offsets) δX and δY to the Cartesian components of the CIP in the GCRS. If these offsets are given by $\delta\psi$ and $\delta\epsilon$, the Cartesian corrections can be deduced from (IERS C03, eq. 23)

$$\begin{aligned} \delta X &= \delta\psi \sin\epsilon_A + (\psi_A \cos\epsilon_0 - \chi_A)\,\delta\epsilon, \\ \delta Y &= \delta\epsilon - (\psi_A \cos\epsilon_0 - \chi_A) \sin\epsilon_A\,\delta\psi. \end{aligned} \quad (9.37)$$

The angles ϵ_A, ψ_A, and χ_A, described in Fig. 9.16, are given below and (IERS C03)

$$\epsilon_0 = 84381''.448. \quad (9.38)$$

By means of the observed offsets δX and δY, the position of the CIP in the GCRS is given by

$$\begin{aligned} X &= X_{\text{IAU}} + \delta X, \\ Y &= Y_{\text{IAU}} + \delta Y. \end{aligned} \quad (9.39)$$

9.3.4 IAU 2000 Precession–Nutation Model

The IAU 2000A P-N model replaces the IAU 1976 precession model (Lieske et al. 1977) and the IAU 1980 nutation model after 1 January 2003. Together with empirical rate corrections to the IAU 1976 precession, the MHB2000 model (MHB: Mathews, Herring, Buffet; Mathews et al. 2002; Herring et al. 2002) forms the basis for the IAU 2000A precession–nutation model. MHB2000 provides a transfer function T that in frequency space relates some nutational amplitude η of a realistic Earth model with that from a rigid Earth η_R. Nutational amplitude for a rigid Earth is taken from REN2000 (Souchay et al. 1999). In the MHB model one has

$$\eta(\sigma) = T(\sigma)\,\eta_R(\sigma) \tag{9.40}$$

with σ as dimensionless frequency, $f = \sigma\Omega$. The transfer function $T(\sigma)$ can be approximated by

$$T(\sigma) = \frac{e_R - \sigma}{e_R + 1} N_0 \left[1 + (1+\sigma)\left(Q_0 + \sum_{\alpha=1}^{4} \frac{Q_\alpha}{\sigma - s_\alpha} \right) \right]. \tag{9.41}$$

The dynamical ellipticity, $e_R = (C - A)/A$, used for the calculation of nutations of a rigid Earth is related with the dynamical flattening $H_d = (C - A)/C = 0.0032737548$ by

$$e_R = \frac{H_d}{1 - H_d} \tag{9.42}$$

and for the Earth $e_R \simeq 0.0032845075$.

Values for N_0, Q_0, Q_a, and for the resonance frequencies s_a are (Mathews et al. 2002, $I = \sqrt{-1}$)

$$\begin{aligned}
N_0 &= 1 + 1.224 \times 10^{-5} \\
s_1 &= 3.11279 \times 10^{-3} + I \cdot (+3.76098 \times 10^{-4}) \\
s_2 &= -1 - 2.31811 \times 10^{-3} + I \cdot (+2.50607 \times 10^{-5}) \\
s_3 &= -1 + 9.73555 \times 10^{-4} + I \cdot (+7.78663 \times 10^{-4}) \\
s_4 &= 4.1324 \times 10^{-4} + I \cdot (+9.28220 \times 10^{-8}) \\
Q_0 &= -1.65291 \times 10^{-1} + I \cdot (+3.18995 \times 10^{-2}) \\
Q_1 &= -9.48081 \times 10^{-1} + I \cdot (+6.78857 \times 10^{-2}) \\
Q_2 &= 4.89324 \times 10^{-2} + I \cdot (+1.61700 \times 10^{-3}) \\
Q_3 &= 2.96114 \times 10^{-4} + I \cdot (-9.56740 \times 10^{-5}) \\
Q_4 &= -1.10856 \times 10^{-5} + I \cdot (-1.22654 \times 10^{-6})
\end{aligned} \tag{9.43}$$

The quantities s_a are four complex resonance frequencies corresponding to Chandler wobble (CW; s_1), retrograde free-core nutation (RFCN; s_2), prograde free-core nutation (PFCN; s_3), and inner core wobble (ICW; s_4).

Let us now discuss the problem of frequencies σ and amplitudes η related with the MHB-transfer function. To this end we will first consider the 18.6-year nutation term with celestial period P_1. It has a celestial frequency of $\omega_1 = 1/P_1$ cpsd. In the terrestrial system there are two frequencies that contribute to this nutation: a retrograde and a prograde one with frequencies

$$\sigma_1^\pm = \pm\omega_1 - 1.$$

For the 18.6 y nutation this gives the terrestrial frequencies in cpsd:

$$\sigma_1^- = -1.0001471; \quad \sigma_1^+ = -0.9998529.$$

Suppose we have the values for $\Delta\epsilon_R$ and $\Delta\psi_R$ for a rigid Earth model like REN2000 (Souchay et al. 1999). Concentrating on the inphase terms only, the retrograde and prograde amplitudes are given by

$$\eta^- = -\frac{1}{2}(\Delta\epsilon + \Delta\psi \sin\epsilon_0); \quad \eta^+ = -\frac{1}{2}(\Delta\epsilon - \Delta\psi \sin\epsilon_0). \tag{9.44}$$

In Mathews and Shapiro (1992) or Defraigne et al. (1995), also out-of-phase terms are considered for the calculation of retrograde and prograde amplitudes. For the 18.6-year nutation, one finds (Mathews et al. 2002)

$$\eta^-_{1,R} = -8050.866; \quad \eta^+_{1,R} = -1177.044$$

in mas. For the real part of the MHB-transfer function one then finds

$$T_1^- = 0.9967503; \quad T_1^+ = 1.0028789$$

from which the nonrigid amplitudes

$$\eta_1^- = -8024.703; \quad \eta_1^+ = -1180.433$$

can be derived. The corresponding values for $\Delta\epsilon$ and $\Delta\psi$ can be obtained from

$$\Delta\epsilon = -(\eta^- + \eta^+)$$
$$\Delta\psi \sin\epsilon_0 = +(\eta^+ - \eta^-) \tag{9.45}$$

with $\sin\epsilon_0 = 0.397776995$. Celestial periods, terrestrial frequencies, and values for the complex transfer function for selected nutations are given in Table 9.2.

9.3 Celestial Intermediate Pole

Table 9.2 Celestial periods (in days), terrestrial frequencies (in cycles per sidereal day), and values for the MHB-transfer function for selected nutations

Cel. period	Terr. frequency	$T(\sigma)$ Re	$T(\sigma)$ Im
−6798.38	−1.0001471	0.9967503	−0.0001692
+6798.38	−0.9998529	1.0028789	0.0001561
−3399.19	−1.0002942	0.9930053	−0.0003593
+3399.19	−0.9997058	1.0054203	0.0003057
−365.26	−1.0027378	1.3203904	−0.0085198
+365.26	−0.9972622	1.0262780	0.0009504
−182.62	−1.0054758	1.0867797	0.0022284
+182.62	−0.9945242	1.0333842	0.0009888
−121.75	−1.0082136	1.0708026	0.0021656
+121.65	−0.9917864	1.0364296	0.0009709
−27.55	−1.0362911	1.0625614	0.0029169
+27.55	−0.9637089	1.0369836	0.0000519
−13.66	−1.0732011	1.0713060	0.0041405
+13.66	−0.9267989	1.0292679	−0.0014364
−9.13	−1.1094930	1.0815098	0.0052826
+9.13	−0.8905070	1.0211961	−0.0030058

Exercise 9.1: Verify the numbers of Table 9.2. Derive amplitudes η_R from the REN2000 model for these nutations, apply the transfer function to get η-values, and finally compute $\Delta\epsilon$ and $\Delta\psi$ for a realistic Earth's model.

Compared with the IAU 1980 nutation model, the new MHB2000 model considers effects from the ocean tides, the atmosphere, the anelasticity of the mantle, and core–mantle couplings. As mentioned above, the FCN is not part of the IAU 2000A model but is, however, monitored by VLBI. In contrast to this, the geodetic nutation is included in the IAU 2000A model. It is simply added to the numbers of the MHB2000 series (Schuh et al. 2003). The IAU 2000A nutation model provides a series for nutation in longitude $\Delta\psi$ and obliquity $\Delta\epsilon$ with respect to the mean ecliptic of date of the form (6.69). Some terms can be found in Table 9.3. The IAU 2000B model presents a truncated version of the full IAU 2000A model with an accuracy of about 1 mas for the years 1950–2050. Software containing the IAU 2000 models can be downloaded from the IERS websites (IERS C03/C10).

Exercise 9.2: Use Maple™ to compute the (complex) transfer function of (9.41). Plot the real part of $T(\sigma)$ for the range $\sigma = -1.010\ldots-0.990$.

The following series for the conventional precession angles ζ_A, θ_A, and z_A have been derived from the precession part of the IAU 2000A model by Capitaine et al. (2003c):

Table 9.3 Largest lunisolar and planetary nutation terms. The units for A, A'', B, and B'' are mas and for A', A''', B', and B''' mas/cen; periods are in days. The fundamental angles Ω, F, D, l, l' are given by (9.50). The mean longitudes (l_{Me}, l_{Ve}, l_E, l_{Ma}, l_{Ju}, l_{Sa}, l_{Ur}, l_{Ne}) and the general precession in longitude p_a are given by (9.51) (from IERS C03, Tab. 5.3)

l	l'	F	D	Ω	Period	A_i	A'_i	B_i	B'_i	A''_i	A'''_i	B''_i	B'''_i
0	0	0	0	1	−6798.383	−17206.4161	−17.4666	9205.2331	0.9086	3.3386	0.0029	1.5377	0.0002
0	0	2	−2	2	182.621	−1317.0906	−0.1675	573.0336	−0.3015	−1.3696	0.0012	−0.4587	−0.0003
0	0	2	0	2	13.661	−227.6413	−0.0234	97.8459	−0.0485	0.2796	0.0002	0.1374	−0.0001
0	0	0	0	2	−3399.192	207.4554	0.0207	−89.7492	0.0470	−0.0698	0.0000	−0.0291	0.0000
0	1	0	0	0	365.260	147.5877	−0.3633	7.3871	−0.0184	1.1817	−0.0015	−0.1924	0.0005
0	1	2	−2	2	121.749	−51.6821	0.1226	22.4386	−0.0677	−0.0524	0.0002	−0.0174	0.0000
1	0	0	0	0	27.555	71.1159	0.0073	−0.6750	0.0000	−0.0872	0.0000	0.0358	0.0000
0	0	2	0	1	13.633	−38.7298	−0.0367	20.0728	0.0018	0.0380	0.0001	0.0318	0.0000
1	0	2	0	2	9.133	−30.1461	−0.0036	12.9025	−0.0063	0.0816	0.0000	0.0367	0.0000
0	−1	2	−2	2	365.225	21.5829	−0.0494	−9.5929	0.0299	0.0111	0.0000	0.0132	−0.0001

l	l'	F	D	Ω	L_{Me}	L_{Ve}	L_E	L_{Ma}	L_{Ju}	L_{Sa}	L_{Ur}	L_{Ne}	p_A	Period	Longitude		Obliquity	
															A_i	A''_i	B_i	B''_i
0	0	0	0	1	0	0	−1	0	−2	5	0	0	0	311921.26	−0.3084	0.5123	0.1647	0.2735
0	0	0	0	2	0	0	0	0	−2	5	0	0	1	311927.52	−0.1444	0.2409	−0.0771	−0.1286
0	0	0	0	0	0	0	0	0	0	0	0	0	2	2957.35	−0.2150	0.0000	0.0932	0.0000
0	0	1	−1	1	0	0	−3	0	0	0	0	0	0	−88082.01	0.1200	0.0598	−0.0641	0.0319
0	−1	0	0	0	0	−8	12	0	0	0	0	0	0	2165.30	−0.1166	0.0000	0.0505	0.0000
0	0	0	0	1	0	0	0	−8	3	0	0	0	2	−651391.30	−0.0462	0.1604	0.0000	0.0000
0	0	0	0	0	0	1	−1	0	0	0	0	0	0	583.92	0.1485	0.0000	0.0000	0.0000
0	0	0	0	0	0	0	0	−16	4	5	0	0	0	34075700.82	0.1440	0.0000	0.0000	0.0000
0	0	0	0	1	0	0	1	0	−1	0	0	0	0	398.88	−0.1223	−0.0026	0.0000	0.0000
0	0	0	0	0	0	0	−1	2	0	0	0	0	0	37883.60	−0.0460	−0.0435	0.0246	−0.0232

9.3 Celestial Intermediate Pole

$$\zeta_A = 2\overset{''}{.}5976176 + 2306\overset{''}{.}0809506\,t + 0\overset{''}{.}3019015\,t^2 + 0\overset{''}{.}0179663\,t^3$$
$$- 0\overset{''}{.}0000327\,t^4 - 0\overset{''}{.}0000002\,t^5,$$

$$\theta_A = 2004\overset{''}{.}1917476\,t - 0\overset{''}{.}4269353\,t^2 - 0\overset{''}{.}0418251\,t^3 \tag{9.46}$$
$$- 0\overset{''}{.}0000601\,t^4 - 0\overset{''}{.}0000001\,t^5,$$

$$z_A = -2\overset{''}{.}5976176 + 2306\overset{''}{.}0803226\,t + 1\overset{''}{.}0947790\,t^2 + 0\overset{''}{.}0182273\,t^3$$
$$+ 0\overset{''}{.}0000470\,t^4 - 0\overset{''}{.}0000003\,t^5.$$

Derived precession angles (Fig. 9.16) are (IERS C03)

$$\psi_A = 5038\overset{''}{.}47875\,t - 1\overset{''}{.}07259\,t^2 - 0\overset{''}{.}001147\,t^3,$$
$$\omega_A = \epsilon_0 - 0\overset{''}{.}02524\,t + 0\overset{''}{.}05127\,t^2 - 0\overset{''}{.}007726\,t^3, \tag{9.47}$$
$$\epsilon_A = \epsilon_0 - 46\overset{''}{.}84024\,t - 0\overset{''}{.}00059\,t^2 + 0\overset{''}{.}001813\,t^3,$$
$$\chi_A = 10\overset{''}{.}5526\,t - 2\overset{''}{.}38064\,t^2 - 0\overset{''}{.}001125\,t^3$$

with
$$t = (\mathrm{JD}_{TT} - \mathrm{J2000.0})/36{,}525.$$

By means of these latter angles, the precession matrix $\mathcal{P}$ can be written as

$$\mathcal{P} = \mathcal{R}_1(-\epsilon_0)\,\mathcal{R}_3(\psi_A)\,\mathcal{R}_1(\omega_A)\,\mathcal{R}_3(-\chi_A). \tag{9.48}$$

In these descriptions the precessional motion is given as a function of TT as time variable. In the original expansions, however, the time variable was TDB. The difference TDB − TT is expensive to compute and corresponds to only about 1.7 mas × sin l'. If TDB is replaced by TT, the resulting error in the precessional quantity ψ_A has an annual period and an amplitude of $2\overset{''}{.}7 \times 10^{-9}$. Thus, at an accuracy level of a few nanoarcseconds, using TT instead of TDB has no consequence for the precession matrix (McCarthy and Petit 2003).

The IAU 2000 precession angles ζ_A, θ_A, and z_A and the appropriate rotation matrix may be calculated with the *AstroRef* package by calling `Precession`.

9.3.5 Fundamental Nutation Angles

Each lunisolar nutation term is characterized by of a set of five integers N_j, defining the corresponding argument (arg) as a linear combination of the fundamental Delaunay angles $F_j = l, l', F, D, \Omega$:

$$\mathrm{arg} = \sum_{j=1}^{5} N_j F_j \tag{9.49}$$

with (when using the IAU 2000A series)

$$F_1 \equiv l = \text{mean anomaly of Moon}$$
$$= 134.96340251° + 1717915923\overset{''}{.}2178\,t + 31\overset{''}{.}8792\,t^2$$
$$+ 0\overset{''}{.}051635\,t^3 - 0\overset{''}{.}00024470\,t^4,$$

$$F_2 \equiv l' = \text{mean anomaly of Sun}$$
$$= 357.52910918° + 129596581\overset{''}{.}0481\,t - 0\overset{''}{.}5532\,t^2$$
$$+ 0\overset{''}{.}000136\,t^3 - 0\overset{''}{.}00001149\,t^4,$$

$$F_3 \equiv F = L - \Omega$$
$$= 93.27209062° + 1739527262\overset{''}{.}8478\,t - 12\overset{''}{.}7512\,t^2 \quad (9.50)$$
$$- 0\overset{''}{.}001037\,t^3 + 0\overset{''}{.}00000417\,t^4,$$

$$F_4 \equiv D = \text{mean elongation of Moon from Sun}$$
$$= 297.85019547° + 1602961601\overset{''}{.}2090\,t - 6\overset{''}{.}3706\,t^2$$
$$+ 0\overset{''}{.}006593\,t^3 - 0\overset{''}{.}00003169\,t^4,$$

$$F_5 \equiv \Omega = \text{mean longitude of ascending node of Moon}$$
$$= 125.04455501° - 6962890\overset{''}{.}5431\,t + 7\overset{''}{.}4722\,t^2$$
$$+ 0\overset{''}{.}007702\,t^3 - 0\overset{''}{.}00005939\,t^4.$$

Here, t is again
$$t = (\text{JD}_{\text{TT}} - \text{J2000.0}) / 36{,}525.$$

The mean longitudes have been derived by Souchay et al. (1999), based on the theory and constants of VSOP82 (Bretagnon 1982) and ELP2000 (Chapront-Touzé and Chapront 1983), as well as the expansion in Simon et al. (1994).

The planetary nutation terms are characterized similarly with additional fundamental angles $F_6, \ldots, F_{14}$. F_6 to F_{13} are the mean longitudes of Mercury to Neptune including the Earth ($l_{\text{Me}}, l_{\text{Ve}}, l_{\text{E}}, l_{\text{Ma}}, l_{\text{Ju}}, l_{\text{Sa}}, l_{\text{Ur}}, l_{\text{Ne}}$), and F_{14} is the general precession in longitude p_a (Kinoshita and Souchay 1990):

$$F_6 \equiv l_{\text{Me}} = 4.402608842 + 2608.7903141574\,t$$
$$F_7 \equiv l_{\text{Ve}} = 3.176146697 + 1021.3285546211\,t$$
$$F_8 \equiv l_{\text{E}} = 1.753470314 + 628.3075849991\,t$$
$$F_9 \equiv l_{\text{Ma}} = 6.203480913 + 334.0612426700\,t$$
$$F_{10} \equiv l_{\text{Ju}} = 0.599546497 + 52.9690962641\,t$$
$$F_{11} \equiv l_{\text{Sa}} = 0.874016757 + 21.3299104960\,t$$

9.3 Celestial Intermediate Pole

$$F_{12} \equiv l_{Ur} = 5.481293872 + 7.4781598567\,t$$
$$F_{13} \equiv l_{Ne} = 5.311886287 + 3.8133035638\,t$$
$$F_{14} \equiv p_a = 0.024381750\,t + 0.00000538691\,t^2 \tag{9.51}$$

with
$$t = (\mathrm{JD_{TT}} - \mathrm{J2000.0})/36{,}525.$$

The nutation angles $\Delta\epsilon$ and $\Delta\psi$ as well as the appropriate rotation matrix may be calculated using the function Nutation from the *AstroRef* package.

9.3.6 Frame Bias Matrix

The Cartesian components of the CIP in the GCRS can be obtained from the matrix

$$\mathcal{M} = \mathcal{N}\mathcal{P}\mathcal{B} \tag{9.52}$$

Here, $\mathcal{P}$ is the precession, $\mathcal{N}$ the nutation matrix, and $\mathcal{B}$ is the so-called frame bias matrix (Capitaine et al. 2003a):

$$\mathcal{B} = \mathcal{R}_1(-\eta_0)\,\mathcal{R}_2(\xi_0)\,\mathcal{R}_3(d\alpha_0). \tag{9.53}$$

This time-independent matrix transforms any GCRS vector into the corresponding vector in a mean system for the epoch J2000.0. The quantities ξ_0 and η_0 describe the position of the CIP in the GCRS at epoch J2000.0 (e.g., IERS C03):

$$\begin{aligned}\xi_0 &= -0\overset{''}{.}0166170 \pm 0\overset{''}{.}0000100\\ \eta_0 &= -0\overset{''}{.}0068192 \pm 0\overset{''}{.}0000100,\end{aligned} \tag{9.54}$$

and $d\alpha_0$ is the right ascension of the mean equinox J2000.0 in the GCRS (Chapront et al. 2002):

$$d\alpha_0 = (-0.01460 \pm 0.00050)''. \tag{9.55}$$

The frame bias matrix is returned from the *AstroRef* package after calling the function Bias.

9.3.7 The Series by Capitaine and Wallace

By means of a series expansion that takes into account frame bias, precession, and nutation, the coordinates X and Y of the CIP in the GCRS are given directly with microarcsecond accuracy (Capitaine and Wallace 2006; Petit and Luzum 2010). The series are based upon the IAU 2000A precession–nutation model and the corresponding pole offsets at epoch J2000.0 with respect to the GCRS pole. They have the form (IERS C03, eqs. 15, 16)

$$\begin{aligned}
X = {} & -0\overset{''}{.}01661699 + 2004\overset{''}{.}19174288\, t - 0\overset{''}{.}42721905\, t^2 \\
& -0\overset{''}{.}19862054\, t^3 - 0\overset{''}{.}00004605\, t^4 + 0\overset{''}{.}00000598\, t^5 \\
& + \sum_i \left[(a_{s,0})_i \sin(\arg_i) + (a_{c,0})_i \cos(\arg_i) \right] \\
& + \sum_i \left[(a_{s,1})_i\, t \sin(\arg_i) + (a_{c,1})_i\, t \cos(\arg_i) \right] \\
& + \sum_i \left[(a_{s,2})_i\, t^2 \sin(\arg_i) + (a_{c,2})_i\, t^2 \cos(\arg_i) \right] \\
& + \sum_i \left[(a_{s,3})_i\, t^3 \sin(\arg_i) + (a_{c,3})_i\, t^3 \cos(\arg_i) \right] \\
& + \sum_i \left[(a_{s,4})_i\, t^4 \sin(\arg_i) + (a_{c,4})_i\, t^4 \cos(\arg_i) \right]
\end{aligned} \quad (9.56)$$

$$\begin{aligned}
Y = {} & -0\overset{''}{.}00695078 - 0\overset{''}{.}02538199\, t - 22\overset{''}{.}40725099\, t^2 \\
& + 0\overset{''}{.}00184228\, t^3 + 0\overset{''}{.}00111306\, t^4 + 0\overset{''}{.}00000099\, t^5 \\
& + \sum_i \left[(b_{c,0})_i \cos(\arg_i) + (b_{s,0})_i \sin(\arg_i) \right] \\
& + \sum_i \left[(b_{c,1})_i\, t \cos(\arg_i) + (b_{s,1})_i\, t \sin(\arg_i) \right] \\
& + \sum_i \left[(b_{c,2})_i\, t^2 \cos(\arg_i) + (b_{s,2})_i\, t^2 \sin(\arg_i) \right] \\
& + \sum_i \left[(b_{c,3})_i\, t^3 \cos(\arg_i) + (b_{s,3})_i\, t^3 \sin(\arg_i) \right] \\
& + \sum_i \left[(b_{c,4})_i\, t^4 \cos(\arg_i) + (b_{s,4})_i\, t^4 \sin(\arg_i) \right].
\end{aligned} \quad (9.57)$$

The full IAU 2000/2006 series for X and Y are available electronically on the IERS Conventions Center website. The largest nonpolynomial terms in the expansions for $X(t)$ and $Y(t)$ can be found in Table 9.4.

9.3.8 Motion of the CIP in the ITRS

Terrestrial pole coordinates x_p and y_p are published by the IERS. To these the components of the ocean tides and lunisolar terms with periods of less than 2 days in the GCRS have to be added:

$$(x_p, y_p) = (x, y)_{\text{IERS}} + (\Delta x, \Delta y)_{\text{OT}} + (\Delta x, \Delta y)_{\text{LIB}}. \quad (9.58)$$

9.3 Celestial Intermediate Pole

Table 9.4 The largest nonpolynomial terms in the expansions (9.56) for $X(t)$ (*upper panel*) and (9.57) for $Y(t)$ (*lower panel*) (units: µas). Expressions for the fundamental arguments are given by (9.50) and (9.51) (IERS C03, Tables 5.2a and 5.2b)

i	$(a_{s,0})_i$	$(a_{c,0})_i$	l	l'	F	D	Ω	L_{Me}	L_{Ve}	L_E	L_{Ma}	L_J	L_{Sa}	L_U	L_{Ne}	p_A
1	−6844318.44	1328.67	0	0	0	0	1	0	0	0	0	0	0	0	0	0
2	−523908.04	−544.76	0	0	2	−2	2	0	0	0	0	0	0	0	0	0
3	−90552.22	111.23	0	0	2	0	2	0	0	0	0	0	0	0	0	0
4	82168.76	−27.64	0	0	0	0	2	0	0	0	0	0	0	0	0	0
5	58707.02	470.05	0	1	0	0	0	0	0	0	0	0	0	0	0	0
...																

i	$(a_{s,1})_i$	$(a_{c,1})_i$	l	l'	F	D	Ω	L_{Me}	L_{Ve}	L_E	L_{Ma}	L_J	L_{Sa}	L_U	L_{Ne}	p_A
1307	−3328.48	205833.15	0	0	0	0	1	0	0	0	0	0	0	0	0	0
1308	197.53	12814.01	0	0	2	−2	2	0	0	0	0	0	0	0	0	0
1309	41.19	2187.91	0	0	2	0	2	0	0	0	0	0	0	0	0	0
...																

i	$(b_{s,0})_i$	$(b_{c,0})_i$	l	l'	F	D	Ω	L_{Me}	L_{Ve}	L_E	L_{Ma}	L_J	L_{Sa}	L_U	L_{Ne}	p_A
1	1538.18	9205236.26	0	0	0	0	1	0	0	0	0	0	0	0	0	0
2	−458.66	573033.42	0	0	2	−2	2	0	0	0	0	0	0	0	0	0
3	137.41	97846.69	0	0	2	0	2	0	0	0	0	0	0	0	0	0
4	−29.05	−89618.24	0	0	0	0	2	0	0	0	0	0	0	0	0	0
5	−17.40	22438.42	0	1	2	−2	2	0	0	0	0	0	0	0	0	0
...																

i	$(b_{s,1})_i$	$(b_{c,1})_i$	l	l'	F	D	Ω	L_{Me}	L_{Ve}	L_E	L_{Ma}	L_J	L_{Sa}	L_U	L_{Ne}	p_A
963	153041.82	878.89	0	0	0	0	1	0	0	0	0	0	0	0	0	0
964	11714.49	−289.32	0	0	2	−2	2	0	0	0	0	0	0	0	0	0
965	2024.68	−50.99	0	0	2	0	2	0	0	0	0	0	0	0	0	0
...																

Table 9.5 Daily and subdaily variations in polar motion due to ocean tides

Tide	Argument						Doodson number	Period (days)	x_p		y_p	
	χ	l	l'	F	D	Ω			sin	cos	sin	cos
2Q1	1	−2	0	−2	0	−2	125.755	1.1669259	0.3	3.4	−3.4	0.3
σ1	1	0	0	−2	−2	−2	127.555	1.1603495	0.5	4.2	−4.1	0.5
Q1	1	−1	0	−2	0	−2	135.655	1.1195148	6.2	26.3	−26.3	6.2
RO1	1	1	0	−2	−2	−2	137.455	1.1134606	1.3	5.0	−5.0	1.3
O1	1	0	0	−2	0	−2	145.555	1.0758059	48.8	132.9	−132.9	48.8
M1	1	−1	0	0	0	0	155.655	1.0347187	−4.5	−9.6	9.6	−4.5
P1	1	0	0	−2	2	−2	163.555	1.0027454	26.1	51.2	−51.2	26.1
K1	1	0	0	0	0	0	165.555	0.9972695	−77.5	−151.7	151.7	−77.5
	1	0	0	0	0	−1	165.565	0.9971233	−10.5	−20.6	20.6	−10.5
J1	1	1	0	0	0	0	175.455	0.9624365	−3.5	−7.3	7.3	−3.5
2N2	2	−2	0	−2	0	−2	235.755	0.5377239	−6.1	−1.6	3.1	3.4
μ2	2	0	0	−2	−2	−2	237.555	0.5363232	−7.6	−2.0	3.4	4.2
N2	2	−1	0	−2	0	−2	245.655	0.5274312	−56.9	−12.9	11.1	32.9
ν2	2	1	0	−2	−2	−2	247.455	0.5260835	−11.0	−2.4	1.9	6.4
M2	2	0	0	−2	0	−2	255.555	0.5175251	−330.2	−27.0	37.6	195.9
L2	2	1	0	−2	0	−2	265.455	0.5079842	9.4	−1.4	−1.9	−5.6
T2	2	0	−1	−2	2	−2	272.556	0.5006854	−8.5	3.5	3.3	5.1
S2	2	0	0	−2	2	−2	273.555	0.5000000	−144.1	63.6	59.2	86.6
K2	2	0	0	0	0	0	275.555	0.4986348	−38.5	19.1	17.7	23.1
	2	0	0	0	0	−1	275.565	0.4985982	−11.4	5.8	5.3	6.9

Only the largest terms are given. The unit is μas; χ denotes GMST $+ \pi$ (from IERS C03/C10, Tab. 8.2)

IERS bulletins A and B provide $(x, y)_{\text{IERS}}$. The terms $(\Delta x, \Delta y)_{\text{OT}}$ and $(\Delta x, \Delta y)_{\text{LIB}}$ describe daily and subdaily variations of the pole coordinates due to ocean tides and librations, respectively.

The IERS C10 values of $(\Delta x, \Delta y)_{\text{OT}}$ are based upon a model by Ray et al. (1994). This model starts with a numerical calculation of global tidal heights (Schwiderski 1980). This Schwiderski model is known to be fairly reliable since it was constrained by more than 2000 coastal, island, and bottom pressure measurements. From the global tidal height fields (leading to the matter terms, i.e., changes in the moment of inertia tensor), the tidal currents (leading to the motion terms, i.e., relative angular momentum terms) were derived. Computer subroutines for the calculation of the oceanic tidal terms are available from IERS websites (http://www.iers.org). Table 9.5 shows amplitudes and arguments for selected daily and subdaily ocean tide terms.

According to the definition of the CIP, astronomically forced librations terms $(\Delta x, \Delta y)_{\text{LIB}}$ with periods of less than 2 days are considered to be polar motion. Some of these terms are shown in Table 9.6. Such forced librations terms, formerly called *nutation terms* in polar motion, are diurnal and semidiurnal tidal terms that

9.3 Celestial Intermediate Pole

Table 9.6 Polar motion coefficients of sin(arg) and cos(arg) in $(\Delta x, \Delta y)_{\text{LIB}}$ due to the tidal potential of degree n in units of 1 µas

	Argument						Period	x_p		y_p	
n	χ	l	l'	F	D	Ω	PM	sin	cos	sin	cos
4	0	0	0	0	0	−1	6798.384	0.0	0.6	−0.1	−0.1
3	0	−1	0	1	0	2	6159.136	1.5	0.0	−0.2	0.1
3	0	−1	0	1	0	1	3231.496	−28.5	−0.2	3.4	−3.9
3	0	−1	0	1	0	0	2190.350	−4.7	−0.1	0.6	−0.9
3	0	1	1	−1	0	0	438.360	−0.7	0.2	−0.2	−0.7
3	0	1	1	−1	0	−1	411.807	1.0	0.3	−0.3	1.0
3	0	0	0	1	−1	1	365.242	1.2	0.2	−0.2	1.4
3	0	1	0	1	−2	1	193.560	1.3	0.4	−0.2	2.9
3	0	0	0	1	0	2	27.432	−0.1	−0.2	0.0	−1.7
3	0	0	0	1	0	1	27.322	0.9	4.0	−0.1	32.4
3	0	0	0	1	0	0	27.212	0.1	0.6	0.0	5.1
3	0	−1	0	1	2	1	14.698	0.0	0.1	0.0	0.6
3	0	1	0	1	0	1	13.719	−0.1	0.3	0.0	2.7
3	0	0	0	3	0	3	9.107	−0.1	0.1	0.0	0.9
3	0	0	0	3	0	2	9.095	−0.1	0.1	0.0	0.6
2	1	−1	0	−2	0	−1	1.11970	−0.4	0.3	−0.3	−0.4
2	1	−1	0	−2	0	−2	1.11951	−2.3	1.3	−1.3	−2.3
2	1	1	0	−2	−2	−2	1.11346	−0.4	0.3	−0.3	−0.4
2	1	0	0	−2	0	−1	1.07598	−2.1	1.2	−1.2	−2.1
2	1	0	0	−2	0	−2	1.07581	−11.4	6.5	−6.5	−11.4
2	1	−1	0	0	0	0	1.03472	0.8	−0.5	0.5	0.8
2	1	0	0	−2	2	−2	1.00275	−4.8	2.7	−2.7	−4.8
2	1	0	0	0	0	0	0.99727	14.3	−8.2	8.2	14.3
2	1	0	0	0	0	−1	0.99712	1.9	−1.1	1.1	1.9
2	1	1	0	0	0	0	0.96244	0.8	−0.4	0.4	0.8
			Rate of secular polar motion in (µas/year) due to the zero frequency tide								
4	0	0	0	0	0	0			−3.80		−4.3

Only terms with amplitudes larger than 0.5 µas are given. $\chi = \text{GMST} + \pi$ (from: IERS C10, Tab. 5.1a)

result from the external gravitational torque due to the nonaxisymmetric matter distribution of the Earth, i.e., the nonzonal parts of the Earth's potential coefficients. Such terms have been calculated for a rigid Earth by Bretagnon et al. (1997), Folgueira et al. (1998a,b), Souchay et al. (1999), Roosbeek (1999), Bizouard et al. (2000, 2001), Getino et al. (2001), and Escapa et al. (2002, 2003). Several authors have extended these calculations to a realistic Earth's model: Brzeziński (2001, 2002), Brzeziński and Capitaine (2003), and Mathews and Bretagnon (2003). For more details, see the IERS C03/C10.

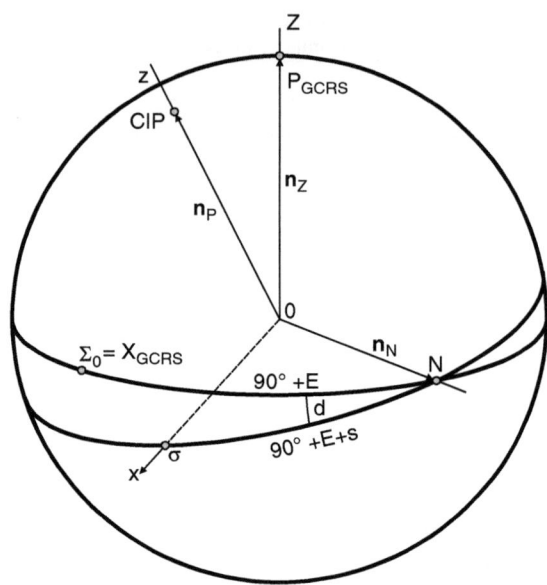

Fig. 9.17 The quantity s defining the location of the CIO

9.4 CIO, TIO, ERA, and GAST

9.4.1 Celestial Intermediate Origin

The CIP defines the z-axis of the astronomical intermediate system, replacing the IRP in the classical system. The x-axis in the classical system is given geometrically, by the equinox of date, that depends upon the complicated motion of the Earth about the Sun. In contrast, the x-axis in the new system, namely, the celestial intermediate origin (CIO), is a point that is defined kinematically and has no connection with the ecliptic.

Let (X, Y, Z) denote the spatial GCRS coordinates and (x, y, z) those of the intermediate system. The x-axis on the CIP equator will be denoted by σ (Fig. 9.17), while N is the node where the CIP equator and the GCRS equator intersect. The angle between N and σ along the moving CIP equator is denoted by $90° + E + s$, where $90° + E$ is the angle between the GCRS x-axis and N (Fig. 9.17; e.g., McCarthy and Petit 2003). The position of σ at time t is shown in Fig. 9.18.

The CIO (σ) on the intermediate equator is defined such that the intermediate system's angular velocity vector at no time has a component in the direction of the CIP, i.e., the point σ never moves *along* the CIP equator. This *nonrotating origin* condition (NRO condition; Guinot 1979; Aoki and Kinoshita 1983) together with a suitably chosen orientation at epoch J2000.0 defines the CIO.

9.4 CIO, TIO, ERA, and GAST

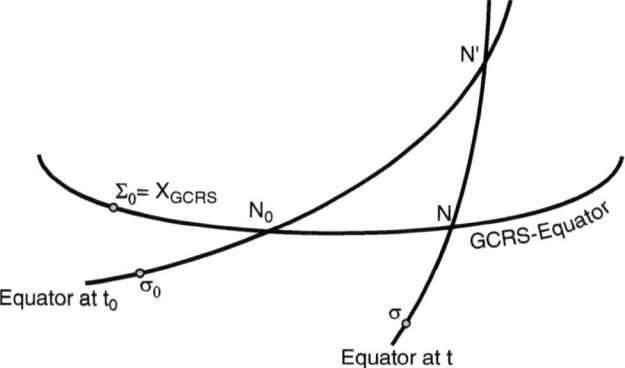

Fig. 9.18 The position of σ at time t

In the following we will describe the unit vector in direction of the CIP by its spherical coordinates (d, E) and the CIO by an angle s such that the node N in Fig. 9.17 on the CIP equator has longitude $90° + E + s$. The corresponding angle in the GCRS is $90° + E$. (In other words, the quantity s is the difference between the GCRS and intermediate right ascension of the node N.) Accordingly, the transformation from the GCRS into the intermediate system is given by

$$\mathcal{R}_3(-(90° + E + s))\,\mathcal{R}_1(d)\,\mathcal{R}_3(90° + E). \tag{9.59}$$

The first rotation is carried out about the Z-axis of the GCRS with unit vector $\mathbf{n}_Z$, the second about the nodal direction given by $\mathbf{n}_N$, and the third about the CIP with unit vector $\mathbf{n}_P$. From this, it follows that the instantaneous angular velocity of the intermediate system in the GCRS is given by

$$\mathbf{\Omega} = \dot{E}\mathbf{n}_Z + \dot{d}\mathbf{n}_N - (\dot{E} + \dot{s})\mathbf{n}_P. \tag{9.60}$$

Projection of this angular velocity onto $\mathbf{n}_P$ leads to ($\mathbf{n}_Z \cdot \mathbf{n}_P = \cos d$)

$$\mathbf{\Omega} \cdot \mathbf{n}_P = \dot{E}(\cos d - 1) - \dot{s}, \tag{9.61}$$

which should vanish by definition, i.e., the CIO obeys the following dynamical equation:

$$s(t) = \int_{t_0}^{t} (\cos d - 1)\dot{E}\,dt - (\sigma_0 N_0 - \Sigma_0 N_0). \tag{9.62}$$

From $X = \sin d \cos E, Y = \sin d \sin E, Z = \cos d$ for the Cartesian GCRS coordinates of the CIP, one finds

$$X\dot{Y} - Y\dot{X} = \dot{E}(1 - \cos^2 d), \tag{9.63}$$

Table 9.7 Series for $s(t)$ compatible with the IAU 2000A precession–nutation model: all terms exceeding 0.5 μas during the interval 1975–2025, unit μas; from: IERS C03, Tab. 5.2c

$$s(t) = -XY/2 + 94 + 3808.35\,t - 119.94\,t^2 - 72574.09\,t^3 + \sum_k C_k \sin\alpha_k$$
$$+1.71\,t\,\sin\Omega + 3.57\,t\,\cos 2\Omega + 743.53\,t^2 \sin\Omega + 56.91\,t^2 \sin(2F - 2D + 2\Omega)$$
$$+9.84\,t^2 \sin(2F + 2\Omega) - 8.85\,t^2 \sin 2\Omega$$

Argument α_k	Amplitude C_k
Ω	−2640.73
2Ω	−63.53
$2F - 2D + 3\Omega$	−11.75
$2F - 2D + \Omega$	−11.21
$2F - 2D + 2\Omega$	+4.57
$2F + 3\Omega$	−2.02
$2F + \Omega$	−1.98
3Ω	+1.72
$l' + \Omega$	+1.41
$l' - \Omega$	+1.26
$l + \Omega$	+0.63
$l - \Omega$	+0.63

so that the dynamical equation for s can also be written in the form (e.g., Capitaine et al. 1986)

$$s = -\int_{t_0}^{t} \frac{1}{1+Z}\left(X\frac{dY}{dt} - Y\frac{dX}{dt}\right) dt - (\sigma_0 N_0 - \Sigma_0 N_0). \tag{9.64}$$

The original convention

$$\sigma_0 N_0 = \Sigma_0 N_0 \tag{9.65}$$

completes the definition of the CIO, so that we have

$$s = -\int_{t_0}^{t} \frac{1}{1+Z}\left(X\frac{dY}{dt} - Y\frac{dX}{dt}\right) dt. \tag{9.66}$$

To ensure continuity with the quasi-classical transformation, this convention was slightly modified later (see IERS C10).

One possibility for calculating s without recourse to numerical integration is to use a suitable Poisson series. Although a series giving s directly can be constructed, it turns out that the quantity $s + XY/2$ is a smoother curve and consequently requires fewer terms; a series can be found in Capitaine and Wallace (2006). See also Table 9.7.

Figure 9.19 depicts the quantity s for the years between 1800 and 2200 (see also Fukushima 2001).

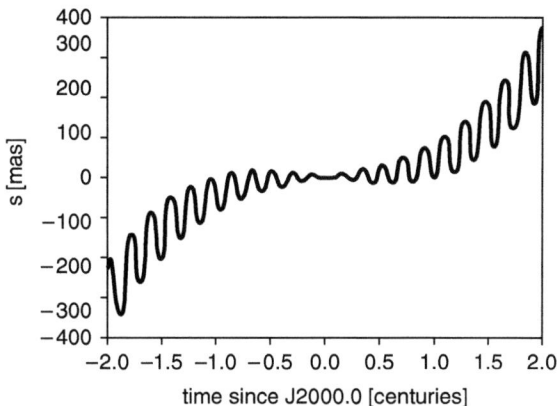

Fig. 9.19 The quantity s between 1800 and 2200

9.4.2 Terrestrial Intermediate Origin

By analogy with the CIO, defined in the GCRS and located by the quantity s, an intermediate origin in the ITRS is defined, called the terrestrial intermediate origin (TIO), located by a quantity s'. From (9.66), we obtain by substitution $X \to x_p, Y \to y_p$ and $Z \simeq 1$

$$s' = (1/2) \int_{t_0}^{t} \left(x_p \dot{y}_p - \dot{x}_p y_p \right) dt. \tag{9.67}$$

To a good approximation, one finds (Lambert and Bizouard 2002)

$$s' = -47.0 \,\mu\text{as} \times t. \tag{9.68}$$

The quantity s' is returned by the *AstroRef* function `TioPosition`.

9.4.3 Earth Rotation Angle θ and GAST

The earth rotation angle (ERA) θ is the angle between CIO (σ) and TIO (ϖ) at time T on the CIP equator (Fig. 9.20). Its classical counterpart, GAST, is the angle between the equinox (Υ) and the TIO.

For θ one finds (Capitaine et al. 2000)

$$\theta(T) = 2\pi \left(0.7790572732640 + 1.00273781191135448\, T \right), \tag{9.69}$$

with

$$T = \text{JD}_{\text{UT1}} - \text{J2000.0}.$$

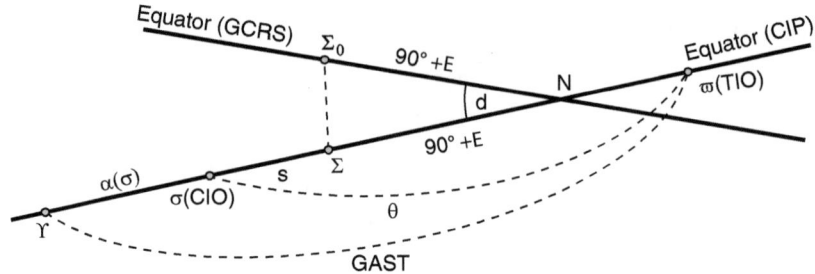

Fig. 9.20 The earth rotation angle (ERA) θ and GAST

The linear relation between θ (ERA) and UT1 is independent of the IAU 2000A model. In contrast, each successive precession model requires its own expression for GMST; the IAU 2000A GMST formula, written out using θ as a basis, reads (Capitaine et al. 2003b)

$$\text{GMST} = 0\overset{''}{.}014506 + \theta + 4612\overset{''}{.}15739966\,t + 1\overset{''}{.}39667721\,t^2$$
$$- 0\overset{''}{.}00009344\,t^3 + 0\overset{''}{.}00001882\,t^4. \quad (9.70)$$

The constant terms in GMST and s were chosen to satisfy the requirement that for 1 January 2003, 0^h TT = JD 2 452 640.5, all three extant ways of computing the rotational orientation of the Earth (the GMST/equinox method, both pre- and post-2000, and the new ERA/CIO method) agreed. In order to determine these numbers, a nominal TAI − UT1 value of 32.3 s at the transition date was adopted, and both polar motion and free-core nutation were neglected (Capitaine et al. 2003b).

The numerical expression for GAST compatible with the IAU 2000A precession–nutation is written in the form (Capitaine et al. 2003b)

$$\text{GAST} = \text{GMST} + \Delta\psi \cos\epsilon_A + \text{CT}, \quad (9.71)$$

with

$$\text{CT} = \sum_k \left(C'_{s,k} \sin\alpha_k + C'_{c,k} \cos\alpha_k\right) - 0\overset{''}{.}00000087\,t \sin\Omega \quad (9.72)$$

and

$$t = (\text{JD}_{\text{TT}} - \text{J2000.0})/36{,}525.$$

The *complementary terms*, CT, are small corrections that have to be added to the classical (kinematical) equation of the equinoxes $\Delta\psi \cos\epsilon_A$ in order to reach microarcsecond accuracy; the largest has an amplitude of about 2.6 mas. Terms larger than 0.5 µas are listed in Table 9.8.

Values for θ, GMST, or GAST may be determined with the *AstroRef* package by calling the functions `EarthRotAngle`, `GMST`, or `GAST`, respectively.

Table 9.8 Complementary terms in GAST

Argument α_k	Amplitude $C'_{s,k}$
Ω	2640.96
2Ω	63.52
$2F - 2D + 3\Omega$	11.75
$2F - 2D + \Omega$	11.21
$2F - 2D + 2\Omega$	−4.55
$2F + 3\Omega$	2.02
$2F + \Omega$	1.98
3Ω	−1.72
$l' + \Omega$	−1.41
$l' - \Omega$	−1.26
$l + \Omega$	−0.63
$l - \Omega$	−0.63

Only terms exceeding 0.5 μas are shown (the $C'_{c,k}$ parts are smaller and therefore have been neglected). Amplitudes in μas

9.5 Transformations from the GCRS to the ITRS

The transformation between the GCRS and the ITRS can be realized in two different ways. First, by using the relations given above, a quasi-classical transformation can be constructed. This equinox-based method starts with the frame bias, then applies precession and nutation using the IAU 2000A model, and takes Earth rotation into account using the GAST(t, UT1). Finally, two rotations about the pole coordinates account for polar motion. The second method uses the CIP coordinates (X, Y) to deal directly with frame bias, precession, and nutation, then the quantity s to locate the CIO, and finally the ERA θ(UT1) and the transformations for polar motion.

9.5.1 Quasi-Classical Equinox-Based Transformation

The quasi-classical relation between a vector $\mathbf{X}_I$ in the GCRS and the same vector $\mathbf{X}_T$ in the ITRS takes the form

$$\mathbf{X}_T = \mathcal{W}(t)\, \mathcal{R}_3(\text{GAST}; t, \text{UT1})\, \mathcal{N}(t)\, \mathcal{P}(t)\, \mathcal{B}\, \mathbf{X}_I. \tag{9.73}$$

Definitions of the components of (9.73) are as follows:

- The parameter t, also used in the following relations, is defined by

$$t = (\text{JD}_{TT} - \text{J2000.0}) / 36{,}525.$$

- $\mathcal{B}$ denotes the frame bias matrix that was already discussed above,

$$\mathcal{B} = \mathcal{R}_1(-\eta_0)\, \mathcal{R}_2(\xi_0)\, \mathcal{R}_3(d\alpha_0).$$

- $\mathcal{P}$ is the precession matrix. It is either given by

$$\mathcal{P}(t) = \mathcal{R}_3(-z_A)\,\mathcal{R}_2(+\theta_A)\,\mathcal{R}_3(-\zeta_A)$$

or, equivalently, by

$$\mathcal{P}(t) = \mathcal{R}_3(\chi_A)\,\mathcal{R}_1(-\omega_A)\,\mathcal{R}_3(-\psi_A)\,\mathcal{R}_1(\epsilon_0).$$

- $\mathcal{N}$ is the nutation matrix, given by

$$\mathcal{N}(t) = \mathcal{R}_1(-\epsilon)\,\mathcal{R}_3(-\Delta\psi)\,\mathcal{R}_1(\epsilon_A).$$

Here, $\epsilon = \epsilon_A + \Delta\epsilon$ and

$$\epsilon_A = 23°26'21''.448 - 46''.84024\,t - 0''.00059\,t^2 + 0''.001813\,t^3$$

describes the obliquity of the mean equator with respect to the mean ecliptic of date.

- The matrix $\mathcal{W}$ finally describes polar motion. It is given by

$$\mathcal{W}(t) = \mathcal{R}_1(-y_p)\,\mathcal{R}_2(-x_p)\,\mathcal{R}_3(s'), \tag{9.74}$$

where x_p and y_p are the pole coordinates of the CIP in the ITRS.

9.5.2 CIO-Based Transformation

In the CIO-based transformation, the relation between the GCRS and the ITRS is written in the form

$$\mathbf{X}_T = \mathcal{W}(t)\,\mathcal{R}_3(\theta;\mathrm{UT1})\,\mathcal{Q}(t)\,\mathbf{X}_I. \tag{9.75}$$

Definitions of the components of (9.75) are as follows:

- The time variable t is again

$$t = (\mathrm{JD}_{TT} - \mathrm{J2000.0})\,/\,36525.$$

- The matrix $\mathcal{Q}$ describes the frame bias as well as precession and nutation. It is given by

$$\mathcal{Q}(t) = \mathcal{R}_3(-s)\,\mathcal{R}_3(-E)\,\mathcal{R}_2(d)\,\mathcal{R}_3(E). \tag{9.76}$$

Written out, the matrix $\mathcal{Q}$ takes the form

$$\mathcal{Q}(t) = \mathcal{R}_3(-s) \cdot \begin{bmatrix} 1+\cos^2 E(\cos d - 1) & (\cos d - 1)\sin E \cos E & -\sin d \cos E \\ (\cos d - 1)\sin E \cos E & 1+\sin^2 E(\cos d - 1) & -\sin d \sin E \\ \sin d \cos E & \sin d \sin E & \cos d \end{bmatrix}.$$

9.5 Transformations from the GCRS to the ITRS

The components of $\mathcal{Q}$ can also expressed in terms of the Cartesian coordinates (X, Y, Z) of the geocentric unit vector towards the CIP

$$\mathcal{Q}(t) = \mathcal{R}_3(-s) \cdot \begin{bmatrix} 1 - aX^2 & -aXY & -X \\ -aXY & 1 - aY^2 & -Y \\ X & Y & 1 - a\left(X^2 + Y^2\right) \end{bmatrix} \quad (9.77)$$

with

$$a = (1 + Z)^{-1}.$$

The matrix $\mathcal{Q}$ may be calculated using the function NutationPrecessionBias of the *AstroRef* package.
- $\mathcal{W}$ is again the polar motion matrix (9.74).

The CIO-based transformations between the celestial and the terrestrial coordinate system are supplied by the *AstroRef* package with the functions CRStoTRS and TRStoCRS.

Note that the figures of the IAU 2000 model given in last sections are based on the IERS conventions 2003. With the IERS conventions of 2010, some of them changed slightly. However, the structures of the formulae remained the same. See IERS C10 for details.

Exercise 9.3: Implement the classical and the modern transformations from the GCRS to the ITRS and vice versa with Maple™ by making use of the corresponding functions from the *AstroRef* package (see Appendix B and package's help in Maple™)! For 1 January 2011, 12^h UTC, transform the GCRS vector $\mathbf{X}_I = (5, 3, 6)$ into the ITRS vector using the classical and the modern transformation! Check if both ways agree at the level of $1\,\mu$as! (Assure setting Maple™ to at least 25 significant digits!)

Chapter 10
Astronomical Software - Yearbooks

10.1 Software Implementations

Many of the algorithms set out in this book are provided for in readily accessible software packages. Indeed, the combination of the availability of such software and the complicated algorithms needed in modern high-precision work is starting to render obsolete some parts of the traditional yearbooks such as the *Astronomical Almanac*. This is further compounded by the wide availability of computer-controlled telescopes, which perform onerous calculations *on the fly* that were a daily chore for observers in former times.

Two packages in particular can be regarded as authoritative: SOFA (from the IAU) and NOVAS (from USNO). It should be noted that neither is an executable application: they are both subroutine libraries used by software developers when writing programs that need to deal with astronomical reference frames and timescales.

10.1.1 SOFA

SOFA stands for *Standards of Fundamental Astronomy* and is an initiative of the International Astronomical Union (IAU) (Division 1, Fundamental Astronomy). A panel of astronomy and software experts, established in 1997, develops software libraries and makes them publicly available at http://www.iausofa.org/. The libraries comprise source code, available (at the time of writing) in Fortran and ANSI C variants and downloadable onto all common platforms such as Linux, Macintosh, and MS-Windows. There are no platform dependencies, such as system calls, and the software can be used on any computer with appropriate compilers.

There are two groups of subroutines. One, the "vector-matrix library," supports operations used for handling positions, velocities, and rotations and also deals with spherical angles, including conversions to and from sexagesimal format. The other is the main "astronomy library." This handles calendars, timescale transformations, Earth rotation angle and sidereal time, ephemerides, precession, nutation, polar motion, star space motion, stellar catalog conversions, and geodetic-geocentric transformations. Some of the subroutines implement standard models that are defined by IAU resolution and, de facto, constitute an authoritative restatement of those standards. But the majority are utility functions, such as for constructing the usual rotation matrices, but calling the *canonical* functions as required for things like evaluating nutation series.

SOFA also provides "cookbook" documentation that offers a tutorial guide to using key capabilities of the libraries. At present, one of these covers Earth attitude (precession, nutation, Earth rotation, polar motion) and the other timescales and calendars.

10.1.2 NOVAS

The US Naval Observatory (USNO) has produced the software package NOVAS. The acronym stands for *Naval Observatory Vector Astrometry Subroutines*. NOVAS is available in ANSI Fortran and C. It is the basis for the numbers in the Astronomical Almanacs. The package can be found under http://www.usno.navy.mil/USNO/astronomical-applications/software-products/novas.

10.2 Astronomical Yearbooks

To help with calculations concerning the relationship between observable or apparent places of astronomical objects and catalog places (usually mean places with respect to the equator and equinox of a given epoch T_0), astronomical yearbooks can be useful. Such yearbooks contain a selection of astronomical data for the year in question. The French "Connaissance des temps ou des mouvements célestes" has the longest tradition of all yearbooks that are still available today. It appeared for the first time back in 1679. Today, the "Connaissance des temps" is issued by IMCCE.

Likely, the two most important yearbooks are "The Astronomical Almanac" (US States Naval Observatory, Washington, together with Her Majesty's Nautical Almanac Office, Taunton, UK) and the Russian yearbook edited by the Institute of Applied Astronomy (IPA) in St. Petersburg. Electronic versions of such yearbooks are available commercially. The *Multiyear Interactive Computer Almanac* (MICA), the interactive computer version of Astronomical Almanac, is provided by the US Naval Observatory. A similar Russian software package (AE) can be obtained from IPA. In the past at least, the annual Apparent Places of Fundamental Stars (APFS) played an important role.

10.2.1 APFS

The astronomical yearbook APFS is a compendium of apparent (and mean) places of 1,535 FK5 stars and was published from 1941 until 2003, when publication ceased. Until the 1959 edition, it was produced by HMNAO in the UK, after that by the *Astronomisches Rechen-Institut* (ARI) in Heidelberg, Germany. Besides apparent places, it contains tables of Besselian day numbers for the computation of annual aberration, short periodic nutations, sidereal time at 0^h UTC, the conversion of mean solar time into sidereal time, and diurnal aberration.

10.2.2 Astronomical Almanac

The *Astronomical Almanac* is produced jointly by the US Naval Observatory (Washington) and HM Nautical Almanac Office in the UK. Its predecessor (from 1960 until 1981) was the *Astronomical Ephemeris*. The publication is a descendant of the American yearbook *The American Ephemeris and Nautical Almanac* published since 1855 and its British counterpart *The Nautical Almanac and Astronomical Ephemeris* published since 1767. The contents of the 2006 edition illustrate its breadth:

- A: Phenomena: Lunar phases, occultations, observability of planets, Moon and Sun rise and set times, twilight, eclipses, etc.
- B: Timescales and coordinate systems: The various timescales and coordinate systems and how they are related, universal time, sidereal time, position and velocity of the Earth, Besselian day numbers, nutation terms in longitude and obliquity, matrix elements for precession and nutation, etc.
- C: Sun: Coordinates, distance and apparent diameter, equation of time, etc.
- D: Moon: Lunar phases, orbital elements, libration angles, etc.
- E: Major planets: Osculating orbital elements of Mercury, Venus, Earth, Mars, Jupiter, Saturn, Uranus, Neptune, and Pluto
- F: Satellites of the planets: Ephemerides and phenomena of the satellites of Mars, Jupiter, Saturn, Uranus, Neptune, and Pluto
- G: Minor planets and comets: Orbital elements, opposition dates, geocentric coordinates, etc., for asteroids and comets
- H: Stars and stellar systems: J2006.5 star positions, J2000.0 positions of radio sources, etc.
- J–N: Observatories, tables, and data: Calendar data, the IAU system of astronomical constants, explanations, glossary, and index

Chapter 11
Astronomical Constants

11.1 Natural Constants

The following part is based upon Soffel and Klioner (2008). Astronomical constants appear when the dynamics of an astronomical system is under discussion. From a fundamental point of view, the dynamics of any physical system can be described by means of just a few fundamental physical interactions: gravity, electromagnetism, and the weak or the strong force. These interactions are described by means of certain fields (the metric field g in the case of gravity; a vector potential A in the case of electromagnetism) that obey certain field equations: Einstein's equations (EIN) in the case of general relativity (GRT) or Maxwell's equations (MAX) in the case of electromagnetism. These fundamental laws of nature contain certain natural constants describing certain properties of the interaction such as strength, propagation velocities, etc. In GRT the Newtonian gravitational constant G plays a central role; in electromagnetism and special relativity (SRT), it is the vacuum speed of light c. In the microscopic world, Planck's reduced constant $\hbar$ is of similar importance. As is well known, the numerical values of three natural constants can be chosen arbitrarily (i.e., by convention), thereby fixing the basic physical units. Geometrized units are a theoretical well-known choice, where $G = \hbar = c = 1$; the unit of time then is the Planck time $T_P = (\hbar G/c^5)^{1/2} = 5.4 \times 10^{-44}$ s, the unit of length is the Planck length $L_P = cT_P = 1.6 \times 10^{-35}$ m, and the unit of mass is $(\hbar c/G)^{1/2} = 2.2 \times 10^{-8}$ kg. For practical purposes, such units obviously are quite inconvenient.

Historically, the basic physical units for time (the second), length (the meter), and mass (the kilogram) have been chosen by means of physical prototypes or properties of astronomical bodies. Before 1956, e.g., the second was defined as the fraction of 1/86 400 of a mean solar day and then until 1867 as 1/31 556 925.8747 of the tropical year 1900. The meter in 1793 was defined as a fraction of 10^{-7} of the Earth's quadrant passing through Paris, and in 1889, it was given by the international prototype of platinum-iridium rod kept at *Bureau International des Poids et Mesures* (BIPM), Paris. Still, the actual definition of the kilogram is through the platinum-iridium prototype of mass, also kept at BIPM. Such prototypes clearly have several disadvantages: precise copies have to be manufactured; they might change their properties due to interactions with the environment. There are indications that copies of the kilogram prototype became heavier in the course of time; mass differences of up to 50 μg have been reported. Possibly the prototype has lost mass because of cleaning procedures.

For these reasons, one tries to define the basic units through natural constants. This has been done for the second and the meter; soon, this goal will also be realized for the kilogram, e.g., by counting the number of atoms of a macroscopic silicon sphere (the Avogadro method; see, e.g., Becker and Gläser 2001).

11.2 Defined and Measurable Natural Constants

All basic constants, defined or measurable, form a system of units. The international system of units is the SI (from French *Système International d'Unités*).

Today, the SI second is defined as "duration of 9 192 631 770 periods of the radiation corresponding to the transition between the two hyperfine levels of the ground state of the cesium-133 atom." The SI meter is defined as "length of the path traveled by light in vacuum during a time interval of 1/299 792 458 of a second." Note that these two definitions fix the vacuum speed of light once and forever; it has become a *defined* natural constant. In contrast to this, the value for G is not defined; it is a *measurable* natural constant.

Clearly multiples of the SI second, SI meter, or SI kilogram can be introduced for convenience without any problems. The astronomical unit, e.g., might be defined by a fixed value in terms of the SI meter rather than by relation to a certain ephemeris.

11.3 Problems Related with Natural Units

The definition of a natural unit related with a defined natural constant is related with a corresponding fundamental law of nature (e.g., special relativity). In case a violation of that law will be detected in the future, the definition might have to be abandoned because of serious problems. Consider, e.g., the definition of the meter and the isotropy of space. Using an old definition of the meter, the famous experiment by Michelson and Morley from 1887 showed that the speed of light is independent of the propagation direction in space (isotropy of space). This fundamental experiment has been repeated over and over again. In recent time, Müller et al. (2003) compared the resonant frequencies of two orthogonal cryogenic resonators subject to Earth's rotation over the period of 1 year with the result that a possible upper limit for $\Delta c/c$ is of order 3×10^{-15}. Suppose that 1 day a violation of the isotropy of space will be found. In that case with the present definition of the meter, the length of a meter stick would depend upon its orientation in space, really an unpleasant situation.

11.4 Body Constants and Framework

The two astronomical constants G and c are related with GRT and Maxwell's theory of electromagnetism. In the latter case the properties of astronomically interesting light rays follow from Maxwell's theory in the limit of geometrical optics. On the other hand, exact GRT is too complex to treat solar system problems. Not even the constants describing certain aspects of astronomical bodies could be defined. These constants will be called *body constants*; examples for such body constants are the mass of a body, its potential coefficients (mass multipole moments), its intrinsic angular momentum (spin), etc. Consider, e.g., the mass of an astronomical body which in Newton's theory of gravity is well defined and given as integral over the density. In relativity, because of the mass-energy equivalence ($E = mc^2$), all kinds of energy contribute to the inertial or gravitational mass even the gravitational field itself. Now, GRT is a nonlinear theory, and in principle, one is not able to separate the gravitational field of a body (A) from that of another body (B). For that reason, one resorts to an approximation to GRT or a whole class of metric theories of gravity if one allows for a violation of Einstein's theory of gravity. Such an approximation will be called a *framework*. For solar system applications, usually, one employs the (first) post-Newtonian framework (a weak field, slow-motion approximation to GRT) or the parametrized post-Newtonian framework (PPN) with a suitable choice of coordinates (e.g., harmonic coordinates). This framework might contain additional constants such as the PPN parameters β, γ, α_1, etc. whose numerical values are related with tests of GRT (deviations from $\beta = \gamma = (\alpha_1 + 1) = 1$ indicate a violation of GRT at the PN level). As is well known (e.g., Damour et al. 1991), body constants can be defined in the basic (post-Newtonian) framework. For $\beta = \gamma = 1; \alpha_1 = 0$, the mass of a body E (Earth) can be defined in the local co-moving system (GCRS) with coordinates $(T, \mathbf{X})$ as (e.g., Damour et al. 1991)

$$M_E(T) = \int_E d^3\Sigma + \frac{1}{6c^2}\frac{d^2}{dT^2}\left(\int_E d^3 X \mathbf{X}^2 \Sigma\right) - \frac{4}{3c^2}\frac{d}{dT}\left(\int_E d^3 X X^a \Sigma^a\right),$$

where Σ and Σ^a are the gravitational (energy) mass density and mass current in the GCRS. Though this theoretical post-Newtonian expression for the mass of the Earth looks quite complicated, it appears as a parameter in the gravitational potential W_E outside the Earth in the simple, quasi-Newtonian form:

$$W_E = \frac{GM_E}{R} + \cdots.$$

Body constants will be time dependent in general; this time dependence together with their realistic errors has to be indicated explicitly. Major sources for such a time dependence are dust accretion or energy loss and solar wind in case of our Sun. For our Sun, with luminosity $L = 4 \times 10^{33}$ erg/s, the fractional mass variation is of order $\dot{M}_S/M_S \sim 10^{-13}$ per year.

11.5 Initial Values and Model

Bodies with their body constants plus initial conditions appear in a dynamical model, e.g., for the motion of the gravitational N-body problem (ephemeris equations). Such a model might involve a variety of interactions (not only gravitational) and might contain additional constants describing certain features of them (e.g., a lag angle to describe the tidal friction in the Earth–Moon system). Examples are the basic equations for the DE, INPOP, or EPM ephemerides. Note that if we start with a certain model and add another interaction (e.g., we consider potential coefficients of higher order for a certain body of the model), we basically face *another model*. Initial values are intimately related with the underlying model. If the values of certain initial values are discussed, the full underlying model has to be specified in some way or another.

In contrast to this, one expects the body constants to have some well-defined (time-dependent) values within a certain framework. Clearly, given a certain data set, different models will imply different values for them with certain errors. However, if realistic errors are given, these values should be compatible with each other.

To ensure consistency of several models, realistic errors should be given. To this end, correlations have to be studied, possibly different models or even different branches of science have to be consulted, etc. This implies that chasing after the current best estimate of a constant is highly problematic as long as a realistic estimate of the error is not given.

11.6 Current Best Estimates

According to IAU 2009 Resolution B2, current best estimates for astronomical constants should continuously be updated. Table 11.1 shows a list of astronomical constants and their current best estimates, based on the IAU 2009 system of astronomical constants (Luzum et al. 2011). Table 11.2 (Folkner et al. 2009) shows parameters of major asteroids.

11.6 Current Best Estimates

Table 11.1 Astronomical constants and their current best estimates

Constant	Value	Reference
	Natural defining constant	
c	2.99792458×10^8 m/s	CODATA (2006)
	Natural measurable constant	
G	$6.67428(67) \times 10^{-11}$ m^3 kg^{-1} s^{-2}	CODATA (2006)
	Auxiliary defining constants	
k	$1.720209895 \times 10^{-2}$	IAU (1976)
L_G	$6.969290134 \times 10^{-10}$	IAU 2000 resolution
L_B	$1.550519768 \times 10^{-8}$	IAU 2000 resolution
	Body constants	
M_M/M_E	$1.23000371(4) \times 10^{-2}$	Standish (2006)
M_S/M_{Me}	$6.0236(3) \times 10^6$	Anderson et al. (1987)
M_S/M_{Ve}	$4.08523719(8) \times 10^5$	Konopliv et al. (1999)
M_S/M_{Ma}	$3.09870359(2) \times 10^6$	Konopliv et al. (2006)
M_S/M_{Ju}	$1.047348644(17) \times 10^3$	Jacobson et al. (2000)
M_S/M_{Sa}	$3.4979018(1) \times 10^3$	Jacobson et al. (2006)
M_S/M_{Ur}	$2.290298(3) \times 10^4$	Jacobson et al. (1992)
M_S/M_{Ne}	$1.941226(3) \times 10^4$	Jacobson (2009)
M_S/M_{Pl}	$1.36566(28) \times 10^8$	Tholen et al. (2008)
M_S/M_{Eris}	$1.191(14) \times 10^8$	Brown and Schaller (2007)
M_{Ceres}/M_S	$4.72(3) \times 10^{-10}$	Pitjeva and Standish (2007)
M_{Pallas}/M_S	$1.03(2) \times 10^{-10}$	Pitjeva and Standish (2007)
M_{Vesta}/M_S	$1.35(2) \times 10^{-10}$	Pitjeva and Standish (2007)

Table 11.2 Parameters of major asteroids, radii R in km, taxonomic class TC, product GM of the gravitational constant and the mass in km^3/s^2, and density ρ in g/cm^3; from Folkner et al. (2009)

ID	Name	R	TC	GM	ρ	ID	Name	R	TC	GM	ρ
1	Ceres[a]	474.0	G	62.178	2.1	63	Ausonia	51.6	S	0.102	2.7
2	Pallas	266.0	B	13.402	2.5	65	Cybele	118.6	C	0.694	1.5
3	Juno	117.0	Sk	1.536	3.4	69	Hesperia	69.1	M	0.414	4.5
4	Vesta	265.0	V	17.630	3.4	78	Diana	60.3	C	0.085	1.4
5	Astraea	59.5	S	0.159	2.7	94	Aurora	102.4	C	0.414	1.4
6	Hebe	92.6	S	0.605	2.7	97	Klotho	41.4	M	0.089	4.5
7	Iris	99.9	S	0.796	2.9	98	Ianthe	52.5	C	0.055	1.4
8	Flora	67.9	S	0.236	2.7	105	Artemis	59.5	C	0.088	1.5
9	Metis	95.0	S	0.567	2.4	111	Ate	67.3	C	0.116	1.4
10	Hygiea	203.6	C	5.364	2.3	135	Hertha	39.6	M	0.078	4.5
11	Parthenope	77.7	S	0.356	2.7	139	Juewa	78.3	C	0.188	1.4
13	Egeria	103.8	C	0.412	1.3	145	Adeona	75.6	C	0.151	1.3
14	Irene	76.0	S	0.348	2.8	187	Lamberta	65.6	C	0.105	1.3
15	Eunomia	127.7	S	1.638	2.8	192	Nausikaa	51.6	S	0.107	2.8
16	Psyche	126.6	M	2.233	3.9	194	Prokne	84.2	C	0.182	1.1
18	Melpomene	70.3	S	0.267	2.7	216	Kleopatra	62.0	M	0.299	4.5

(continued)

Table 11.2 (continued)

ID	Name	R	TC	GM	ρ	ID	Name	R	TC	GM	ρ
19	Fortuna	100.0	Ch	0.463	1.7	230	Athamantis	54.5	S	0.126	2.8
20	Massalia	72.8	S	0.291	2.7	324	Bamberga	114.5	CP	0.661	1.6
21	Lutetia	47.9	M	0.139	4.5	337	Devosa	29.6	M	0.033	4.5
22	Kalliope	90.5	M	0.491	2.4	344	Desiderata	66.1	C	0.114	1.4
23	Thalia	53.8	S	0.129	3.0	354	Eleonora	77.6	Sl	0.327	2.5
24	Themis	99.0	C	0.403	1.5	372	Palma	94.3	C	0.355	1.5
25	Phocaea	37.6	S	0.040	2.7	405	Thia	62.5	C	0.092	1.4
27	Euterpe	48.0	S	0.084	2.7	409	Aspasia	80.8	C	0.216	1.5
28	Bellona	60.5	S	0.165	2.7	419	Aurelia	64.5	C	0.102	1.4
29	Amphitrite	106.1	S	0.906	2.7	451	Patientia	112.5	C	0.610	1.5
30	Urania	49.8	S	0.095	2.7	488	Kreusa	75.1	C	0.164	1.4
31	Euphrosyne	128.0	C	1.139	1.9	511	Davida	163.0	C	1.638	1.4
41	Daphne	87.0	Ch	0.527	2.9	532	Herculina	111.1	S	0.886	2.3
42	Isis	50.1	S	0.092	2.6	554	Peraga	47.9	C	0.044	1.4
45	Eugenia	107.3	C	0.397	1.2	654	Zelinda	63.7	Ch	0.090	1.2
51	Nemausa	73.9	C	0.144	1.3	704	Interamnia	158.3	C	2.464	2.2
52	Europa	151.3	C	1.354	1.4	747	Winchester	85.9	C	0.196	1.1
60	Echo	30.1	S	0.021	2.7						

[a] Ceres is now designated a dwarf planet

References

Allan DW, Hellwig H, Kartaschoff P, Vanier J, Vig J, Winkler GMR, Yannoni NF (1988) Standard terminology for fundamental frequency and time metrology. In: Proceedings of the 42nd Annual IFCS, Baltimore, MD, 1988, pp 419–425

Anderson JD, Colombo G, Esposito PB, Lau EL, Trager GB (1987) The mass, gravity field, and ephemeris of Mercury. Icarus 71:337–349

Aoki S, Kinoshita H (1983) Note on the relation between the equinox and Guinot's non-rotating origin. Celestial Mech Dyn Astron 29:335–360

Aoki S, Guinot B, Kaplan GH, Kinoshita H, McCarthy DD, Seidelmann PK (1982) The new definition of Universal time. A&A 105:359–361

Arias EF (2010) Current and future realizations of coordinate time scales. In: Klioner SA, Seidelman PK, Soffel MH (eds) Relativity in fundamental astronomy. Proceedings of of IAU Symposium No. 261, 2009. Cambridge University Press, Cambridge, pp 95–101

Artz T, Bockmann S, Nothnagel A, Steigenberger P (2010) Sub-diurnal variations in the Earth's rotation from continuous VLBI campaigns. J Geophys Res 115:B05404, 16

Audoin C, Guinot B (2001) The measurement of Tim: Time, frequency and the atomic clock. Cambridge University Press, Cambridge

Barbieri C, Capaccioli M, Ganz R, Pinto G (1972) Accurate positions of the Planet Pluto in the years 1969–1970. Astron J 77:521–522

Barbieri C, Capaccioli M, Pinto G (1975) Accurate positions of the Planet Pluto in the years 1971–1974. Astron J 80:412–414

Barbieri C, Pinocchio L, Capaccioli M, Pinto G, Schoenmaker AA (1979) Accurate positions of the Planet Pluto from 1974 to 1978. Astron J 84:1890–1893

Barbieri C, Benacchio L, Capaccioli M, Gemmo AG (1988) Accurate positions of the Planet Pluto from 1979 to 1987. Astron J 96:396–399

Bay Z (1947) Reflection of microwaves from the moon. Hung Phys Acta 1:1–22

Becker P, Gläser M (2001) Kilogramm und Mol: SI-Basiseinheiten für Masse und Stoffmenge. Physik in unserer Zeit 32:254–259

Bertotti B, Iess L, Tortora P (2003) A test of general relativity using radio links with the Cassini spacecraft. Nature 425:374–376

Bizouard C, Folgueira M, Souchay J (2000) Comparison of the short periodic rigid Earth nutation series. In: Dick S, McCarthy D, Luzum B (eds) Proceedings of IAU Colloquium 178, Publications of the Astron. Soc. Pac. Conf. Ser., vol 208, pp 613–617

Bizouard C, Folgueira M, Souchay J (2001) Short periodic nutations: comparison between series and influence on polar motion. In: Capitaine N (ed) Proceedings of the Journées 2000 - Systèmes de Référence spatio-temporels, Paris, 2001, pp 260–265

Black GJ, Campbell DB, Harmon JK (2010) Radar measurements of Mercury's north pole at 70 cm wavelength. Icarus 209:224–229

Böhm J, Schuh H (2004) Vienna mapping functions in VLBI analyses, Geophys. Res. Lett., 31, L01603

Böhm J, Böhm S, Nilsson T, Pany A, Plank L, Spicakova H, Teke K, Schuh H (2011) The new Vienna VLBI Software VieVS. Geodesy for Planet Earth. In: Kenyon S, Pacino MC, Marti U (eds) Proceedings of the 2009 IAG Symposium, Buenos Aires, Argentina, 31 August 2011–4 September 2009. International Association of Geodesy Symposia Series, vol 136, ISBN 978-3-642-20337-4

Bretagnon P (1982) Theory for the motion of all the planets – the VSOP82 solution. A&A 114:278

Bretagnon P (1984) Amélioration des theories planétaire analytiques. Celestial Mech Dyn Astron 34:193–201

Bretagnon P, Francou G (1988) Planetary theories in rectangular and spherical variables, VSOP87 solutions. A&A 202:309

Bretagnon P, Rocher P, Simon J-L (1997) Theory of the rotation of the rigid Earth. A&A 319(1):305–317

Brown ME, Schaller EL (2007) The Mass of Dwarf Planet Eris. Science 316:1585

Brzeziński A (2001) Diurnal and sub-diurnal terms in nutation: A simple theoretical model for a non-rigid Earth. In: Capitaine N (ed) Proceedings of the Journées 2000 - Systèmes de Référence spatio-temporels, Paris, 2001, pp 243–251

Brzeziński A (2002) Circular 2, IAU Comission 19 WG "Precession–Nutation"

Brzeziński A, Capitaine N (2003) Lunisolar perturbations in Earth rotation due to the triaxial figure of the Earth: Geophysical aspects. In: Capitaine N (ed) Proceedings of the Journées 2001 – Systèmes de Référence spatio-temporels, Paris, 2003, pp 51–58

Capitaine N (1990) The Celestial pole coordinates. Celestial Mech Dyn Astron 48:127–143

Capitaine N, Gontier A-M (1993) Accurate procedure for deriving UT1 at a submilliarcsecond accuracy from Greenwich Sidereal Time or from the stellar angle. A&A 275:645–650

Capitaine N, Wallace PT (2006) High precision methods for locating the Celestial intermediate pole and origin. A&A 450:855–872

Capitaine N, Guinot B, Souchay J (1986) A non-rotating origin on the instantaneous equator: Definition, properties and use. Celestial Mech Dyn Astron 39:283–307

Capitaine N, Guinot B, McCarthy D (2000) Definition of the Celestial ephemeris Origin and of UT1 in the International Celestial Reference Frame. A&A 355:398–405

Capitaine N, Chapront J, Lambert S, Wallace P (2003a) Expressions for the Celestial intermediate pole and Celestial ephemeris origin consistent with the IAU 2000A precession–nutation model. A&A 400:1145–1154

Capitaine N, Wallace P, McCarthy D (2003b) Expressions to implement the IAU-2000 definition of UT1. A&A 406:1135–1149

Capitaine N, Wallace P, Chapront J (2003c) Expressions for IAU-2000 precession quantities. A&A 412:567–586

Carter WE, Robertson DS, MacKay JR (1985) Geodetic Radio interferometric surveying: Applications and results. J Geophys Res 90:4577–4587

Chapront-Touzé M, Chapront J (1983) The lunar ephemeris ELP 2000. A&A 124:50–62

Chapront J, Chapront-Touzé M, Francou G (2002) An new determination of lunar orbital parameters, precession constant and tidal acceleration from LLR measurements. A&A 387: 700–709

Charlot P (1990) Radio-source structure in astrometric and geodetic very long baseline interferometry. Astron J 99:1309–1326

CODATA (2006), Mohr P, Taylor B, Newell D (2008) CODATA recommended values of the physical constants: 2006. Rev Mod Phys 80:633–730

Cohen CJ, Hubbard EC, Oesterwinter C (1967) New Orbit for Pluto and analysis of differential corrections. Astron J 8:973–988

Cooperstock F, Faraoni V, Vollick D (1998) The influence of the cosmological expansion on local systems. Ap J 503:61–66

Damour T, Soffel M, Xu C (1991) General-relativistic Celestial mechanics I. Phys Rev D 43:3273

Damour T, Soffel M, Xu C (1994) General-relativistic Celestial mechanics IV. Phys Rev D 49:618
Defraigne P, Dehant V, Paquet P (1995) Link between the retrograde-prograde nutations and nutations in obliquity and longitude. Celestial Mech Dyn Astron 62:363–376
DeMets C, Gordon RG, Argus DF, Stein S (1994) Effect of the recent revisions to the geomagnetic reversal time scale on estimates of current plate motions. Geophys Res Lett 21:2191–2194
de Sitter S (1916) On Einstein's theory of gravitation and its astronomical consequences. Mon Not Roy Astron Soc 77:155–184
Dravins D, Lindegren L, Madsen S (1999) Astrometric radial velocities, I. Non-spectroscopic methods for measuring stellar radial velocity. A&A 348:1040–1051
Drewes H (1998) Combination of VLBI, SLR and GPS determined station velocities for actual plate kinematic and crustal deformation models. In: Feissel M (ed) Geodynamics, IAG Symposia. Springer, Berlin
Drewes H (2006) The actual plate kinematic and crustal deformation model (APKIM 2005) as basis for a non-rotating ITRS. In: Drewes H (ed) IAG Symposia, vol 134: Geodetic reference frames. IAG Symposium Munich, 9–14 October 2006. Spinger, Berlin, pp 95–100
Drewes H, Angermann D (2001) The actual plate kinematic and crustal deformation model 2000 (APKIM2000) as a geodetic reference system. AIG 2001 Scientific Assembly, Budapest, 2–8 September 2001. www.dgfi.badw.de/fileadmin/DOC/DS_APKIM.pdf
Escapa A, Getino J, Ferrándiz JM (2002) Indirect effect of the triaxiality in the Hamiltonian theory for the rigid Earth nutations. A&A 389(3):1047–1054
Escapa A, Getino J, Ferrándiz JM (2003) Influence of the triaxiality of the non-rigid Earth on the J2 forced nutations. In: Capitaine N (ed) Proceedings of the Journées 2001 - Systèmes de Référence spatio-temporels, Paris, 2003
Eubanks TM, Matsakis DN, Martin JO, Archinal BA, McCarthy DD, Klioner SA, Shapiro S, Shapiro II (1997) Advances in Solar System Tests of Gravity. American Physical Society, APS/AAPT Joint Meeting, abstract K11.05, 1105
Fairhead L, Bretagnon P (1990) An analytical formula for the time transformation TB-TT. A&A 229:240–247
Fedorov EP (1963) Nutation and forced motion of the Earth's pole. Pergamon Press, Oxford
Fedorov PN, Akhmetov VS, Bobylev VV, Gontcharov GA (2010) The XPM catalogue as a realization of the ICRS in optical and near-infrared ranges of wavelengthsm. Mon Not Roy Astron Soc 415:665–672
Fey A, Charlot P (1997) VLBA observations of Radio reference frame sources, II. Astrometric suitability based on observed structure. Ap J Suppl 111:95
Fey A, Gordon D, Jacobs C (eds) (2009) IERS Technical Notes, No 35, BKG, Frankfurt/Main
Fienga A, Manche H, Laskar J, Gastineau M (2008) INPOP06: A new numerical planetary ephemeris. A&A 477:315–327
Fienga A, Laskar J, Morley T, Manche H, Kuchynka P, Le Poncin-Lafitte C, Budnik F, Gastineau M, Somenzi L (2009) INPOP08, a 4-D planetary ephemeris: From asteroid and time-scale computations to ESA Mars Express and Venus Express contributions. A&A 507:1675–1686
Fienga A, Manche H, Kuchynka P, Laskar J, Gastineau M (2010) Planetary and Lunar Ephemerides INPOP10A. In: Capitaine N (ed) Proceedings of the Journées 2010 - Systèmes de Référence spatio-temporels, Paris, 2010
Fienga A, Laskar J, Kuchynka P, Manche H, Desvignes G, Gastineau M, Cognard I, Theureau G (2011) The INPOP10a planetary ephemeris and its applications in fundamental physics. Celestial Mech Dyn Astron 111(3):363–385
Folgueira M, Souchay J, Kinoshita H (1998a) Effects on the nutation of the non-zonal harmonics of third degree. Celestial Mech Dyn Astron 69(4):373–402
Folgueira M, Souchay J, Kinoshita H (1998b) Effects on the nutation of C4m and S4m harmonics. Celestial Mech Dyn Astron 70(3):147–157
Folgueira M, Bizouard C, Souchay J (2001) Diurnal and subdiurnal luni-solar nutations: Comparisons and effects. Celestial Mech Dyn Astron 81:191–217

Folkner W (2011) Uncertainties in the JPL Planetary Ephemeris, in: Proc. of the Journées 2010 - Systèmes de Référence spatio-temporels, N. Capitaine (ed.), Paris

Folkner W et al (2010) Recent developments in planetary ephemeris observations, JPL. Journées 2010 - Systèmes de réference spatio-temporels, Paris, 20–22 September 2010. http://syrte.obspm.fr/journees2010/powerpoint/folkner.pdf

Folkner WM, Standish EM, Williams JG, Boggs DH (2007) The planetary and lunar ephemerides DE418. JPL Interoffice Memorandum 343R-07-005. ftp://naif.jpl.nasa.gov/pub/naif/generic_kernels/spk/planets/a_old_versions/de418_announcement.pdf. Accessed 2 Aug 2007

Folkner WM, Williams JG, Boggs DH (2009) The planetary and lunar ephemerides DE421. Interoffice Memorandum, 343.R-08-003; see also: IPN Progress Report 42–178, August 15, 2009

Francou G, Simon JL (2011) New analytical planetary theories VSOP2010. In: Capitaine N (ed) Proceedings of the Journées 2010 - Systèmes de Référence spatio-temporels, Paris, 2011, pp 85–86

Fricke W, Schwan H, Lederle T, Bastian U, Bien R, Burkhardt G, du Mont B, Hering R, Jährling R, JahreißH, Röser S, Schwerdtfeger H-M, Walter HG (1988) Fifth Fundamental Catalogue (FK5), Part 1. The Basic Fundamental Stars. Veröffentlichungen Astronomisches Rechen-Institut Heidelberg, No. 32, Verlag G. Braun, Karlsruhe

Froeschlé M, Mignard F, Arenou F (1997) Determination of the PPN Parameter gamma with the HIPPARCOS Data. In: Proceedings of the ESA Symposium Hipparcos - Venice '97, 13–16 May, Venice, Italy, ESA SP-402 (July 1997), pp 49–52

Fukushima T (2001) Global rotation of the nonrotating origin. Astron J 122:482–486

Gemmo AG, Barbieri C (1994) Astrometry of Pluto from 1969 to 1989. Icarus 108:174–179

Getino J, Ferrándiz JM, Escapa A (2001) Hamiltonian theory for the non-rigid Earth: Semi-diurnal terms. A&A 370(1):330–341

Gripp AE, Gordon RG (1990) Current plate velocities relative to the hotspots incorporating the NUVEL-1 global plate motion model. Geophys Res Lett 17:1109–1112

Gripp AE, Gordon RG (2002) Young tracks of hotspots and current plate velocities. Geophys J Int 150:321–361

Gubanov V, Rusinov Y, Surkis I, Kurdubov S, Shabun C (2004) Project: Global analysis of 1979–2004 VLBI data. In: Vandenberg NV, Baver KD (eds) International Service for Geodesy and Astrometry, 2004 General Metting Proceedings, Ottawa, Canada, NASA/CP-2204-212255, pp 315–319

Guinot B (1979) Basic problems in the kinematics of the rotation of the Earth. In: McCarthy D, Pilkington J (eds) Time and the Earth's rotation. Reidel Publishing Company, Dordrecht, pp 7–18

Hambly NC, Irwin MJ, MacGillivray HT, Irwin MJ, MacGillivray HT (2001) The SuperCOSMOS Sky survey, Paper I: Introduction and description. Mon Not Roy Astr Soc 326:1279, 1295

Hambly N, Read M, Mann R, Sutorius E, Bond I, MacGillivray H, Williams P, Lawrence A (2004) The SuperCOSMOS Science archive, astronomical data analysis software and systems (ADASS) XIII. In: Ochsenbein F, Allen MG, Egret D (eds) Proceedings of the conference held 12–15 October, 2003, Strasbourg, France; ASP Conference Proceedings, vol 314, San Francisco. Astronomical Society of the Pacific, p 137

Herring TA (1992) Modelling atmospheric delays of space geodetic data. In: DeMunk JC, Spoelstra TA (eds) Symposium on refraction of transatmospheric signals in Geodesy. Netherlands Geodetic Comission Series No. 36, pp 157–164

Herring T, Mathews P, Buffett B (2002) Modelling of nutation-precession: Very long baseline interferometry results. J Geophys Res 107:B4

Hilton J, Hohenkerk C (2011) A comparison of the high accuracy planetary ephemerides DE421, EPM2008, and INPOP08. In: Capitaine N (ed) Proceedings of the Journées 2010 - Systèmes de Référence spatio-temporels, Paris, 2011, pp 77–80

Hofmann F, Müller J, Biskupek L (2010) Lunar laser ranging test of the Nordtvedt parameter and a possible variation in the gravitational constant. A&A 522:L5

Høg E, Fabricius C, Makarov VV, Urban S, Corbin T, Wycoff G, Bastian U, Schwekendiek P, Wicenec A (2000) The Tycho-2 catalogue of the 2.5 million brightest stars. A&A 355:L27–L30

Höling B (1990) Ein Lasergyroskop zur Messung der Erdrotation, Dissertation, Universität Tübingen

Hurst RB, Stedman GE, Schreiber KU, Thirkettle RJ, Graham RD, Rabeendran N, Wells J-PR (2009) Experiments with a 834 m^2 ring laser interferometer. J Appl Phys 105:113115

Jacobson RA (2009) The orbits of the Neptunian satellites and the orientation of the pole of Neptune. Astron J 137:4322–4329

Jacobson RA, Campbell J, Taylor AH, Synott SP (1992) The masses of Uranus and its major satellites from voyager tracking data and Earth-based Uranian satellite data. Astron J 103(6):2068–2078

Jacobson RA, Haw RJ, McElrath TP, Antreasian PG (2000) A Comprehensive Orbit Reconstruction for the Galileo Prime Mission in the J2000 System. J Astronaut Sci 48(4):495–516

Jacobson RA, Antreasian PG, Bordi JJ, Criddle KE, Ionasescu R, Jones JB, Mackenzie RA, Meek MC, Parcher D, Pelletier FJ, Owen WM Jr, Roth DC, Roundhill IM, Stauch JR (2006) The gravity field of the Saturnian system from satellite observations and spacecraft tracking data. Astron J 132:2520

Jeffreys H (1963) Preface to Fedorov (1963)

Jensen KS (1979) Accurate astrometric position of Pluto. A&A Suppl 36:395–398

Jones DL, Fomalont E, Dhawan V, Romney J, Folkner W, Lanyi G, Border J (2011) Very Long Baseline Array Astrometric Observations of the Cassini Spacecraft at Saturn, Astron J **141** 29:1–10

Jones R, Tryon P (1987) Continuous time series models for unequally spaced data applied to modeling atomic clocks. SIAM J Sci Stat Comput 8(1):71–81

Jurgens RF (1982) Earth-based radar studies of Planetary surfaces and Atmospheres. IEEE Trans Geosci Rem Sens 20:293–305

Kaplan G (2005) USNO Circular No. 179, The IAU Resolutions on Astronomical Reference Systems, Time Scales, and Earth Rotation Models, Explanation and Implementation. US Naval Observatory, Washington. http://aa.usno.navy.mil/publications/docs/Circular_179.pdf. Accessed 20 Oct 2005

Kinoshita H, Souchay J (1990) The theory of nutation for the rigid Earth model at the second order. Celestial Mech Dyn Astron 48:187

Klemola AR, Harlan EA (1982) Astrometric observations of the outer Planets and minor Planets: 1980–1982. Astron J 87:1242–1243

Klemola AR, Harlan EA (1984) Astrometric observations of the outer Planets and minor Planets: 1982–1983. Astron J 89:879–881

Klemola AR, Harlan EA (1986) Astrometric observations of the outer Planets and minor Planets: 1984–1985. Astron J 92:195–198

Klioner S (2000) Relativity in modern astrometry and celestial mechanics – Overview. In: Johnston KJ, McCarthy DD, Luzum BJ, Kaplan GH (eds) Towards models and constants for sub-microarcsecond astrometry. Proceedings of the IAU Colloquium 180, Washington, DC, pp 265–274

Klioner S (2003a) A practical relativistic model for microarcsecond astrometry in space. Astron J 125(3):1580–1597

Klioner S (2003b) Proposal for the representation of the astrometric parameters, available from the Gaia document archive. http://www.rssd.esa.int/llink/livelink

Klioner S, Kopeikin S (1992) Microarcsecond astrometry in Space: Relativistic effects and reduction of observations. Astron J 104(2):897–914

Klioner S, Soffel M (2004) Refining the relativistic model for GAIA: cosmiligical effects in the BCRS. In: Proceedings of the GAIA meeting, 4–7 October 2004, Paris (ESA SP-576), 2004, pp 305–308

Konopliv AS, Banerdt W, Sjogren W (1999) Venus gravity: 180th degree and order model. Icarus 139:3–18

Konopliv AS, Yoder CF, Standish EM, Yuan DN, Sjogren WL (2006) A global solution for the Mars static and seasonal gravity, Mars orientation, Phobos and Deimos masses, and Mars ephemeris. Icarus 182:23–50

Konopliv AS, Asmar SW, Folkner WM, Karatekin O, Nunes DC, Smrekar SE, Yoder CF, Zuber MT (2011) Mars high resolution gravity fields from MRO, Mars seasonal gravity, and other dynamical parameters. Icarus 211:401–428

Kopeikin S, Efroimsky M, Kaplan G (2011) Relativistic celestial mechanics of the solar system. Wiley-VCH, New York

Kovalevsky J (1995) Modern astrometry. Springer, Berlin

Kovalevsky J, Lindegren L, Perryman MAC, Hemenway PD, Johnston KJ, Kislyuk VS, Lestrade JF, Morrison LV, Platais I, Röser S, Schilbach E, Tucholke HJ, de Vegt C, Vondrák J, Arias F, Gontier AM, Arenou F, Brosche P, Florkowski DR, Garrington ST, Preston RA, Ron C, Rybka SP, Scholtz RD, Zacharias N (1997) The Hipparcos catalogue as a realisation of the extragalactic reference system. A&A 323:620–633

Kovalevsky J, Seidelmann PK (2004) Fundamentals of astrometry. Cambridge University Press, Cambridge

Krasinsky GA, Pitjeva EV, Sveshnikov ML, Chunaeva LI (1993) The motion of major planets from observations 1769–1988 and some astronomical constants. Celestial Mech Dyn Astron 55:1–23

Kreemer C, Holt WE, Haines AJ (2003) An integrated global model of present-day plate motions and plate boundary deformation. Geophys J Int 154:8–34

Kurdubov S (2007) QUASAR software in IAA EOP service: Global solution and Daily SINEX. In: Böhm J, Pany A, Schuh H (eds) Proceedings of the 18th European VLBI for Geodesy and Astrometry working meeting. Geowissenschaftliche Mitteilungen, Heft Nr. 79, pp 79–81

Lambert S, Bizouard C (2002) Positioning the terrestrial ephemeris origin in the international terrestrial frame. A&A 394:317–321

Lambert SB, Le Poncin-Lafitte C (2011) Improved determination of γ by VLBI. A&A 529:id.A70

Laskar J, Robutel P, Joutel F, Gastineau M, Correia ACM, Levrard B (2004) A long term numerical solution for the insolation quantities of the Earth. A&A 428:261–285

Laskar J, Fienga A, Gastineau M, Manche H (2011) La2010: A new orbital solution for the long term motion of the Earth. A&A 532:A89

Lebach DE, Corey BE, Shapiro II, Ratner MI, Webber JC, Rogers AE, Davis JL, Herring TA (1995) Measurement of the solar gravitational deflection of radio waves using very-long-baseline interferometry. Phys Rev Lett 75:1439–1442

Lefevre H (1993) The fiber-optic gyroscope. Artech House Inc, London

Lieske J, Lederle T, Fricke W, Morando B (1977) Expression for the precession quantities based upon the IAU (1976) system of astronomical constants. A&A 58:1–16

Luzum B, Capitaine N, Fienga A, Folkner W, Fukushima T, Hilton J, Hohenkerk C, Krasinsky G, Petit G, Pitjeva E, Soffel M, Wallace P (2011) The IAU 2009 system of astronomical constants: The report of the IAU working group on numerical standards for fundamental Astronomy. Celestial Mech Dyn Astron 110:293–304

Macek WM, Davies DTM Jr (1963) Rotation rate sensing with travelling-wave ring lasers. Appl Phys Lett 2(3):67–68

Manche H, Fienga A, Laskar J, Bouquillon S, Francou G, Gastineau M (2010) LLR residuals of INPOP10A and constraints on Post-Newtonian parameters. In: Capitaine N (ed) Proceedings of the Journées 2010 - Systèmes de Référence spatio-temporels, Paris, 2010, pp 65–68

Mashhoon B (1985) Gravitational effects of rotating masses. Found Phys 15:497–515

Mathews P, Bretagnon P (2003) Polar motion equivalent to high frequency nutations for a nonrigid earth with anelastic mantle. A&A 400:1113–1128

Mathews P, Shapiro II (1992) Nutations of the Earth. Ann Rev Earth Planet Sci 20:469–500

Mathews P, Herring T, Buffet B (2002) Modelling of nutation-precession: New nutation series for nonrigid Earth and insights into the Earth's interior. J Geophys Res 107:B4

McCarthy D, Petit G (2003) IERS Conventions 2003, IERS Technical Note No. 32, Verlag des Bundesamtes für Kartographie und Geodäsie, Frankfurt am Main, 2004

McCarthy DD (1996) IERS Conventions (1996), IERS Technical Note No. 21, Paris, Central Bureau of IERS, Observatoire de Paris, 1996

McCarthy DD, Capitaine N (2002) Practical Consequences of Resolution B1.6 IAU2000 Precession-Nutation Model, Resolution B1.7 Definition of Celestial Intermediate Pole, Resolution B1.8 Definition and Use of Celestial and Terrestrial Ephemeris Origin, IERS Technical Note No. 28

Meeus J (1999) Astronomical algorithms, 2nd edn. Willmann-Bell, Richmond

Michelson AA, Gale HG (1925) The effect of the Earth's rotation on the velocity of light. Ap J 61:140–145

Milonni P, Eberly J (1988) Lasers. Wiley, New York

Mofenson J (1946) Radar echoes from the moon. Electronics 19:92–98

Moisson X, Bretagnon P (2001) Analytical Planetary solution VSOP2000. Celestial Mech Dyn Astron 80:205–213

Monet DG, Levine SE, Canzian B, Ables HD, Bird AR, Dahn CC, Guetter HH, Harris HC, Henden AA, Leggett SK, Levison HF, Luginbuhl CB, Martini J, Monet AKB, Munn JA, Pier JR, Rhodes AR, Riepe B, Sell S, Stone RC, Vrba FJ, Walker RL, Westerhout G, Brucato RJ, Reid IN, Schoening W, Hartley M, Read MA, Tritton SB (2003) The USNO-B catalog. Astron J 125:984–993

Mueller II (1969) Spherical and practical astronomy (as applied to geodesy). Frederick Ungar Publishing Co., New York

Müller J, Biskupek L (2007) Variations of the gravitational constant from Lunar Laser Ranging data. Classical Quant Grav 24:4533–4538

Müller J, Nordtvedt K (1998) Lunar laser ranging and the equivalence principle signal. Phys Rev D 58:062001

Müller J, Schneider M, Soffel M, Ruder H (1991) Testing Einstein's theory of gravity by analyzing lunar laser ranging data. Ap J Lett 101:382

Müller H, Herrmann S, Braxmaier C, Schiller S, Peters A (2003) Modern Michelson-Morley experiment using cryogenic optical resonators. Phys Rev Lett 91:020401

Müller J, Williams JG, Turyshev SG (2008) Lunar laser ranging contributions to relativity and Geodesy. In: Dittus H, Lämmerzahl C, Turyshev SG (eds) Lasers, clocks and drag-free control: Exploration of relativistic gravity in space. Astrophysics and Space Science Library, vol 349, pp 457–472

Murphy T (2011) private communication

Murphy T, Strasburg J, Stubbs C, Adelberger E, Angle J, Nordtvedt K, Williams J, Dickey J, Gillespie (2000) The Apache point observatory lunar laser-ranging operation (APOLLO). In: Proceedings of the 12th International Workshop on Laser Ranging, held in Matera, Italy, 13–17 November 2000

Murphy T Jr, Adelberger EG, Battat JBR, Hoyle CD, Johnson NH, McMillan RJ, Michelsen EL, Stubbs CW, Swanson HE (2010) Laser ranging to the lost Lunokhod 1 reflector, arXiv:1009.5720v2. http://arxiv.org/abs/1009.5720v2

Newhall XX, Standish EM, Williams JG (1983) DE102: A numerical integrated ephemeris of the Moon and planets spanning forty-four centuries. A&A 125:150–167

Niell AE (1996) Global mapping functions for the atmosphere delay at radio wavelength. J Geophys Res 100:3227–3246

Niell AE (2000) Improved atmospheric mapping functions for VLBI and GPS'. Earth Planets Space 52:699–702

Nilsson T, Böhm J, Schuh H (2010) Sub-diurnal Earth rotation variations observed by VLBI. Artif Satellites 45(2):49–55. doi:10.2478/v10018-010-0005-8

Nordtvedt K (1968) Testing relativity with laser ranging to the moon. Phys Rev 170(5):1186–1187

Nordtvedt K (1995) The relativistic orbit observables in lunar laser ranging. Icarus 114:51–62

Oppolzer T (1886) Traité de détermination des orbites. Gauthiers-Villars, Paris

Petit G, Luzum B (eds) (2010) IERS conventions, IERS Technical Note No. 36

Pettengill GH, Dyce RB, Campbell DB (1967) Radar measurements at 70 CM of Venus and Mercury. Astron J 72:330–337

Pitjeva EV (2001) Modern numerical ephemerides of planets and the importance of ranging observations for their creation. Celestial Mech Dyn Astron 80:249–271

Pitjeva EV (2005) High-precision ephemerides of Planets – EPM and determinations of some astronomical constants. Solar Syst Res 39(3):176–186

Pitjeva EV (2008) Recent models of planet motion and fundamental constants determined from position observations of planets and spacecraft. In: Capitaine N (ed) Proceedings of the Journées 2007 - Systèmes de Référence spatio-temporels, Paris, 2008, pp 65–69

Pitjeva EV (2009) Ephemerides EPM2008: The updated model, constants, data. In: Soffel M, Capitaine N (eds) Proceedings of the Journées 2008 – Systèmes de Référence spatio-temporels. Lohrmann-Observatorium and Observatoire de Paris, pp 57–60

Pitjeva EV (2010) EPM ephemerides and relativit. In: Klioner S, Seidelmann PK, Soffel M (eds) Proceedings of IAU Symp. No. 261/relativity in fundamental astronomy: Dynamics, reference frame, and data analysis. Cambridge University Press, Cambridge, pp 170–178

Pitjeva EV (2010) Influence of trans-neptunian objects on motion of major planets and limitation on the total TNO mass from planet and spacecraft. In: Lazzaro D, Prialnik D, Schulz R, Fernandez JA (eds) Proceedings of IAU Symp. No. 263/Icy bodies of the solar system. Cambridge University Press, Cambridge, pp 93–97

Pitjeva EV, Standish EM (2007) private communication

Pitjeva EV, Standish EM (2009) Proposals for the masses of the three largest asteroids, the Moon-Earth mass ratio and the Astronomical unit. Celestial Mech Dyn Astron 103(4):365–372

Pitjeva EV, Bratseva OA, Panfilov VE (2010) EPM – Ephemerides of Planets and the Moon of IAA RAS: Their model, accuracy, availability. In: Capitaine N (ed) Proceedings of the Journées 2010 – Systèmes de Référence spatio-temporels, Paris, 2010

Prawirodirdjo L, Bock Y (2004) Instantaneous global plate motion model from 12 years of continuous GPS observations. J Geophys Res 109:B08405, 15

Rapaport M, Teixeira R, Le Campion JF, Ducourant C, Camargo JI, Benevides-Soares P (2002) Astrometry of Pluto and Saturn with the CCD Meridian Instruments of Bordeaux and Valinhos. A&A 383:1054–1061

Ray R, Steinberg D, Chao B, Cartwright D (1994) Diurnal and semidiurnal variations in the Earth's rotation rate induced by oceanic tides. Science 264:830–832

Riehle F (2001) In: Lämmerzahl C, Everitt CW, Hehl F (eds) Gyros, clocks, interferometers...: Testing relativistic gravity in space. Springer, Berlin

Robertson DS, Carter WE (1984) Relativistic deflection of radio signals in the solar gravitational field measured with VLBI. Nature 310:572

Roosbeek F (1999) Diurnal and sub-diurnal terms in RDAN97 series. Celestial Mech Dyn Astron 74(4):243–252

Röser S, Demleitner M, Schilbach E (2010) The PPMXL catalog of positions and proper motions on the ICRS. Astron J 139:2440–2447

Rylkov VP, Vityazev VV, Dementieva AA (1995) Pluto: An analysis of photographic positions obtained with the Pulkovo normal astrograph in 1930–1992. Astron Astrophy Trans 6:251–281

Saastamoinen J (1972) Introduction to the practical computation of astronomical refraction. Bull Geod 106:383

Sagnac G (1913) L'éther lumineux démontré par l'effect du vent relatif d'éther dans un interféromètre en rotation uniforme. Comptes Rendus Acad Sci (Paris) 157:708–710

Schreiber KU, Stedman GE, Klügel T (2003) Earth tide and tilt detection by a ring laser gyroscope. J Geophys Res 108(B):2. doi:10.1029/2001JB000569

Schreiber KU, Klügel T, Velikoseltsev A, Schlüter W, Stedman GE, Wells J-PR (2009) The large ring laser G for continuous Earth rotation monitoring. Pure Appl Geophys (PAGEOPH) 166:148

Schreiber KU, Klügel T, Wells J-PR, Hurst RB, Gebauer A (2011) How to detect the Chandler and the annual Wobble of the Earth with a large ring laser Gyroscope. Phys Rev Lett 107(17):173904. doi:10.1103/PhysRevLett.107.173904

von Schuh H, Dill R, Greiner-Mai H, Kutterer H, Müller J, Nothnagel A, Richter B, Rothacher M, Schreiber U, Soffel M (2003) Erdrotation und globale dynamische Prozesse. Mitteilungen des Bundesamtes für Kartographie und Geodäsie (ISSN 1436–3445), Band 32, erarbeitet innerhalb des DFG-Forschungsvorhabens "Rotation der Erde". Bundesamtes für Kartographie und Geodäsie, Frankfurt am Main. ISBN 3-89888-883-5, 2003, IV + 118 pp

Schwiderski EW (1980) On charting global ocean tides. Rev Geophys Space Phys 18:243–268

Seeber G, Böder V, Goldan H-J, Schmitz M, Wübbena G (1996) Precise DGPS positioning in marine and airborne applications. In: Beutler u.a. GPS trends in precise terrestrial, airborne, and spaceborne applications, IAG Symposium No. 113. Springer, Berlin

Seidelmann PK (ed) (1992) Explanetory supplement to the astronomical almanac. University Science Books, Mill Valley

Sella GF, Dixon TH, Mao A (2002) REVEL: A model for recent plate velocities from space geodesy. J Geophys Res 107:2081, 30

Shapiro II, Reasenberg RD, Chandler JF, Babcock RW (1988) Measurement of the de Sitter precession of the Moon: A relativistic three-body effect. Phys Rev Lett 61:2643–2646

Sharai SG, Budnikova NA (1969) Theory of the motion of the Planet Pluto. Trans Inst Theor Astron 10:1–173; published as NASA Technical Translation F-491

Shelus P (2001) Lunar laser ranging: Glorious past and a bright future. Surv Geophy 22:517–535

Simon J-L, Bretagnon P, Chapront J, Chapront-Touzé M, Francou G, Laskar J (1994) Numerical expressions for precession formulae and mean elements for the Moon and Planets A&A 282:663–683

Skrutskie MF, Cutri RM, Stiening R, Weinberg MD, Schneider S, Carpenter JM, Beichman C, Capps R, Chester T, Elias J, Huchra J, Liebert J, Lonsdale C, Monet DG, Price S, Seitzer P, Jarrett T, Kirkpatrick JD, Gizis JE, Howard E, Evans T, Fowler J, Fullmer L, Hurt R, Light R, Kopan EL, Marsh KA, McCallon HL, Tam R, Van Dyk S, Wheelock S (2006) The two Micron All Sky Survey (2MASS). Astron J 131:1163

Sobel D, Andrewes JH (2003) The illustrated longitude, the true story of a Lone Genius who solved the greatest scientific problem of his time. Walker & Company, St. Louis

Soffel M (1989) Relativity in astrometry, celestial mechanics and geodesy. Springer, Berlin

Soffel M, Klioner S (2004) The BCRS ans the large scale structure of the universe. In: Finkelstein A, Capitaine N (eds) Proceedings of the Journées 2003 - Systèmes de Référence spatio-temporels, St.Petersburg, 2004, pp 297–301

Soffel M, Klioner S (2008) On astronomical constants. In: Capitaine N (ed) Proceedings of the Journées 2007 – Systèmes de Référence spatio-temporels, Paris Observatory, 2008, pp 58–60

Soffel M, Ruder H, Schneider M (1986) The dominant relativistic terms in the lunar theory. A&A 157:357–364

Soffel M, Klioner SA, Petit G, Wolf P, Kopeikin SM, Bretagnon P, Brumberg VA, Capitaine N, Damour T, Fukushima T, Guinot B, Huang T-Y, Lindegren L, Ma C, Nordtvedt K, Ries JC, Seidelmann PK, Vokrouhlický D, Will CM, Xu C (2003) The IAU 2000 resolutions for astrometry, celestial mechanics, and metrology in the relativistic framework: Explanatory supplement. Astron J 126:2687–2706

Souchay J, Loysel B, Kinoshita H, Folgueira M (1999) Corrections and new developments in rigid Earth nutation theory, III. Final tables REN-2000 including crossed-nutation and spin-orbit coupling effects. A&A Suppl Ser 135:111–131

Spagna A, Lattanzi MG, McLean B, Bucciarelli B, Carollo D, Drimmel R, Greene G, Morbidelli R, Pannunzio R, Sarasso M, Smart R, Volpicelli A (2006) The guide star catalog, II. Properties of the GSC 2.3 release. Memorie della Societa Astronomica Italiana 77:1166

Standish EM Jr (1982) Orientation of the JPL ephemerides, DE200/LE200, to the dynamical equinox of J2000. A&A 114:297–302

Standish EM Jr (1990a) The orservational basis for JPL's DE200, planetary ephemerides of the Astronomical Almana. A&A 233:252–271

Standish EM Jr (1990b) An approximation to the inter planet ephemeris errors in JPL's DE200. A&A 233:272–274

Standish EM Jr (1998) JPL Planetary and Lunar Ephemerides, DE405/LE405, Interoffice Memorandum, 312.F-98-048

Standish EM Jr (2003) JPL planetary ephemeris DE410, Interoffice Memorandum, 312.N-03-109
Standish EM Jr (2006) JPL Planetary, DE414, Interoffice Memorandum, 343R-06-002
Standish EM Jr, Newhall XX, Williams JG, Folkner WF (1995) JPL planetary and lunar ephemerides, DE403/LE403, JPL IOM, 314, 10
Stedman GE (1997) Ring-laser tests of fundamental physics and geophysics. Rep Prog Phys 60:615–688
Stone RC, Monet DG, Monet AK, Harris FH, Ables HD, Dahn CC, Canzian B, Guetter HH, Harris HC, Henden AA, Levine SE, Luginbuhl CB, Munn JA, Pier JR, Vrba FJ, Walker RL (2003) Upgrades to the flagstaff astrometric scanning transit telescope: A fully automated telescope for astrometry. Astron J 126:2060–2080
Tholen DJ, Buie MW, Grundy W, Elliott G (2008) Masses of Nix and Hydra. Astron J 135(3): 777–784
Titov O (2004) Construction of a Celestial Coordinate Reference Frame from VLBI Data. Astron Rep 48(11):941–948
Titov O, Tesmer V, Böhm J (2004) OCCAM v. 6.0 software for VLBI data analysis. In: Vandenberg NV, Baver KD (eds) International VLBI Service for Geodesy and Astrometry, International Service for Geodesy and Astrometry, 2004 General Metting Proceedings, Ottawa, Canada, NASA/CP-2204-212255, pp 267–271
Turyshev S, Williams J, Nordtvedt K, Shao M, Murphy T (2004) 35 years of testing relativistic gravity: Where do we go from here? In: Proceedings of 302. WE-Heraeus-Seminar: Astrophysics, clocks and fundamental constants, 16–18 June 2003. Springer Lecture Notes Phys., vol 648, pp 301–320
Urban SE, Corbin TE, Wycoff GL, Makarov VV, Høg E, Fabricius C (2001) The AC2000.2 Catalogue, Washington DC: U.S. Naval Observatory. Copenhagen University Observatory, Copenhagen
van Leeuwen F (2007) Hipparcos, the New Reduction of the Raw Data, Astrophysics and Space Science Library, vol 350, Springer, Berlin
Walter HG, Sovers OJ (2000) Astrometry of fundamental catalogues. Springer, Berlin
Webb H (1946) Project Diana: Army radar contacts the moon. Sky Telesc 54:3–6
Weinberg S (1972) Gravitation and cosmology: Principles and applications of the general theory of relativity. Wiley, New York
Wielen R, Schwan H, Dettbarn C, Lenhardt H, Jahreiß H, Jährling R (1999) Sixth catalogue of fundamental Stars (FK6), Part I. Basic fundamental stars with direct solutions. Veröff. Astron. Rechen-Institut Heidelberg, no. 35
Will C (1993) Theory and experiment in gravitational physics. Cambridge University Press, Cambridge
Williams J, Newhall XX, Dickey J (1996) Relativity parameters determined from lunar laser ranging. Phys Rev D 53:6730–6739
Williams J, Turyshev S, Boggs D (2004) Progress in Lunar Laser Ranging Tests of Relativistic Gravity. Phys Rev Lett 93:261101
Williams JG, Turyshev SG, Boggs DH (2009) LLR tests of the equivalence principle with the Earth and Moon, Int J Mod Phys D 18:1129–1175
Zacharias N, Urban SE, Zacharias MI, Hall DM, Wycoff GL, Rafferty TJ, Germain ME, Holdenried ER, Pohlman JW, Gauss FS, Monet DG, Winter L (2000) The First US Naval Observatory CCD Astrograph Catalog. Astron J 120:2131–2147
Zacharias N, Urban SE, Zacharias MI, Wycoff GL, Hall DM, Monet DG, Rafferty TJ (2004) The Second US Naval Observatory CCD Astrograph Catalog (UCAC2). Astron J 127: 3043–3059
Zacharias N, Finch C, Girard T, Hambly N, Wycoff G, Zacharias MI, Castillo D, Corbin T, DiVittorio M, Dutta S, Gaume R, Gauss S, Germain M, Hall D, Hartkopf W, Hsu D, Holdenried E, Makarov V, Martinez M, Mason B, Monet D, Rafferty T, Rhodes A, Siemers T, Smith D, Tilleman T, Urban S, Wieder G, Winter L, Young A (2010) The Third US Naval Observatory CCD Astrograph Catalog (UCAC3). Astron J 139:2184–2199

Zacharias N et al (2012) UCAC4, in preparation
Zappala V, de Sanctis G, Ferreri W (1980) Astrometric observations of Pluto from 1973 to 1979. A&A Suppl 41:29–31
Zappala V, de Sanctis G, Ferreri W (1983) Astrometric positions of Pluto from 1980 to 1982. A&A Suppl 51:385–387

List of Acronyms

AC	Astrographic Catalog
APFS	Apparent Places of Fundamental Stars; a German yearbook
APOLLO	Apache Point Observatory Lunar Laser-ranging Operation
AU	Astronomical Unit
BCRS	Barycentric Celestial Reference System
BDRS	Barycentric Dynamical Reference System
BNM	Bureau National de Métrologie (Paris)
CET	Central European Time
CEST	Central European Summer Time
CCD	Charge-Coupled Device
CIO	Celestial Intermediate Origin
CIP	Celestial Intermediate Pole
CMO	Calculated Minus Observed
cpsd	Cycles per sidereal day
DCF77	German longwave time signal radio station
CRS	Celestial Reference System
DE	Development Ephemeris; numerical solar system ephemerides of Jet Propulsion Laboratory (JPL)
DGPS	Differential GPS
DORIS	Doppler Orbitography by Radiopositioning Integrated on Satellite
EDM	Electromagnetic Distance Measurement
EIH	Einstein–Infeld–Hoffmann
EOP	Earth Orientation Parameter
EPM	Ephemerides of Planets and the Moon, the Russian numerical ephemeris
Eq.E.	Equation of Equinoxes
Eq.T.	Equation of Time
ERA	Earth Rotation Angle
FK	Fundamentalkatalog, fundamental catalog
Gaia	Planned astrometric space mission

GAST	Greenwich Apparent Sidereal Time
GALILEO	European satellite navigation system
GCRS	Geocentric Celestial Reference System
GLONASS	Globalnaja Nawigazionnaja Sputnikowaja Sistema; Russian satellite navigation system
GMT	Greenwich Mean Time
GMST	Greenwich Mean Sidereal Time
GNSS	Global Navigation Satellite System
GPS	Global Positioning System
GRT	General Relativity Theory
GSC	Guide Star Catalogue; stellar catalog for the attitude control of the Hubble Space Telescope
Hipparcos	High-Precision Parallax Collecting Satellite; astrometric satellite and space mission
IAU	International Astronomical Union
ICRS	International Celestial Reference System
IEN	Istituto Elettrotecnico Nazionale (Turin, Italy)
IERS	International Earth Rotation and Reference Systems Service
IERS C03	IERS Conventions 2003 (McCarthy and Petit 2003)
IERS C10	IERS Conventions 2010 (Petit and Luzum 2010)
ILS	International Latitude Service
IMCCE	Institut Mécanique Céleste et de Calcul des Éphémérides
INPOP	Intégrateur Numérique Planétaire de l'Observatoire de Paris, the French numerical ephemeris
IRIS	International Radio Interferometric Surveying
IRP	Instantaneous Rotation Pole
ITRS	International Terrestrial Reference System
IVS	International VLBI Service for Geodesy and Astrometry
J2000.0	The present reference epoch (1 January 2000, 12:00)
JD	Julian Date
JPL	Jet Propulsion Laboratory (Pasadena, USA)
LAGEOS	Laser Geodynamics Satellite
LAST	Local Apparent Sidereal Time
LLR	Lunar Laser Ranging
LMST	Local Mean Sidereal Time
LORAN-C	Long Range Navigation; former navigation system
MERIT	Monitor Earth Rotation and Intercompare the Techniques of observation and analysis
MEX	Mars Express
MGS	Mars Global Surveyor
MJD	Modified Julian Date
MTT	Mapping Temperature Test (Herring 1992)
MRO	Mars Reconnaissance Orbiter
MT$_S$	Mean Solar Time

NICT	National Institute of Information and Communication Technology (Tokyo, Japan)
NIST	National Institute of Standards and Technology (Boulder, Colorado, USA)
NOVAS	Astronomical software developed at USNO
NMF	Niell Mapping Functions (Niell 1996)
NNR	No-Net-Rotation condition
NPL	National Physical Laboratory (Middlesex)
OCA	Observatoire de la Côte d'Azur
PFCN	Prograde Free Core Nutation
PFS	Primary Frequency Standard
POLARIS	Polar Motion Analysis by Radio Interferometric Sampling
PPM	Positions and Proper Motions; a stellar catalog
PPN	Parametrized Post-Newtonian
PTB	Physikalisch Technische Bundesanstalt (Brunswick, Germany)
RFCN	Retrograde Free Core Nutation
SI	Système International d'unités; international system of units
SLR	Satellite Laser Ranging
SOFA	Software for Astronomy (IAU)
TAI	Temps Atomique International; International Atomic Time
TCB	Temps-Coordonnée Barycentrique; Barycentric Coordinate Time
TCG	Temps-Coordonnée Geocentrique; Geocentric Coordinate Time
TDB	Temps Dynamique Barycentrique; Barycentric Dynamical Time
T_{eph}	Barycentric Ephemeris Time
TT	Temps Terrestre; Terrestrial Time
TT_S	True Solar Time
TIO	Terrestrial Intermediate Origin
TRS	Terrestrial Reference System
UCAC	USNO CCD Astrograph Catalog
USNO	US Naval Observatory (Washington)
USNO-A	A stellar catalog from USNO
UT	Universal Time
UT1	UT corrected for polar motion effects
UTC	Universal Time Coordinated
VLBI	Very Long Baseline Interferometry
VMF	Vienna Mapping Function
VSOP	Variations Séculaire des Orbites Planétaires; French semianalytical solar system ephemerides

Appendix A: Solutions to Exercises

Exercise 2.1 on page 30

An observer on the platform will see the tennis ball following a zigzag path because the velocity vectors of the ball and the train will add up. The overall velocity v_G of the ball that this observer will measure therefore is given by

$$v_G^2 = v^2 + v_T^2 .$$

Let us consider only the situation when the tennis ball moves from the bottom to the top of the compartment. As seen from the train station, the ball moves a distance $v\Delta t/2$ along the railway tracks, a distance L upwards, and a total distance of $v_G \Delta t/2$. From the Pythagorean theorem we therefore get

$$L^2 + \left(\frac{v\Delta t}{2}\right)^2 = \left(\frac{v_G \Delta t}{2}\right)^2$$

or

$$\Delta t = \frac{2L}{\sqrt{v_G^2 - v^2}} = \frac{2L}{v_T} = \Delta \tau .$$

Exercise 2.2 on page 31

Let d denote the difference between the proper time interval $\Delta \tau$ and the corresponding coordinate time interval Δt_c,

$$d = \Delta \tau - \Delta t_c .$$

Since $\Delta \tau = \Delta t_c \sqrt{1 - v^2/c^2}$ we get

$$d = \Delta t_c \left(\sqrt{1 - v^2/c^2} - 1\right) .$$

Using the given values d becomes -5 ns.

Exercise 2.3 on page 32
From
$$\mathbf{e}_x \cdot \mathbf{e}_{y'} = \cos\alpha; \qquad \mathbf{e}_y \cdot \mathbf{e}_{y'} = \sin\alpha$$
we get for some arbitrary vector **v**:
$$\mathbf{v} = x\mathbf{e}_x + y\mathbf{e}_y = x'\mathbf{e}_{x'} + y'\mathbf{e}_{y'} = (x' + y'\cos\alpha)\mathbf{e}_x + y'\sin\alpha\,\mathbf{e}_y$$
or
$$x = x' + y'\cos\alpha; \qquad y = y'\sin\alpha.$$
From this we find
$$ds^2 = dx^2 + dy^2 = dx'^2 + dy'^2 + 2\cos\alpha\, dx'dy',$$
or $g_{11} = g_{22} = 1$, $g_{12} = g_{21} = \cos\alpha$. Note that the components of the metric tensor are simply given by the usual scalar product of the basis vectors, i.e., $g_{ij} = \mathbf{e}_i \cdot \mathbf{e}_j$.

Exercise 2.4 on page 33
We start with the differential relations
$$c\,dT = \gamma(c\,dt - (v/c)\,dx); \qquad dX = \gamma(dx - v\,dt).$$
From this we get
$$dS^2 = -\gamma^2(c^2 dt^2 - 2v\,dxdt - (v/c)^2 dx^2) + \gamma^2(dx^2 - 2v\,dxdt + v^2 dt^2)$$
$$= -\gamma^2 c^2 dt^2(1 - (v/c)^2) + \gamma^2 dx^2(1 - (v/x)^2) = -c^2 dt^2 + dx^2 = ds^2.$$

Exercise 2.5 on page 36
The clock on Mount Everest runs faster, and after 1 year, its display will show 10 μs more than the clock in Brunswick.

Exercise 2.6 on page 43

Date	JD	MJD
30.07.1853 12^h	2 398 065.0	−1 935.5
21.04.1937 12^h	2 428 645.0	28 644.5
01.02.1977 12^h	2 443 176.0	43 175.5
01.01.2000 12^h	2 451 545.0	51 544.5
20.05.2006 12^h	2 453 876.0	53 875.5
01.06.2011 12^h	2 455 714.0	55 713.5

Appendix A: Solutions to Exercises

Exercise 2.7 on page 43

At first, calculate the JD value for today, 0^h, and divide it by 7. The remainder depends upon the day of the week. As can be concluded from the remainder of today's date is 0 for Monday, is 1 for Tuesday, etc.

For the date in question, the remainder is 4, and we find that Edmund Halley observed the eclipse on a Friday.

Exercise 2.9 on page 43

The solar eclipse happened at JD = 1 507 900. If your result differs by 366 days, keep in mind that there never was the year 0 BC.

Exercise 2.8 on page 43

JD	Date
2 451 545.00	01.01.2000 12^h
2 450 313.25	17.08.1996 18^h
2 450 026.75	05.11.1995 06^h

Exercise 3.1 on page 59

First it is useful to write

$$g_{00} = -\exp(-2w/c^2) + \mathcal{O}_5 ; \qquad g^{00} = -\exp(+2w/c^2) + \mathcal{O}_5$$

since $\exp(\pm 2w/c^2) = 1 \pm 2w/c^2 + 2w^2/c^4 + \mathcal{O}_6$.

To check the orthonormality condition

$$g^{\alpha\sigma} g_{\sigma\beta} = \delta_{\alpha\beta}$$

we have to distinguish three cases: (1) $\alpha = \beta = 0$, (2) $\alpha = 0, \beta = i$ (or $\alpha = i, \beta = 0$), and (3) $\alpha = i, \beta = j$. We get

$$g^{0\sigma} g_{\sigma 0} = g^{00} g_{00} + g^{0i} g_{0i} = g^{00} g_{00} + \mathcal{O}_5 = 1 + \mathcal{O}_5$$

which can be seen immediately from the exponential relations above. Now,

$$g^{i\sigma} g_{0\sigma} = g^{i0} g_{00} + g^{ij} g_{0j} = -g^{i0} + g_{0i} + \mathcal{O}_5 = \mathcal{O}_5$$

and

$$g^{i\sigma} g_{j\sigma} = g^{i0} g_{j0} + g^{ik} g_{jk} = \delta_{ik}\delta_{jk} + \mathcal{O}_4 = \delta_{ij} + \mathcal{O}_4$$

as was to be shown.

According to the definition (3.13), e.g., for the Christoffel symbol Γ^0_{00}, we get

$$\Gamma^0_{00} = \frac{1}{2}g^{00}(g_{00,0} + g_{00,0} - g_{00,0})$$
$$+ \frac{1}{2}g^{0i}(g_{0i,0} + g_{0i,0} - g_{00,i})$$
$$= -\frac{1}{2}g_{00,0} + \mathcal{O}_5$$
$$= -\frac{w_{,0}}{c^2} + \mathcal{O}_5.$$

Remember that $f_{,0} = \partial f/\partial x^0 = (1/c)\partial f/\partial t$. Let us compute one more Christoffel symbol explicitly:

$$\Gamma^0_{0i} = \frac{1}{2}g^{0\sigma}(g_{\sigma 0,i} + g_{\sigma i,0} - g_{0i,\sigma})$$
$$= \frac{1}{2}g^{00}(g_{00,i} + g_{0i,0} - g_{0i,0})$$
$$+ \frac{1}{2}g^{0j}(g_{j0,i} + g_{ji,0} - g_{0i,j})$$
$$= \frac{1}{2}g^{00}g_{00,i} + \mathcal{O}_6$$
$$= \frac{1}{2}\exp(+2w/c^2)(\exp(-2w/c^2))_{,i} + \mathcal{O}_6$$
$$= -\frac{w_{,i}}{c^2} + \mathcal{O}_6.$$

The other Christoffel symbols can be found in a similar way.

Exercise 3.2 on page 60

The EIH equations can be derived from the geodesic equation (3.16) in the external metric with gravitational potentials

$$\overline{w} = \sum_{B \neq A} w_B; \qquad \overline{w}_i = \sum_{B \neq A} w_i^B.$$

Inserting the Christoffel symbols that have to be evaluated at the position of body A, i.e., $\mathbf{z}_A$, on the right hand side of (3.16), explicitly we get

$$\frac{d^2 z_A^i}{dt^2} = \overline{w}_{,i} - 4\frac{\overline{w}\,\overline{w}_{,i}}{c^2} + \frac{4}{c^2}\overline{w}_{i,t}$$
$$+ \frac{8}{c^2}\overline{w}_{[i,j]}v_A^j - \frac{2}{c^2}v_A^i \overline{w}_{,t}$$
$$- \frac{v_A^i v_A^j}{c^2}(\delta_{ij}\overline{w}_{,k} + \delta_{ik})\overline{w}_{,j} - \delta_{jk}\overline{w}_{,i}$$

Appendix A: Solutions to Exercises

$$-\frac{1}{c^2} v_A^i \overline{w}_{,t}$$

$$-\frac{2}{c^2} v_A^i v_A^j \overline{w}_{,j}$$

or

$$\frac{d^2 z_A^i}{dt^2} = \left(1 - \frac{4}{c^2}\overline{w} + \frac{1}{c^2}\mathbf{v}_A^2\right)\partial_i\overline{w} + \frac{4}{c^2}(\partial_i\overline{w}_j - \partial_j\overline{w}_i)v_A^j - \frac{1}{c^2}(3\partial_t\overline{w} + 4v_A^j\partial_j\overline{w})v_A^i.$$

Inserting (3.7) for w_B and (3.8) w_i^B in the expressions for $\overline{w}$ and $\overline{w}_i$, we finally end up with the Einstein–Infeld–Hoffmann equations of motion.

Exercise 4.2 on page 78

Geocentric position and velocity of the Sun for 1 January 2000, 0^h TT in rectangular equatorial coordinates. Values are in AU for (X, Y, Z) and AU/d for $(\dot{X}, \dot{Y}, \dot{Z})$.

Ephemeris	X $\dot{X}$	Y $\dot{Y}$	Z $\dot{Z}$
INPOP10	+0.16852462	−0.88884290	−0.38535607
	+0.01723395	+0.00275949	+0.00119635
DE405	+0.16852462	−0.88884291	−0.38535606
	+0.01723395	+0.00275949	+0.00119635

Exercise 4.2 on page 78

Position (AU) and velocity (AU/d) of Mercury, Venus, Mars, Jupiter, and Saturn for 1 January 2000, 0^h TT in rectangular equatorial coordinates computed with DE405.

Planet	x $\dot{x}$	y $\dot{y}$	z $\dot{z}$
Mercury	−0.14786722	−0.40062848	−0.19891423
	+0.02117425	−0.00551550	−0.00514110
Venus	−0.72576936	−0.03966770	+0.02790150
	+0.00051891	−0.01851510	−0.00836218
Mars	+1.38322192	−0.00814943	−0.04104003
	+0.00075330	+0.01380716	+0.00631275
Jupiter	+3.99632068	+2.73099373	+1.07327447
	−0.00455810	+0.00587801	+0.00263057
Saturn	+6.40141806	+6.17024926	+2.27303005
	−0.00428574	+0.00352277	+0.00163934

Exercise 4.3 on page 78

In a first step the position of Uranus $\mathbf{x}_U$ and the initial position of Earth $\mathbf{x}_E^{(0)}$ are calculated. Using DE405, we find

$$\mathbf{x}_U = \begin{bmatrix} 20.06830431 \\ 0.65363583 \\ 0.00243779 \end{bmatrix} \text{AU} \qquad \mathbf{x}_E^{(0)} = \begin{bmatrix} 0.50799975 \\ -0.80566552 \\ -0.34926479 \end{bmatrix} \text{AU}$$

leading to an initial distance of $d^{(0)} = 19.61781765$ AU. Since the speed of light is $c = 173.14463268$ AU/day, we get an initial light travel time $\tau^{(0)} = 0.11330$ days. But throughout this interval, the Earth moved. To get a better estimate for the light travel time, we add $\tau^{(0)}$ to the initial Julian date and determine a new position

$$\mathbf{x}_E^{(1)} = \begin{bmatrix} 0.50965169 \\ -0.80477095 \\ -0.34887693 \end{bmatrix} \text{AU}$$

for the Earth and a new distance $d^{(1)} = 19.61609707$ AU. We compare the new light travel time $\tau^{(1)} = 0.11329$ days with our initial result and get a difference of 0.9 s. A more accurate solution would require more of such iterative steps. In our case we are already done. Finally, we add $\tau^{(1)}$ to our initial Julian date and convert it to UTC. The light ray is observed at Earth 23 July 2011, $17^h 14^m 46^s$ UTC.

Exercise 4.4 on page 78

Using, e.g., GetDE405Ephemeris from the *AstroRef* package, we find coordinates for Sun, Venus, and Earth. Then we derive the vectors Venus–Sun and Venus–Earth. Finally we calculate the included angle and get $\alpha = 25°$ which means that more than 95% are illuminated.

Exercise 5.1 on page 110

$$\mathcal{R}_1(2.7) = \begin{bmatrix} 1.00 & 0.00 & 0.00 \\ 0.00 & -0.90 & 0.43 \\ 0.00 & -0.43 & -0.90 \end{bmatrix}$$

$$\mathcal{R}_2(2.7) = \begin{bmatrix} -0.90 & 0.00 & -0.43 \\ 0.00 & 1.00 & 0.00 \\ 0.43 & 0.00 & -0.90 \end{bmatrix}$$

$$\mathcal{R}_3(2.7) = \begin{bmatrix} -0.90 & 0.43 & 0.00 \\ -0.43 & -0.90 & 0.00 \\ 0.00 & 0.00 & 1.00 \end{bmatrix}$$

Appendix A: Solutions to Exercises

Exercise 5.2 on page 110

We get

$$\mathcal{R}_1(2.7) \cdot \mathcal{R}_2(3.1) = \begin{bmatrix} -1.00 & 0.00 & -0.04 \\ 0.02 & -0.90 & -0.43 \\ -0.04 & -0.43 & 0.90 \end{bmatrix}$$

and

$$\mathcal{R}_2(3.1) \cdot \mathcal{R}_1(2.7) = \begin{bmatrix} -1.00 & 0.02 & 0.04 \\ 0.00 & -0.90 & 0.43 \\ 0.04 & 0.43 & 0.90 \end{bmatrix}.$$

Comparing the results, we find that the order of rotational operations is important.

Exercise 5.3 on page 110

$$\mathcal{R}_1(-2.7) = \begin{bmatrix} 1.00 & 0.00 & 0.00 \\ 0.00 & -0.90 & -0.43 \\ 0.00 & 0.43 & -0.90 \end{bmatrix}$$

Rotating backwards inverts the rotational matrix. Since rotational matrices are of orthogonal structure, it is possible to perform the inversion by simply transposing the matrix.

Exercise 5.4 on page 111

Using the *AstroRef* function the transformation can easily be confirmed.

Exercise 5.5 on page 111

The rotation from the horizon system to the equatorial system of the first kind is done by

$$\mathcal{R}^{(h,\delta)}_{(a,A)} = \mathcal{R}_2(\Phi - 90°)\, \mathcal{R}_3(180°)$$

The mirroring operation for the y-axis is carried out by multiplication with

$$\mathcal{M} = \begin{pmatrix} 1 & 0 & 0 \\ 0 & -1 & 0 \\ 0 & 0 & 1 \end{pmatrix}.$$

The rotation between the two equatorial system then reads

$$\mathcal{R}^{(\alpha,\delta)}_{(h,\delta)} = \mathcal{R}_3(-\text{LAST})\, \mathcal{M}.$$

The last rotation from the equatorial system of the second kind to the ecliptical system is given by

$$\mathcal{R}^{(\lambda,\beta)}_{(\alpha,\delta)} = \mathcal{R}_3(\epsilon).$$

Altogether we get

$$\mathcal{R}^{(\lambda,\beta)}_{(a,A)} = \mathcal{R}^{(\lambda,\beta)}_{(\alpha,\delta)} \mathcal{R}^{(\alpha,\delta)}_{(h,\delta)} \mathcal{R}^{(h,\delta)}_{(a,A)}$$
$$= \mathcal{R}_3(\epsilon) \mathcal{R}_3(-\text{LAST}) \mathcal{M} \mathcal{R}_2(\Phi - 90°) \mathcal{R}_3(180°).$$

After expressing spherical coordinates as Cartesian vectors, we may make use of the transformation and get

$$\cos\beta \cos\lambda = -\cos a \cos A \sin\Phi \cos\text{LAST}$$
$$- \cos a \sin A \sin\text{LAST}$$
$$+ \sin a \cos\Phi \cos\text{LAST}$$
$$\cos\beta \sin\lambda = -\cos a \cos A \sin\Phi \sin\text{LAST} \cos\epsilon$$
$$+ \cos a \cos A \cos\Phi \sin\epsilon$$
$$+ \cos a \sin A \cos\text{LAST} \cos\epsilon$$
$$+ \sin a \cos\Phi \sin\text{LAST} \cos\epsilon$$
$$+ \sin a \sin\Phi \sin\epsilon$$
$$\sin\beta = -\sin a \cos\Phi \sin\text{LAST} \sin\epsilon$$
$$+ \cos a \cos A \sin\Phi \sin\text{LAST} \sin\epsilon$$
$$+ \cos a \cos A \cos\Phi \cos\epsilon$$
$$- \cos a \sin A \cos\text{LAST} \sin\epsilon$$
$$+ \sin a \sin\Phi \cos\epsilon$$

And finally the Cartesian coordinates are converted back to spherical ones.

Exercise 5.6 on page 113

Scale	Value
GMST	$18^h 40^m 46\!\!\!.^s 55$
GAST	$18^h 40^m 45\!\!\!.^s 69$
LAST	$19^h 35^m 40\!\!\!.^s 99$

Exercise 5.7 on page 113

First convert CEST to UTC by adding two hours, and then convert it to TT. The difference GAST − GMST = −0.304 s which is the equation of equinoxes.

Exercise 5.8 on page 113

From the linear term in (5.23) we see that a sidereal day is shorter than a solar day and that indeed there may be one solar day for which GMST becomes zero two times. Sidereal time becomes zero when the vernal equinox passes the meridian. For the day in question this should happen once shortly after midnight and again shortly before next midnight. That means the vernal equinox should pass the meridian around midnight. This happens when the Sun is close to the autumnal equinox, i.e., when autumn begins.

For the determination of that day in 2010 we start with GMST for 1 January 2010, 0^h, using (5.23) or *AstroRef*, respectively. We get $GMST_0 = 100.2616°$. From the linear term in (5.23), we see that next midnight GMST would be $\Delta = 0.9856°$ larger. To answer the question we have to find how many days it will take until GMST reaches 360°,

$$n = \text{floor}\left(\frac{360° - \text{GMST}_0}{\Delta}\right) = 263.$$

The event happened on the 263rd day of 2010. With the function `CalendarDate` from the *AstroRef* package, it is easy to find 21 September 2010 to be this day. We may control the result by calculating GMST for 0^h this as well as for the next day. If we are right the first value should lie shortly before 360° and the next value shortly after 0°.

Exercise 6.1 on page 123

The refraction corrections ζ_S (Saastamoinen) and ζ_I (numerical integration) in arcsec for given zenith distances z in degrees are

z	ζ_S	ζ_I
10	10.5	10.5
20	21.6	21.6
30	34.3	34.3
40	49.9	49.9
50	70.8	70.8
60	102.6	102.7
70	162.0	162.1

Exercise 6.2 on page 123

Holding all other parameters at a time, plots may be made showing the change of refraction when the free parameter changes. To see large effects we should set the zenith distance to a rather large value, like 65°. Now let the meteorological parameters take all meaningful values, i.e., pressure $p = 980\ldots1,050$ hPa, temperature $T = -30\ldots+60°C$, and humidity $H_{rel} = 0\ldots100\%$. We find that under large zenith distances, temperature acts on refraction by about $0.5''/°C$, pressure by about $0.1''/$hPa, and humidity by about $0.001''/\%$. This means that astronomical refraction is not very sensitive to humidity, quite sensitive to pressure, and most sensitive to temperature.

Exercise 6.3 on page 123

Since close to the horizon Saastamoinens model is not valid, the numerical integration model (`RefrNumInt`) has to be used. For a zenith distance of 90°, we get a refraction correction of approximately $30'$, i.e., the same as the angular diameter of the Sun. Since refraction seems to "lift" objects, if there was no air, the sunset would just be over. By the way, the same is valid for the Moon since its angular diameter is also around $30'$.

Exercise 6.4 on page 128

The annual and the geocentric parallax of Sirius are

$$\Pi = 0''\!.38 \quad \text{and}$$

$$\Pi_G = 16\,\mu\text{as}.$$

The stellar constants are

$$c = -0.204362356$$
$$d = 1.023928371$$
$$c' = 0.697299282 \quad \text{and}$$
$$d' = 0.056296901.$$

For the determination of the annual parallax corrections for right ascension and declination, the X- and Y-components of the unit vector from the Earth to the solar systems barycenter are needed. Using Maple^{TM} with the *AstroRef* package function `GetDE405Ephemeris`, these quantities read

$$X = 0.187682870 \quad \text{and}$$
$$Y = -0.901096240.$$

The corrections are

$$(\Delta\alpha)_P = -0\overset{''}{.}00 \quad \text{and}$$
$$(\Delta\delta)_P = -0\overset{''}{.}24.$$

Exercise 6.5 on page 129

Referring to the geometry of the problem (see Fig. 6.8, right), we define a 2D coordinate system with the x-axis pointing into the direction of the car (apex) and the y-axis pointing up. Within this coordinate system, the velocity vectors of the car and the rain, respectively, are

$$\mathbf{v}_{car} = v_{car}(1,0)^T \quad \text{and}$$
$$\mathbf{v}_{rain}(\theta) = v_{rain}(-\cos\theta, -\sin\theta)^T.$$

For the driver the vector $\mathbf{v}_{app}$ pointing into the direction from where the raindrops seem to come from is then given by

$$\mathbf{v}_{app}(\theta) = \mathbf{v}_{car} - \mathbf{v}_{rain}(\theta)$$

and the angle between the cars direction and the apparent one becomes

$$\theta' = \arccos\left(\frac{\mathbf{v}_{app}(\theta)\mathbf{v}_{car}}{|\mathbf{v}_{app}(\theta)||\mathbf{v}_{car}|}\right),$$

or if we replace the vectors by their components, we get

$$\theta' = \arccos\left(\frac{v_{car} + v_{rain}\cos\theta}{\sqrt{v_{car}^2 + v_{rain}^2 + 2v_{car}v_{rain}\cos\theta}}\right).$$

The dependency between θ' and θ as well as between $\Delta\theta$ and θ can be seen from the following graphs (panels a and b are for $v_{rain} = 5\,\text{m/s}$, and panels c and d are for $v_{rain} = 500\,\text{m/s}$):

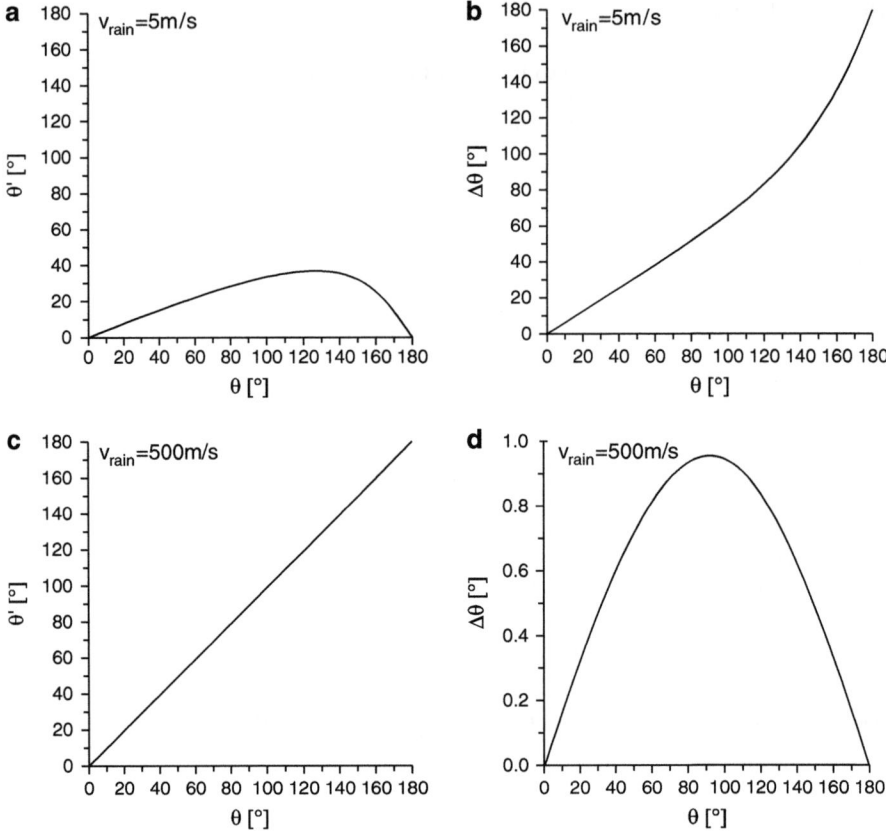

In panel *d* we may find the constant of aberration for the fast raindrops, which is the amplitude of the curve.

Exercise 6.6 on page 130

We have

$$P^\mu_\nu \bar{e}^\nu_{(0)} = \left(\delta^\mu_\nu + \eta_{\nu\alpha} \bar{e}^\mu_{(0)} \bar{e}^\alpha_{(0)}\right) \bar{e}^\nu_{(0)} = \bar{e}^\mu_{(0)}(1 + \eta_{\nu\alpha} \bar{e}^\nu_{(0)} \bar{e}^\alpha_{(0)}) = 0.$$

Furthermore,

$$\bar{k}^\mu = P^\mu_\nu k^\nu = (\delta^\mu_\nu + \eta_{\nu\alpha} \bar{e}^\mu_{(0)} \bar{e}^\alpha_{(0)}) k^\nu = k^\mu - \gamma\sigma \bar{e}^\mu_{(0)}$$

since

$$\eta_{\nu\alpha} k^\nu \bar{e}^\alpha_{(0)} = -k^0 \bar{e}^0_{(0)} + k^i \bar{e}^i_{(0)} = -\gamma - \gamma\beta^i m^i = -\gamma\sigma.$$

Appendix A: Solutions to Exercises

Then,

$$\eta_{\mu\nu}\bar{k}^\mu\bar{k}^\nu = \eta_{\mu\nu}k^\mu k^\nu - \gamma\sigma\eta_{\mu\nu}\bar{e}^\mu_{(0)}k^\nu - \gamma\sigma\eta_{\mu\nu}\bar{e}^\nu_{(0)}k^\mu + \gamma^2\sigma^2\eta_{\mu\nu}\bar{e}^\mu_{(0)}\bar{e}^\nu_{(0)}\,.$$

Here, on the right hand side, the first term vanishes, the following two terms give $+\gamma^2\sigma^2$ each, and the last term equals $-\gamma^2\sigma^2$, so that the result reads $\gamma^2\sigma^2$. Finally, the aberration formula from special relativity, (6.37), results from

$$m'_i = -\eta_{\mu\nu}\bar{e}^\mu_{(i)}\frac{\bar{k}^\nu}{|\bar{k}^\nu|}$$

since

$$|\bar{k}^\nu| = \gamma(1+\boldsymbol{\beta}\cdot\mathbf{m})$$

and

$$-\eta_{\mu\nu}\bar{e}^\mu_{(i)}\bar{k}^\nu = \bar{e}^0_{(i)}\bar{k}^0 - \bar{e}^j_{(i)}\bar{k}^j = m^i + \gamma\beta^i + (\gamma-1)(\boldsymbol{\beta}\cdot\mathbf{m})\beta^i/\beta^2\,.$$

Exercise 6.7 on page 132

Using the *AstroRef* function `GetDE405Ephemeris`, we find for the relevant barycentric velocity components of the Earth:

$$\dot{x} = -0.013075871\,\text{AU/day} \quad \text{and}$$
$$\dot{y} = -0.010605160\,\text{AU/day}\,.$$

Divided by c, we get the Besselian day numbers $C = -12''\!.63$ and $D = +15''\!.58$.

Exercise 6.8 on page 138

First, define a vector $\mathbf{x}(300\,\text{BC})$ pointing to the vernal equinox at 300 BC, i.e., its coordinates are $\mathbf{x}(300\,\text{BC}) = (1,0,0)^T$. The *AstroRef* function `Precession` returns the precession matrix from the reference epoch J2000.0 to the epoch given by the argument. The resulting precession matrix then reads

$$\mathcal{P}(300\,\text{BC} \to \text{now}) = \mathcal{P}(\text{J2000.0} \to 300\,\text{BC})^T \mathcal{P}(\text{J2000.0} \to \text{now})\,.$$

The transposition for the first matrix has to be used because we would like to rotate into the opposite direction. Using the rotation on our vector leads to the new vector $\mathbf{x}(\text{now})$ at our time. If we take 2020 for now, we get

$$\mathbf{x}(2020) = \mathcal{P}(300\,\text{BC} \to 2020)\,\mathbf{x}(300\,\text{BC})$$
$$= (0.846, 0.488, 0.215)^T$$

To find out how far the point moved along the ecliptic, we have to convert these equatorial coordinates into ecliptical ones by rotating about the obliquity. Then we simply determine the ecliptic longitude, and we find out that since 300 BC the vernal equinox moved about 32°, which is a little more than one constellation.

Exercise 6.9 on page 143

Nutation angles in arcsec for the years 2000 until 2010. Comparison between nutation theories IAU 1980 and IAU 2000:

	IAU 1980		IAU 2000	
Year	$\Delta\epsilon$	$\Delta\psi$	$\Delta\epsilon$	$\Delta\psi$
2000	−5.773808	−13.923385	−5.769398	−13.931996
2001	−2.865308	−16.148713	−2.860283	−16.156738
2002	+0.093224	−16.427536	+0.099281	−16.433752
2003	+3.018217	−15.251265	+3.025117	−15.255709
2004	+5.727859	−12.176918	+5.735401	−12.179354
2005	+7.617029	−7.430528	+7.624256	−7.427924
2006	+8.404020	−1.901142	+8.410699	−1.898316
2007	+8.326185	+3.416456	+8.332733	+3.420336
2008	+7.480013	+8.682903	+7.485901	+8.688444
2009	+5.599195	+13.371435	+5.604509	+13.378356
2010	+2.839919	+16.531999	+2.844154	+16.539379

Exercise 6.10 on page 143

For the aimed accuracy we only have to consider the first 5 lines from Table 6.1. If all routines were done right, you will get a plot like this:

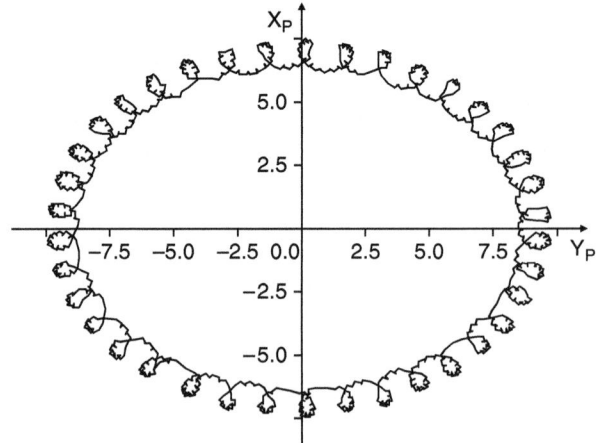

Appendix A: Solutions to Exercises

Exercise 6.11 on page 144
Apparent places of the Sun for 1 January, 12^h TT, for various years:

	AppPlaceSunFast		IMCCE	
Year	α_{app}	δ_{app}	α_{app}	δ_{app}
1600	$18^h45^m40\overset{s}{.}30$	$-23°04'35\overset{''}{.}7$	$18^h45^m40\overset{s}{.}29$	$-23°04'35\overset{''}{.}7$
1800	$18^h47^m35\overset{s}{.}54$	$-23°00'58\overset{''}{.}0$	$18^h47^m35\overset{s}{.}55$	$-23°00'57\overset{''}{.}9$
1900	$18^h46^m23\overset{s}{.}61$	$-23°01'22\overset{''}{.}9$	$18^h46^m23\overset{s}{.}61$	$-23°01'23\overset{''}{.}0$
2000	$18^h45^m06\overset{s}{.}60$	$-23°01'57\overset{''}{.}1$	$18^h45^m06\overset{s}{.}62$	$-23°01'57\overset{''}{.}0$

Exercise 6.12 on page 144
Apparent places for Moon and planets for 1 March 1980, 0^h UTC:

Body	α_{app}	δ_{app}
Moon	$10^h12^m35\overset{s}{.}66$	$+11°14'59\overset{''}{.}0$
Mercury	$23^h18^m51\overset{s}{.}03$	$-00°40'06\overset{''}{.}3$
Venus	$01^h25^m37\overset{s}{.}54$	$+09°45'15\overset{''}{.}7$
Mars	$10^h29^m37\overset{s}{.}55$	$+14°05'26\overset{''}{.}2$
Jupiter	$10^h27^m47\overset{s}{.}95$	$+11°02'55\overset{''}{.}9$
Saturn	$11^h44^m13\overset{s}{.}57$	$+04°18'46\overset{''}{.}3$
Uranus	$15^h33^m13\overset{s}{.}56$	$-18°52'34\overset{''}{.}7$
Neptune	$17^h27^m45\overset{s}{.}96$	$-21°51'52\overset{''}{.}1$

Exercise 6.13 on page 151
Using the Taylor expansion $(1 + x)^{-1/2} = 1 - (1/2)x + (3/8)x^2 + \cdots$, we get to second order in Π

$$\eta = 1 + \Pi \cdot \mathbf{l} - \frac{1}{2}\Pi^2 + \frac{3}{2}(\Pi \cdot \mathbf{l})^2,$$

so that

$$\mathbf{k} = \eta(-\mathbf{l} + \Pi)$$
$$= -\mathbf{l} + \Pi - \mathbf{l}(\Pi \cdot \mathbf{l}) + \Pi(\Pi \cdot \mathbf{l}) + \frac{1}{2}\mathbf{l}\Pi^2 - \frac{3}{2}\mathbf{l}(\Pi \cdot \mathbf{l})^2.$$

Using

$$\boldsymbol{\pi} = \mathbf{l} \times (\Pi \times \mathbf{l}) = \Pi - \mathbf{l}(\Pi) \cdot \mathbf{l}$$

and

$$|\boldsymbol{\pi}|^2 = \Pi^2 - (\Pi \cdot \mathbf{l})^2$$

we see that **k** can indeed be written in the form

$$k = -l\left(1 - \frac{1}{2}|\pi|^2\right) + \pi(1 + l \cdot \Pi).$$

Exercise 9.2 on page 217

The MHB transfer function for $\sigma = -1.010\ldots -0.990$. The RFCN resonance can clearly be seen.

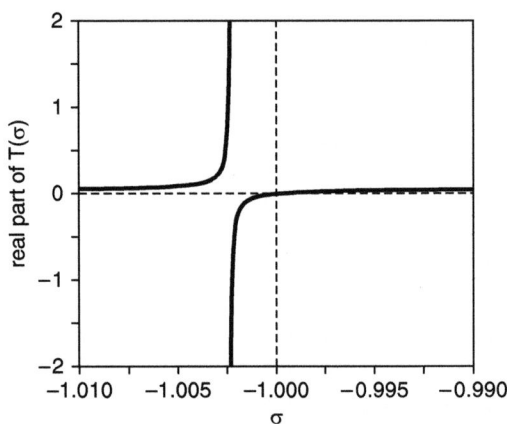

Exercise 9.3 on page 233

Using the classical way, the vector $\mathbf{X}_I$ transforms into

$$\mathbf{X}_T^{(c)} = \begin{pmatrix} -2.025491618199 \\ 5.461798337244 \\ 6.005509356240 \end{pmatrix}.$$

With the modern transformation, $\mathbf{X}_I$ transforms into

$$\mathbf{X}_T^{(m)} = \begin{pmatrix} -2.025491618186 \\ 5.461798337239 \\ 6.005509356249 \end{pmatrix}.$$

The angle between these two vectors is 0.4 µas.

Appendix B: Description of the AstroRef Package

B.1 General Information

The *AstroRef* package has been written in the computer algebra language Maple^{TM}. It provides routines for a lot of issues treated in this book. The reader is encouraged to make use of it to check relations, to use it for the exercises, or to create own functions. The package has been developed in Maple^{TM} 11. It should be clear that these Maple^{TM} routines do not aim at competing with the SOFA or NOVAS routines. They mainly serve didactical purposes.

In order to give accurate results, some of the functions, e.g., the routine which returns polar motion coordinates, need input from real observations. If the computer is connected to the Internet, the functions will try to download the appropriate data on the fly. If there is no access to the web, the user should supply the data files himself.

Large data files are needed to compute numbers from the DE405 solar system ephemeris. These files are not included in the package. The first time when one of the functions that involve DE405 data is invoked, *AstroRef* will try to download the necessary files from the Internet and store them locally. Afterwards the DE405 data will be taken from the local storage.

B.2 Download

The *AstroRef* package can be downloaded from

http://astro.geo.tu-dresden.de/astroref/.

The zip archive contains two files. The routines are included in AstroRef.mla and the associated help pages in AstroRef.hdb.

B.3 Installing and Using the Package

There are two optional ways to get access to the *AstroRef* routines. Users which own administrative rights on their computer should proceed as follows:

1. Locate the library directory of Maple™ by entering the command
   > libname.
2. Uncompress both files of the package into the library directory.
3. Load the package within Maple™ by
   > with(AstroRef).

Users which do not own administrative rights should proceed as follows:

1. Uncompress both files of the package into an arbitrary directory <dir>.
2. Load the package by the following two Maple™ commands (replace <dir> by the full path from (1):
   > libname := "<dir>", libname;
   > with(AstroRef);

If the files were copied into the current working directory, "<dir>" may simply be replaced by "./". Note that the help pages of *AstroRef* will be included into the conventional Maple™ help after the package was loaded.

B.4 Function Reference

B.4.1 Overview

The following functions are implemented in the *AstroRef* package:

Auxiliary Functions

AngleMod	RotMatrix	ConvertLength
VectorToSpherical	SphericalToVector	Sexagesimal
SexagesimalString		

Time Functions

JDDate	CalendarDate	ConvertTime
GetLeapSeconds	GetDeltaUT1	TephOffset
GMST	GAST	EarthRotAngle

CRS and TRS

CRStoTRS	TRStoCRS	Precession
Precession1977	Nutation	Nutation1980
Bias	TioPosition	GetPoleCoords
Obliquity	NutationPrecessionBias	

Refraction
`RefrNumInt`   `RefrSaas`

Apparent Places
`AppPlaceSolarSystem` `AppPlaceSunFast`   `AppPlaceStar`
`EarthObserver`   `CatalogToVector`   `ProperMotion`
`LightTravelTime`   `LightDeflectionSun` `Aberration`

Ephemerides
`GetDE405Ephemeris`   `GetVSOPEphemeris`

B.4.2 Auxiliary Functions

AngleMod

> Syntax:
>   `AngleMod(`α, ϕ`)`
> Arguments
> α arbitrary angle
> ϕ upper limit of the basic period [0,ϕ) in the selected angle system
> Return format:
> β
> Return values:
> β angle α within basic period
> Downloads
>   no

The function `AngleMod` redirects the given angle α to the basic interval [0,ϕ) of the angle system.

RotMatrix

> Syntax:
>   RotMatrix(i,α)
>
> Arguments:
>   i   rotation axis index
> α rotation angle [rad]
>
> Return format:
> $\mathcal{R}_i(\alpha)$
>
> Return values:
> $\mathcal{R}_i(\alpha)$ rotation matrix (3 × 3)
>
> Downloads:
>   no

The function RotMatrix generates a 3-by-3 rotation matrix about the axis i for the given angle α. The axis indices are 1 for the x-, 2 for the y-, and 3 for the z-axis, respectively.

ConvertLength

> Syntax:
>   ConvertLength(L_O, TL_O, TL_T)
>
> Arguments:
> L_O length in the original unit
> TL_O type of original length unit
> TL_T type of target length unit
>
> Return format:
> L_T
>
> Return values:
> L_T length in the target unit
>
> Downloads:
>   no

The function ConvertLength converts the length L_O given in units defined by the string TL_O into the length L_T given in units defined by the string TL_T. Possible values for the strings L_O and L_T are:

- "AU" for astronomical units,
- "M" for meters,
- "KM" for kilometers,

- "LY" for light-years,
- "KLY" for kilo light-years (thousand light-years),
- "MLY" for mega light-years (million light-years),
- "GLY" for giga light-years (billion light-years),
- "PC" for parsec (parallax seconds),
- "KPC" for kiloparsec (thousand parsecs),
- "MPC" for megaparsec (million parsecs), and
- "GPC" for gigaparsec (billion parsecs).

VectorToSpherical

Syntax:
  VectorToSpherical(v)

Arguments:
  v Cartesian row vector (3 × 1)

Return format:
  λ, ϕ, R

Return values:
  λ primary angle of the point [rad]
  ϕ secondary angle of the point [rad]
  R distance of the point from the coordinate origin

Downloads:
  no

The function VectorToSpherical converts Cartesian to spherical coordinates. The primary angle λ is counted in the reference plane, starting from the reference direction up to the direction of the point. The secondary angle ϕ is counted perpendicular to the reference plane, starting from the reference plane up to the point. The distance of the point from the coordinate origin is returned by R. The inverse conversion is carried out by SphericalToVector.

SphericalToVector

Syntax:
SphericalToVector(λ, ϕ)
SphericalToVector(λ, ϕ, R)

Arguments:
- λ primary angle of the point [rad]
- ϕ secondary angle of the point [rad]
- R distance of the point from the coordinate origin

Return format:
v

Return values:
- v Cartesian row vector (3 × 1)

Downloads:
no

The function SphericalToVector converts spherical to Cartesian coordinates. It is the inverse of VectorToSpherical. The meaning of the parameters is equivalent. If the distance R is omitted, unity will be assumed.

Sexagesimal

Syntax:
Sexagesimal(v)
Sexagesimal(v, a)

Arguments:
- v decimal degrees or hours respectively
- a decimal places for arcseconds or seconds respectively

Return format:
d, m, s

Return values:
- d degrees or hours respectively
- m arcminutes or minutes respectively
- s arcseconds or seconds respectively

Downloads:
no

The function Sexagesimal converts an angle from decimal into sexagesimal format. Alternatively, this function may be used for converting decimal hours into sexagesimal format. If no accuracy is given, the function takes the maximum accuracy for arcseconds or seconds, respectively. The integer value a sets the number of decimal places for rounding the arcsecond or second values depending from if the user gave an angle or a time. Due to round-off, it may happen that full periods are reached, like $24^\text{h}00^\text{m}00^\text{s}$ instead of $00^\text{h}00^\text{m}00^\text{s}$. This effect cannot be suppressed since the function does not know if the user sent degrees or hours. If v is negative, all resulting components will be negative. To get a sexagesimal triple in a printable version, e.g., for data exports into files, use SexagesimalString.

SexagesimalString

Syntax:
  SexagesimalString(v)
  SexagesimalString(v,a)
  SexagesimalString(v,a,f)
  SexagesimalString(v,a,f,p)

Arguments:
  v decimal degrees or hours respectively
  a decimal places for arcseconds or seconds respectively
  f minimum number of digits for degrees or hours respectively
  p Boolean: true writes plus sign if v is positive

Return format:
  s

Return values:
  s string representing sexagesimal value

Downloads:
  no

The function SexagesimalString converts decimal degrees or hours into sexagesimal format, similar to the function Sexagesimal. But here, strings are produced instead of numbers. The meaning of the argument a is similar to the description in Sexagesimal. Arc minutes and full arc seconds or minutes and full seconds, respectively, are always written with two digits, therefore filled with zeros if necessary. For example, if the minute value is 4, the string will be "04". The positive value f defines the number of digits to be filled with zeros if necessary for the degrees or hours part, respectively. The default for f is 1. If the Boolean argument p is set to true, there will be a plus sign for positive values of v. If the argument is omitted or set to false, there will be no plus sign for positive values. Negative values will produce minus signs at any case.

Time Functions

JDDate

> Syntax:
>   JDDate(Y,M,D,H)
>
> Arguments:
>   Y   year
>   M   month of year
>   D   day of month
>   H   hour of day
>
> Return format:
>   JD
>
> Return values:
>   JD   Julian date for the given instant of time
>
> Downloads:
>   no

The function JDDate converts the given calendar date to the Julian date format. The arguments Y, M, and D are integers while H is given as float. For prehistoric dates Y may be negative. Note that for prehistoric dates there is a discrepancy since Y=0 refers to the year 1 BC, Y=-1 to 2 BC and so on. Use the function CalendarDate for inverse conversion.

CalendarDate

> Syntax:
>   CalendarDate(JD)
>
> Arguments:
>   JD   Julian date
>
> Return format:
>   Y,M,D,H
>
> Return values:
>   Y   year
>   M   month of year
>   D   day of month
>   H   hour of day
>
> Downloads:
>   no

The function `CalendarDate` converts the given Julian date into a calendar date. It is the inverse conversion of the function `JDDate`.

ConvertTime

> Syntax:
>   ConvertTime(JD_O, TS_O, TS_T)
>
> Arguments:
> JD_O Julian date in original time system
> TS_O type of original time system
> TS_T type of target time system
>
> Return format:
> JD_T
>
> Return values:
> JD_T Julian date in target time system
>
> Downloads:
>   yes

The function `ConvertTime` converts the Julian date JD_O from the time system defined by the string TS_O into the time system defined by the string TS_T. Possible values for TS_O are:

- "UTC" for universal coordinated time,
- "TAI" for international atomic time, and
- "TT" for terrestrial time.

For the target type TS_T, the strings

- "UTC" for universal coordinated time,
- "TAI" for international atomic time,
- "TT" for terrestrial time,
- "UT1" for universal time (correction level 1),
- "Teph" for ephemeris time (used for calculating ephemeris),
- "TB" as a synonym for "Teph", and
- "TDB" for barycentric dynamical time

are allowed. Internally the function converts the input Julian date into the UTC timescale. In a second step, this UTC date is converted into the output timescale. Since for the conversion between TAI and UTC knowledge of the leap seconds is required, downloads according to the description in `GetLeapSeconds` may be necessary. Furthermore, for the conversion between UTC and UT1, knowledge of ΔUT1 is required. That is why downloads according to `GetDeltaUT1` may be necessary.

GetLeapSeconds

Syntax:
   GetLeapSeconds(JD)

Arguments:
   JD   Julian date UTC

Return format:
   N

Return values:
   N   offset TAI − UTC [s]

Downloads:
   yes

The function GetLeapSeconds returns the offset TAI–UTC for the given Julian date JD. Since the dates for the introduction of leap seconds cannot be predicted, it is necessary to update the switch data files regularly. If the download option of *AstroRef* is active (see variable AstroRef_Download_On in Sect. B.5), the function will try to receive the information about leap seconds from the Internet and store it in a local leap seconds file. The download mechanism is described here: if there is no local file, the download is initiated. If there is a local file, its modification date is compared to the current date. If a potential switch date has passed since the last time the local file has been updated, a new download is initiated. In this way downloads take place only once half a year as long as Maple™ retains access to the local file. Therefore it might be meaningful to change the local storage directory to some defined location instead of using the current working directory which is the default (see variable AstroRef_LocalDir in Sect. B.5).

GetDeltaUT1

Syntax:
   GetDeltaUT1(JD)

Arguments:
   JD   Julian date UTC

Return format:
   ΔUT1

Return values:
   ΔUT1   offset UT1-UTC [s]

Downloads:
   yes

The function `GetDeltaUT1` returns the difference between UT1–UTC, which is determined and published continuously by the International Earth Rotation and Reference Systems Service (IERS). That is why regular updates of the data are required. If the download capabilities of *AstroRef* are switched on (see variable `AstroRef_Download_On` in Sect. B.5), the package will try to receive this data from the Internet and save it to a local file. The download mechanism is described here: if still there is no local file, the download is initiated. If there is a local file, it is checked, whether the data line requested by the user is already included. If the data could not be found or is marked with a "predicted" flag by the IERS (future or very recent dates), the download is initiated. By comparison of the local file date and the current date, it is ensured that downloads happen only once per day.

TephOffset

> Syntax:
>   `TephOffset(JD)`
>
> Arguments:
>   JD   Julian date TT
>
> Return format:
> T_{eph}
>
> Return values:
> T_{eph} offset $T_{eph} - TT$ [ms]
>
> Downloads:
>   no

The function `TephOffset` determines the offset between the ephemeris time and the terrestrial time. Ephemeris time is needed as time argument for `GetDE405Ephemeris` and for `GetVSOPEphemeris`. When `ConvertTime` is called to convert some timescale to T_{eph}, this function will be called internally.

GMST

Syntax:
  GMST (JD)

Arguments:
  JD   Julian date TT

Return format:
  GMST

Return values:
  GMST Greenwich Mean Sidereal Time [rad]

Downloads:
  yes

The function GMST determines the Greenwich mean sidereal time for the given JD according to the IAU 2000 definition. Note that the return value is in radians, not in hours.

GAST

Syntax:
  GAST (JD)

Arguments:
  JD   Julian date TT

Return format:
  GAST

Return values:
  GAST Greenwich Apparent Sidereal Time [rad]

Downloads:
  yes

The function GAST determines Greenwich apparent sidereal time for the given JD according to the IAU 2000 definition. Note that the return value is in radians, not in hours.

EarthRotAngle

Syntax:
   EarthRotAngle(JD)

Arguments:
   JD   Julian date TT

Return format:
   θ

Return values:
   θ Earth Rotation Angle [rad]

Downloads:
   yes

The function EarthRotAngle calculates the Earth rotation angle (ERA) for the given instant of time JD. For convenience, the argument for the function is TT although internally UT1 is needed. The internal conversion between those two timescales requires calling the functions GetLeapSeconds and GetDeltaUT1 which may download data from the Internet. For more information about downloading, see the reference of these functions.

CRS and TRS

CRStoTRS

Syntax:
   CRStoTRS(X_{CRS}, JD)

Arguments:
   X_{CRS} Cartesian row vector at the CRS (3×1)
   JD   Julian date UTC

Return format:
   X_{TRS}

Return values:
   X_{TRS} Cartesian row vector at the TRS (3×1)

Downloads:
   yes

The function CRStoTRS performs all the operations which are necessary to transform the celestial vector X_{CRS} into the terrestrial vector X_{TRS} for the given date JD. The transformation steps are in agreement with the IAU 2000 resolutions and give correct results up to the microarcsecond level. The function uses the modern transformation algorithm. It internally calls GetPoleCoords, GetDeltaUT1, and GetLeapSeconds, which are reliant to regular downloads. See the reference of these functions for more information. The inverse transformation is performed by TRStoCRS.

TRStoCRS

Syntax:
  TRStoCRS ($\mathbf{X}_{TRS}$, JD)

Arguments:
  $\mathbf{X}_{TRS}$ Cartesian row vector at the TRS (3 × 1)
  JD   Julian date UTC

Return format:
  $\mathbf{X}_{CRS}$

Return values:
  $\mathbf{X}_{CRS}$ Cartesian row vector at the CRS (3 × 1)

Downloads:
  yes

The function TRStoCRS performs all the operations which are necessary to transform the terrestrial vector $\mathbf{X}_{TRS}$ into the celestial vector $\mathbf{X}_{CRS}$ for the given date JD. The transformation steps are in agreement with the IAU 2000 resolutions and give correct results up to the microarcsecond level. The function uses the modern transformation algorithm. It internally calls GetPoleCoords, GetDeltaUT1, and GetLeapSeconds, which are reliant to regular downloads. See the reference of these functions for more information. The inverse transformation is performed by CRStoTRS.

Precession

Syntax:
  Precession(JD)

Arguments:
  JD Julian date TT

Return format:
  $\mathcal{P}, \zeta_A, \theta_A, z_A$

Return values:
  $\mathcal{P}$ precession matrix (3 × 3)
  ζ_A precession angle [rad]
  θ_A precession angle [rad]
  z_A precession angle [rad]

Downloads:
  no

The function Precession determines the precession angles as well as the precession matrix according to the IAU 2000 theory of precession. The precession data are valid for a precessional rotation from the reference epoch J2000.0 to the epoch given by the argument JD.

In order to determine the precession data in accordance to the IAU 1977 theory of precession, use the *AstroRef* function Precession1977.

Precession1977

Syntax:
  Precession1977(JD)

Arguments:
  JD Julian date TT

Return format:
  $\mathcal{P}, \zeta_A, \theta_A, z_A$

Return values:
  $\mathcal{P}$ precession matrix (3 × 3)
  ζ_A precession angle [rad]
  θ_A precession angle [rad]
  z_A precession angle [rad]

Downloads:
  no

The function `Precession1977` determines the precession angles as well as the precession matrix according to the IAU 1977 theory of precession. The precession data are valid for a precessional rotation from the reference epoch J2000.0 to the epoch given by the argument JD.

In order to determine the precession data in accordance to the IAU 2000 theory of precession, use the *AstroRef* function `Precession`.

Nutation

> Syntax:
>   `Nutation(JD)`
>
> Arguments:
>   JD   Julian date TT
>
> Return format:
> $\mathcal{N}, \Delta\epsilon, \Delta\psi$
>
> Return values:
> $\mathcal{N}$ nutation matrix (3 × 3)
> $\Delta\epsilon$ nutation in obliquity [rad]
> $\Delta\psi$ nutation in longitude [rad]
>
> Downloads:
>   no

The function `Nutation` determines the nutation angles as well as the nutation matrix for the given epoch JD according to the IAU 2000 theory of nutation.

In order to determine the nutation data in accordance to the IAU 1980 theory of nutation, use the *AstroRef* function `Nutation1980`.

Nutation1980

Syntax:
   Nutation1980(JD)

Arguments:
   JD   Julian date TT

Return format:
   $\mathcal{N}, \Delta\epsilon, \Delta\psi$

Return values:
   $\mathcal{N}$ nutation matrix (3×3)
   $\Delta\epsilon$ nutation in obliquity [rad]
   $\Delta\psi$ nutation in longitude [rad]

Downloads:
   no

The function Nutation1980 determines the nutation angles as well as the nutation matrix for the given epoch JD according to the IAU 1980 theory of nutation.

In order to determine the nutation data in accordance to the IAU 2000 theory of nutation, use the *AstroRef* function Nutation.

Bias

Syntax:
   Bias

Arguments:
   -

Return format:
   $\mathcal{B}$

Return values:
   $\mathcal{B}$ frame bias matrix (3×3)

Return values:
   no

The function Bias returns the frame bias matrix. Since this matrix has constant elements, no arguments are required.

NutationPrecessionBias

> Syntax:
> NutationPrecessionBias(JD)
>
> Arguments:
> JD   Julian date TT
>
> Return format:
> Q
>
> Return values:
> Q   nutation-precession-bias matrix (3 × 3)
>
> Downloads:
> no

The function NutationPrecessionBias calculates the combined nutation–precession–bias matrix for the given instant of time JD. This matrix is used in the modern transformation between celestial and terrestrial reference system. For the transition from the celestial to the terrestrial system, Q may be used directly, while for the inverse transition the inverse of Q has to be used.

GetPoleCoords

> Syntax:
> GetPoleCoords(JD)
>
> Arguments:
> JD   Julian date UTC
>
> Return format:
> x_p, y_p
>
> Return values:
> x_p angular pole coordinate x [rad]
> y_p angular pole coordinate y [rad]
>
> Downloads:
> yes

The function GetPoleCoords returns the pole coordinates, which are determined and published continuously by the International Earth Rotation and Reference Systems Service (IERS). That is why regular updates of the data are required. The download mechanism works equivalent to the one described in GetDeltaUT1.

Obliquity

Syntax:
  Obliquity(JD)

Arguments:
  JD   Julian date TT

Return format:
  ϵ

Return values:
  ϵ obliquity of the ecliptic [rad]

Downloads:
  no

The function Obliquity calculates the obliquity of the ecliptic for the epoch given by JD.

TioPosition

Syntax:
  TioPosition(JD)

Arguments:
  JD   Julian date TT

Return format:
  s'

Return values:
  s' longitude of TIO in ITRS [rad]

Downloads:
  no

The function TioPosition determines the longitude of the terrestrial intermediate origin within the ITRS, i.e., the quantity s' for the given instant of time JD.

Refraction

RefrNumInt

Syntax:
  RefrNumInt $(z', T, p, H_{rel}, \Phi_{geo})$
  RefrNumInt $(z', T, p, H_{rel}, \Phi_{geo}, h_0)$
  RefrNumInt $(z', T, p, H_{rel}, \Phi_{geo}, h_0, \lambda)$

Arguments:
  z' observed zenith distance [rad]
  T air temperature at observing site [°C]
  p atmospheric pressure at observing site [hPa]
  H_{rel} relative humidity at observing site [%]
  Φ_{geo} geographical latitude of observing site [rad]
  h_0 height of observing site above geoid [m]
  λ wavelength of light [μm]

Return format:
  ζ

Return values:
  ζ refraction correction $z - z'$ [rad]

Downloads:
  no

The function RefrNumInt calculates the correction angle $\zeta = z - z'$ of the astronomical refraction, where z is the geometrical zenith distance and z' is the observed zenith distance, which is influenced by refraction. When the arguments h_0 or λ are omitted, the function will assume $h_0 = 0$ m or $\lambda = 0.574$ μm.

This function allows for calculation of correction angles for zenith distances up to the horizon but works slower than the *AstroRef* function RefrSaas.

RefrSaas

Syntax:
  RefrSaas $(z', T, p, H_{\text{rel}})$

Arguments:
  z' observed zenith distance [rad]
  T air temperature at observing site [°C]
  p atmospheric pressure at observing site [hPa]
  H_{rel} relative humidity at observing site [%]

Return format:
  ζ

Return values:
  ζ refraction correction [rad]

Downloads:
  no

The function RefrSaas calculates the correction angle ζ of the astronomical refraction. This correction is the difference between the geometrical and the observed zenith distance, $z - z'$, while the latter is influenced by refraction.

This function works faster than RefrNumInt but is limited to zenith distances less than 70°.

Apparent Places

AppPlaceSolarSystem

> Syntax:
>   AppPlaceSolarSystem(JD,ID)
>   AppPlaceSolarSystem(JD,ID,Λ,Φ)
>   AppPlaceSolarSystem(JD,ID,Λ,Φ,h)
>
> Arguments:
>   JD Julian date UTC
>   ID DE405–ID of solar system body
>   Λ   astronomical longitude (east positive) of observer [rad]
>   Φ   astronomical latitude (north positive) of observer [rad]
>   h   height of observer above reference ellipsoid [m]
>
> Return format:
> α, δ, r
>
> Return values:
> α apparent right ascension [rad]
> δ apparent declination [rad]
>   r distance to the object [AU]
>
> Downloads:
>   yes

The function AppPlaceSolarSystem returns the apparent or even the observed place of the solar system body defined by ID for the given instant of time JD. The ephemeris of the solar system body is taken from DE405 by GetDE405Ephemeris. Therefore, the accepted values for ID are DE405_MERCURY, DE405_VENUS, DE405_EARTH, DE405_MARS, DE405_JUPITER, DE405_SATURN, DE405_URANUS, DE405_NEPTUNE, DE405_PLUTO, DE405_MOON, or DE405_SUN (see GetDE405Ephemeris for details). Note, however, that calling the function for DE405_EARTH is not very meaningful. If no position of the observer is given by Λ and Φ (geocentric), apparent places are returned, otherwise (topocentric) observed places. If only h is omitted, a position on the reference ellipsoid is assumed. See the reference of EarthObserver for details.

The internal call of GetDE405Ephemeris as well as the necessary internal time conversions carried out by ConvertTime may require downloading. See the references of these functions for details.

AppPlaceSunFast

> Syntax:
>   `AppPlaceSunFast(JD)`
>
> Arguments:
>   JD Julian date TT
>
> Return format:
> α, δ
>
> Return values:
> α apparent right ascension of the Sun [rad]
> δ apparent declination of the Sun [rad]
>
> Downloads:
>   no

The function `AppPlaceSunFast` rapidly calculates the apparent place of the Sun based on VSOP87. The algorithm introduced by Meeus (1999) was found to give reasonable results of the order of $1''$ for the time interval from 3000 BC to 3000 AD. Besides the rapid execution compared to the *AstroRef* function `AppPlaceSolarSystem`, no downloads are necessary for this function.

AppPlaceStar

Syntax:
    `AppPlaceStar` ($JD_{Obs}, \alpha_{Cat}, \delta_{Cat}, \Pi, JD_{Cat}, \mu_\alpha, \mu_\delta, \dot{r}, \Lambda, \Phi, h$)
    `AppPlaceStar` ($JD_{Obs}, \alpha_{Cat}, \delta_{Cat}, \Pi, JD_{Cat}, \mu_\alpha, \mu_\delta, \dot{r}, \Lambda, \Phi$)
    `AppPlaceStar` ($JD_{Obs}, \alpha_{Cat}, \delta_{Cat}, \Pi, JD_{Cat}, \mu_\alpha, \mu_\delta, \dot{r}$)

Arguments:
- JD_{Obs} Julian date UTC of observation
- α_{Cat} right ascension from stellar catalog [rad]
- δ_{Cat} declination from stellar catalog [rad]
- Π parallax [rad]
- JD_{Cat} Julian date TDB of catalog coordinates
- μ_α angular rate of change for right ascension (proper motion) [rad/d]
- μ_δ angular rate of change for declination (proper motion) [rad/d]
- $\dot{r}$ rate of change for distance [AU/d]
- Λ astronomical longitude (east positive) of observer [rad]
- Φ astronomical latitude (north positive) of observer [rad]
- h   height of observer above reference ellipsoid [m]

Return format:
    $\alpha, \delta, \mathtt{r}$

Return values:
- α apparent right ascension [rad]
- δ apparent declination [rad]
- $\mathtt{r}$ distance to the object [AU]

Downloads:
    yes

The function `AppPlaceStar` calculates the apparent or even the observed place of an object from a stellar catalog for the given instant of time JD_{Obs}. The coordinates α_{Cat} and δ_{Cat} and the parallax Π which are valid for JD_{Cat} as well as the proper motions μ_α, μ_δ, and $\dot{r}$ may be found in a stellar catalog. If Π or $\dot{r}$ are not included, the user may set these values to zero. See the references of `CatalogToVector` and `ProperMotion` for details.

If no position of the observer is given by Λ and Φ (geocentric), apparent places are returned, otherwise (topocentric) observed places. If only h is omitted, a position on the reference ellipsoid is assumed. See the reference of `EarthObserver` for details. Note that for stars the differences between apparent and observed places are negligibly small for most applications due to their large distances.

The internal call of `GetDE405Ephemeris` as well as the necessary internal time conversions carried out by `ConvertTime` may require downloading. See the references of these functions for details.

EarthObserver

Syntax:
```
EarthObserver(JD)
EarthObserver(JD,Λ,Φ)
EarthObserver(JD,Λ,Φ,h)
```

Arguments:

- JD Julian date TT
- Λ   astronomical longitude (east positive) of topocenter [rad]
- Φ   astronomical latitude (north positive) of topocenter [rad]
- h   height above reference ellipsoid [m]

Return format:

$\mathbf{x}, \dot{\mathbf{x}}$

Return values:

- $\mathbf{x}$ barycentric equatorial position vector of topocenter (3×1) [AU]
- $\dot{\mathbf{x}}$ barycentric equatorial velocity vector of topocenter (3×1) [AU/d]

Downloads:

yes

The function `EarthObserver` calculates the barycentric equatorial position and velocity vector of some topocenter on Earth, given by its longitude, latitude, and height. The reference for the heights is the ellipsoid defined by IAU in 1976. If h is omitted, a position on the surface of the reference ellipsoid is assumed. If only JD is given, the position and velocity of the geocenter will be returned.

Since the function internally calls `GetDE405Ephemeris` to get the barycentric position of the geocenter, downloads may be required. See the function reference of `GetDE405Ephemeris` for details.

CatalogToVector

Syntax:
  CatalogToVector(α, δ)
  CatalogToVector(α, δ, Π)

Arguments:
  α right ascension [rad]
  δ declination [rad]
  Π parallax [rad]

Return format:
  $\mathbf{x}_{\text{Cat}}$

Return values:
  $\mathbf{x}_{\text{Cat}}$ barycentric equatorial position vector (3 × 1) [AU]

Downloads:
  no

The function CatalogToVector transforms an equatorial position given by spherical coordinates (α, δ) and the annual parallax Π into Cartesian equatorial coordinates. The routine is intended to be applied to positions given in stellar catalogs. It works very similar to the *AstroRef* function SphericalToVector except for the special handling of the annual parallax Π, which defines the distance of the star here. If the parallax is not given or zero, the star is placed at a distance of 1 Mpc.

ProperMotion

Syntax:

ProperMotion($\mathbf{x}_{Cat}$, JD_{Obs}, JD_{Cat}, μ_α, μ_δ)

ProperMotion($\mathbf{x}_{Cat}$, JD_{Obs}, JD_{Cat}, μ_α, μ_δ, $\dot{r}$)

Arguments:

- $\mathbf{x}_{Cat}$ vector at JD_{Cat} (3×1) [AU]
- JD_{Obs} Julian date TDB of observation
- JD_{Cat} Julian date TDB of catalog coordinates
- μ_α angular rate of change for right ascension (proper motion) [rad/d]
- μ_δ angular rate of change for declination (proper motion) [rad/d]
- $\dot{r}$ rate of change for distance [AU/d]

Return format:

$\mathbf{x}_{Obs}$

Return values:

- $\mathbf{x}_{Obs}$ vector at JD_{Obs} (3×1) [AU]

Downloads:

no

The function ProperMotion corrects the catalog position of some star for its proper motion or even space motion. In most cases stellar catalogs do not only supply positions of stars but also their rates of change with the time. The catalog positions are valid for JD_{Cat} and are transformed to some moment of observation JD_{Obs}. For current catalogs very often JD_{Cat}= J2000.0 = 2 451 545. The input vector $\mathbf{x}_{Cat}$ may be obtained by calling the function CatalogToVector. Note that some stellar catalogs do not directly supply the rate of change in right ascension for μ_α, but a value $\mu_\alpha \cos\delta$ that is corrected for convergence of right ascension lines in the direction to the poles. For such catalogs, it is important to divide the given values by $\cos\delta$ before calling ProperMotion. Some stellar catalogs do not give a rate of change for the distance $\dot{r}$. If this parameter is omitted, zero will be assumed and only proper motion corrections will be carried out.

LightTravelTime

Syntax:
  LightTravelTime(JD, **x**, ID)

Arguments:
  JD Julian date TT
  x barycentric equatorial position vector of observer (3 × 1) [AU]
  ID DE405–ID of solar system body

Return format:
  τ

Return values:
  τ light travel time [d]

Downloads:
  yes

The function LightTravelTime determines the time interval between sending a light ray from a solar system body defined by ID and receiving the light ray at the position **x** for the given moment of observation JD. The ephemeris of the solar system body is taken from GetDE405Ephemeris. Therefore, the argument ID takes one of the index values DE405_MERCURY, DE405_VENUS, DE405_EARTH, DE405_MARS, DE405_JUPITER, DE405_SATURN, DE405_URANUS, DE405_NEPTUNE, DE405_PLUTO, DE405_MOON, or DE405_SUN (see GetDE405Ephemeris for details). The calculation of light travel times is an iterative procedure. Iteration stops when an accuracy of 10^{-8} days is reached for τ. Note that the radii of the target bodies are not considered in the calculation. As described in the reference for GetDE405Ephemeris, downloads may take place.

LightDeflectionSun

> Syntax:
> LightDeflectionSun($x_{O \to T}$, $x_{S \to T}$, $x_{S \to O}$)
>
> Arguments:
> $x_{O \to T}$ equatorial vector from observer at t_O to target body at t_E (3 × 1) [AU]
> $x_{S \to T}$ equatorial vector from Sun at t_E to target body at t_E (3 × 1) [AU]
> $x_{S \to O}$ equatorial vector from Sun to observer at t_O
>
> Return format:
> $x'_{O \to T}$
>
> Return values:
> $x'_{O \to T}$ same as $x_{O \to T}$, but corrected for light deflection (3 × 1) [AU]
>
> Downloads:
> no

The function LightDeflectionSun is used to correct for the influence of gravitational light deflection induced by the Sun on the observed position of some target body, e.g., a solar system object or a star. For solar system objects, it has to be distinguished between the moment of observation t_O and the moment t_E when the light was emitted from the body. The difference $t_O - t_E$ may be determined by LightTravelTime.

Aberration

> Syntax:
> Aberration($x_{O \to T}$, $\dot{x}$)
>
> Arguments:
> $x_{O \to T}$ equatorial vector from observer to target body (3 × 1) [AU]
> $\dot{x}$ barycentric equatorial velocity vector of observer (3 × 1) [AU/d]
>
> Return format:
> $x'_{O \to T}$
>
> Return values:
> $x'_{O \to T}$ same as $x_{O \to T}$, but corrected for aberration (3 × 1) [AU]
>
> Downloads:
> no

The function Aberration corrects the input vector $x_{O \to T}$ for the influence of aberration induced by the velocity of the observer. The latter may be determined, e.g., by the *AstroRef* function EarthObserver if the observer resides on the Earth.

Ephemeris

GetDE405Ephemeris

Syntax:
   GetDE405Ephemeris(JD)

Arguments:
   JD   Julian date T_{eph}

Return format:
   $\mathbf{x}, \dot{\mathbf{x}}$

Return values:
   $\mathbf{x}$ barycentric equatorial position matrix (11×3) [AU]
   $\dot{\mathbf{x}}$ barycentric equatorial velocity matrix (11×3) [AU/d]

Downloads:
   yes

The function GetDE405Ephemeris may be used to calculate very precise positions and velocities of the solar system objects Sun, Mercury, Venus, Earth, Moon, Mars, Jupiter, Saturn, Uranus, Neptune, and Pluto. The user should note that the input timescale is T_{eph}. To convert a usual time into T_{eph}, use ConvertTime. For DE405 the Julian date JD must lie between the years 1600 and 2200.

For the given instant of time JD, two matrices are returned by this function. In every row, there are the three vector components of one particular solar system body, for position and velocity, respectively. For convenient access to the individual bodies, 11 index variables are defined. These variables read:

- DE405_MERCURY (row 1),
- DE405_VENUS (row 2),
- DE405_EARTH (row 3),
- DE405_MARS (row 4),
- DE405_JUPITER (row 5),
- DE405_SATURN (row 6),
- DE405_URANUS (row 7),
- DE405_NEPTUNE (row 8),
- DE405_PLUTO (row 9),
- DE405_MOON (row 10), and
- DE405_SUN (row 11).

If in Maple™ the function is called by
> Pos,Vel := GetDE405Ephemeris(JD)
it is possible to get access to the position vector of Uranus by
> Pos[DE405_URANUS,1..3]

Note that this returns a line vector which may be transposed to a row vector by usual Maple™ commands.

The function needs rather large data files which are not delivered together with *AstroRef*. Instead, the data files are downloaded by request from remote servers as long as the download capabilities of *AstroRef* are switched on (see variable AstroRef_Download_On in Sect. B.5). The data files are split into portions of 20 years beginning at 1600 and ending at 2200. Each data file has a size of 6 MB. The function will download only the file which is needed to fulfill the current request and store it locally for subsequent usage. Note that when the user writes a new Maple™ source file in another directory, the time-consuming downloads may start again. Therefore, it might be meaningful to change the local storage directory to some defined location instead of using the current working directory which is the default (see variable AstroRef_LocalDir in Sect. B.5).

GetVSOPEphemeris

Syntax:
  GetVSOPEphemeris(JD)

Arguments:
  JD   Julian date T_{eph}

Return format:
  $\mathbf{x}, \dot{\mathbf{x}}$

Return values:
  $\mathbf{x}$ barycentric equatorial position matrix (2×3) [AU]
  $\dot{\mathbf{x}}$ barycentric equatorial velocity matrix (2×3) [AU/d]

Downloads:
  no

The function GetVSOPEphemeris returns the position and the velocity of the Sun and the Earth for the given Julian date JD. The results are based on VSOP87 and are less precise than the functions of GetDE405Ephemeris. But this function works faster and without downloads. The user should note that the input timescale is T_{eph}. To convert a usual time into T_{eph}, use ConvertTime.

Two matrices are returned by GetVSOPEphemeris. In the first row there are the three vector components of the Sun, and in the second row the vector components of the Earth. For convenient access to the rows, two index variables are defined. These variables read:

– VSOP_SUN (row 1) and
– VSOP_EARTH (row 2).

If in Maple™ the function is called by
> Pos,Vel := GetVSOPEphemeris(JD)

it is possible to get access to the velocity vector of the Sun by
> `Vel[VSOP_SUN,1..3]`
Note that this returns a line vector which may be transposed to a row vector by usual Maple™ commands.

B.5 Package Variables

When loading the *AstroRef* package, some package variables are initialized. The user may overwrite some of the variables to adapt the package behavior to his preferences.

AstroRef_FTP_DownloadOn

- Default value: true
- Type: Boolean
- Overwrite: yes

If set to true automatical downloads will be performed by some of the package functions if needed. These downloads include Earth rotation data, leap second definitions, and DE405 ephemeris files.

AstroRef_FTP_VerboseMode

- Default value: true
- Type: Boolean
- Overwrite: yes

When downloads are active, all communication will be displayed as Maple™ output. By setting the value of this variable to false, no output will be produced.

AstroRef_FTP_TimeOut

- Default value: 5
- Type: integer
- Overwrite: yes

FTP communication for downloading data from the Internet is based on sending commands and waiting for the response. By default the package will wait 5 s for the response after sending a command. If the user experiences problems while downloading, it might help to increase the timeout value. On the other hand, if the connection is fast, the downloading procedure could be accelerated by decreasing the timeout value.

AstroRef_FTP_UserName

- Default value: "anonymous"
- Type: string
- Overwrite: yes

Connecting to any FTP server requires a user name. For public servers like the ones used in *AstroRef*, it is common to use an anonymous account. As long as the user does not own a special account, this value should remain unchanged.

AstroRef_FTP_Password

- Default value: "Maple AstroRef Package"
- Type: string
- Overwrite: yes

Even an anonymous FTP account needs some (pseudo) password. Very often the providers ask the users to type in their email address instead of a real password. In that way the providers get some knowledge about who is using their services. However, any other phrase like the one set as default here will work as well. It would be very kind if the user replaces the default value by his email address.

AstroRef_LocalDir

- Default value: "./"
- Type: string
- Overwrite: yes

Any file downloaded from the Internet by *AstroRef* will be stored locally. Subsequent calls of the same procedure will not restart the download process again as long as the data files are valid. By default the files will be saved in the current working directory. If the user prefers another directory, he may overwrite this value. Note that the string has to be ended by slash ("/")!

AstroRef_Local_TAIUTC

- Default value: "tai-utc.dat"
- Type: string
- Overwrite: yes

Leap seconds, the offset between TAI and UTC, are stored in a local file with the name defined by this variable. For more information, see the function reference of `GetLeapSeconds` in Sect. B.4.

AstroRef_Remote_TAIUTC

- Default value: "ftp://astro.geo.tu-dresden.de/astroref/tai-utc.dat"
- Type: string
- Overwrite: yes

Leap seconds, the offset between TAI and UTC, are downloaded from the location defined by this variable. For more information, see the function reference of `GetLeapSeconds` in Sect. B.4.

AstroRef_Local_Finals

- Default value: "finals2000A.all"
- Type: string
- Overwrite: yes

Earth rotation data like polar motion coordinates or the offset between UTC and UT1 are stored in a local file with the name defined by this variable. For more information, see the function reference of `GetPoleCoords` and `GetDeltaUT1` in Sect. B.4.

AstroRef_Remote_Finals

- Default value: "ftp://astro.geo.tu-dresden.de/astroref/finals2000A.all"
- Type: string
- Overwrite: yes

Earth rotation data like polar motion coordinates or the offset between UTC and UT1 are downloaded from the location defined by this variable. For more information, see the function reference of `GetPoleCoords` and `GetDeltaUT1` in Sect. B.4.

AstroRef_Remote_DE405Dir

- Default value: "ftp://astro.geo.tu-dresden.de/astroref/"
- Type: string
- Overwrite: yes

The DE405 data files are downloaded from the location defined by this variable. For more information, see the function reference of `GetDE405Ephemeris` in Sect. B.4.

Index

Aberration, 7, 115, 128, 131, 132, 144, 150, 172, 237, 271, 305
AC. *See* Astrographic catalog (AC)
Almucantar, 98
Analemma, 203
Apex, **128**, 271
APFS. *See* Apparent Places of Fundamental Stars (APFS)
APKIM, 194
Apparent Places of Fundamental Stars (APFS), 144, 236
Astrographic catalog (AC), 170
Astronomical almanac, vi, 235, 237
Astronomical unit, 124
Atomic clock, 3, 4, **11**, 15, 17, 24, 28, 35–37, 45, 160, 163, 185, 203
Atomic time (TAI), 4, **37**, 202, 230, 285, 286, 309
Autocorrelation function, 15
Autocovariance, 15
Azimuth, 6, **101**, 105, 106, 179, 180, 201, 202
Azimuth determination, 180

Barycentric celestial reference system (BCRS), 5, 50, 54, 61, 149–152, **155**, 156, 159
Barycentric coordinate time (TCB), 5, **43**, 61, 78, 159
Barycentric dynamical reference system (BDRS), 5, 50, **61**
Barycentric dynamical time (TDB), 5, **44**, 78, 219, 285, 300, 303
BCRS. *See* Barycentric celestial reference system (BCRS)
BDRS. *See* Barycentric dynamical reference system (BDRS)

Besselian day numbers, **132**, 237, 273
BL Lac objects, 172

CCD astrometry, 61
CCD astronomy, 172
Celestial equator, v, 6, 91, 98, 101, 103, 105, 140, 202
Celestial intermediate origin (CIO), vi, 6, **226**, 229, 231, 232
Celestial intermediate pole (CIP), vi, 6, 202, 208, **211**, 214, 221, 224, 226, 229, 231–233
Celestial meridian, 98, 101, 177
Celestial parallel, 98
Central European Time (CET), **41**, 45, 112, 180
Cesium clock, 3, 17, **21**, 24, 27, 188
CET. *See* Central European Time (CET)
Chandler wobble, 192, **197**, 211, 216
CIO. *See* Celestial intermediate origin (CIO)
CIP. *See* Celestial intermediate pole (CIP)
Constant of aberration, 129, 131, 133, 272
Culmination, 107

Daylight saving time, 41
DE, 5, 44, **70**, 144, 242, 265, 273, 277, 287, 298
Declination, 6, **101**, 105, 122, 124, 128, 131, 133, 134, 139, 153, 170, 173, 203, 270, 298–300, 302, 303
DGPS, 187
DORIS, **188**, 192, 194

Earth orientation parameters (EOP), 6, 10, 67, 163

Earth rotation angle (ERA), **229**, 231, 289
ecliptic, v, 6, 91, 92, 94, **99**, 103, 111, 112, 135, 137–141, 174, 217, 226, 232, 274
   latitude, 103
   longitude, **103**, 139, 274
   meridian, 99
   parallel, 99
   pole, 99, 103
Einstein-Infeld-Hoffmann (EIH), 58, 66, 264
Einstein's theory of gravity (GRT), v, 5, 50, 239
Elevation, 6, **101**, 107, 163
El Niño, 206
EOP. *See* Earth orientation parameters (EOP)
Ephemeris, 5, 43, 44, 49, 50, 52, 60, 61, 66, **70**, 80, 83, 132, 236, 237, 240, 277, 287, 298, 304, 306, 307
EPM, 5, **70**, 242
Equation of equinoxes, **112**, 113, 205, 269
Equation of time, **203**, 237
Equinox, v, 42, **92**, 99, 107, 112, 135, 138–141, 168, 170, 173, 199, 221, 226, 231, 236, 269
ERA. *See* Earth rotation angle (ERA)

Fairhead–Bretagnon series, **44**
Femtosecond comb, 27
Fiber optic gyroscope, 189
Flicker noise, 15
Flicker phase noise, 15
Fluorescence light, 26
Fountain clock, 3, **24**
Frame bias, 174, 211, **221**, 231, 293, 294
Frequency accuracy, 4
Frequency fluctuation, 12
Frequency stability, 4
Fundamental catalog, 5, **169**, 173, 237

Gaia, 6, 51, 80, **166**
GALILEO, 7, 9, **188**
GAST. *See* Greenwich Apparent Sidereal Time (GAST)
Geocentric celestial reference system (GCRS), 4, 35, 50, 61, **156**, 177, 197, 200, 207–209, 211, 214, 222, 226, 227, 229, 231–233, 241
Geocentric coordinate time (TCG), 4, 5, **34**, 43, 78, 159, 193
Geodetic precession, 68, 158
Geoid, 4, 35, 121
Geopotential, 35

Global navigation satellite systems (GNSS), 7, 175, 183, 187, 188
Global positioning system (GPS), 1, 7, 9, 27, 47, **183**, 192, 194
Globalnaja Nawigazionnaja Sputnikowaja Sistema (GLONASS), 7, 187
GMST. *See* Greenwich Mean Sidereal Time (GMST)
GMT, 203
GNSS. *See* Global navigation satellite systems (GNSS)
GPS. *See* Global positioning system (GPS)
Gravitational light deflection, 7, 51, 144, 148, **150**, 305
Gravitational redshift, 35
Greenwich apparent sidereal time (GAST), **111**, 178, 199, 204, 209, 229–231, 268
Greenwich mean sidereal time (GMST), **111**, 178, 199, 204, 209, 229–231, 268
Gregorian calendar, 42
G ring, 191
GRT. *See* Einstein's theory of gravity (GRT)

Hipparcos, 7, 51, **164**, 169, 174
H-maser. *See* Hydrogen maser clock
Hour angle, 6, **102**, 103, 105, 107, 112, 128, 178, 203
Hour circle, 98, **101**, 202
Hydrogen maser clock (H-maser), 3, 22, 37, 188
Hyperfine level, 17, 21

IAU. *See* International astronomical union (IAU)
ICRF. *See* International celestial reference frame (ICRF)
ICRS, 5, 61, 163, 166, **172**
ILS, 193, 197
INPOP, 5, 44, 51, **79**, 242
Instantaneous rotation pole (IRP), v, 6, **208**, 211, 226
International astronomical union (IAU), v, 39, 61, 80, 112, 169, 171, 214, 235, 237
International atomic time (TAI), 4, **37**, 202, 230, 285, 286, 309
International celestial reference frame (ICRF), 69, 72, 170–175
International terrestrial reference system (ITRS), 6, 50, 177, 183, **192**, 199, 207–209, 211, 229, 231–233, 295

Index 313

IRP. *See* Instantaneous rotation pole (IRP)
ITRS. *See* International terrestrial reference system (ITRS)
IVS, 161, 211

Julian calendar, 42
Julian date, **41**, 113, 262, 266, 284, 285

LAGEOS, 181
LAST. *See* Local apparent sidereal time (LAST)
Latitude variations, 197, **201**, 209
Law of sines, 105, 118, 128
Length-of-day variations (LOD variations), 6, 67, 191, **207**, 209
Lines of position, 7, **178**
LLR. *See* Lunar laser ranging (LLR)
LMST. *See* Local mean sidereal time (LMST)
LMT, 203
Local apparent sidereal time (LAST), 107, **111**, 128, 133, 178, 205, 267, 268
Local mean sidereal time (LMST), **112**, 205
LOD variations. *See* Length-of-day variations (LOD Variations)
LORAN-C, 45
Lorentz transformation, **32**, 51, 159, 262
Lunar laser ranging (LLR), 7, 27, 50, **62**, 70, 96, 174

Magnetic sub-level, 19
Mapping function, 163
Marine chronometer, 2
Mean quantities, 140
MERIT, 51
Metric tensor(s), 5, **30**, 31, 32, 53, 55, 57, 130, 155, 156, 159, 262
MHB, 215, 276
Modified Julian date (MJD), **42**, 262

Nautical triangle, 99, **105**
Navigation, 5, 9, 43, 50, 69, 183
NCP. *See* North celestial pole (NCP)
No-net-rotation, 193
Non-rotating origin (NRO), 6, **226**
Nordtvedt parameter, 68
North celestial pole (NCP), 99, **101**, 105, 199
NOVAS, vi, **236**, 277
NRO. *See* Non-rotating origin (NRO)
Nucleus spin, 19

Nutation, 6, 67, 112, 115, 135, 138, **140**, 158, 172, 200, 208, 211, 212, 214, 215, 221, 228, 230, 231, 236, 237, 274, 292–294
NUVEL, 195

Obliquity of the ecliptic, 91, 99, 111, 135, 138, **141**, 203, 295
Optical clock, 4, 24
Optical pumping, 22

Parallactic angle, 105, 139
Parallax, 7, 115, **123**, 150, 172, 270, 300
Parsec, **124**, 281
Pendulum clocks, 3
PFCN. *See* Prograde free core nutation (PFCN)
Phase time, 12
Piezoelectrical effect, 17
Platonic year, 94, 135
Polar motion, 6, 67, 115, 178, **197**, 205, 209, 212, 224, 230, 231, 236, 277, 294
Precession, 6, 67, 94, 112, 115, **135**, 172, 200, 208, 211, 214, 215, 228, 231, 236, 237, 273, 291, 292, 294
Prime vertical plane, 98
Prograde free core nutation (PFCN), 216
Proper motion, 7, 115, **134**, 151, 164, 170, 172, 303
Proper time, 4, **30**, 31, 34, 37, 54, 130, 261

Quartz oscillator, 11, 17
Quasar, 6, 7, 50, 96, 160, 172

Random walk, 15
Refraction, 7, 46, **115**, 269, 296, 297
Retrograde free-core nutation (RFCN), 216, 276
Right ascension, 6, **103**, 107, 122, 124, 128, 131, 133, 134, 139, 140, 153, 172, 178, 205, 221, 270, 298–300, 302, 303
Ring laser, 191
Rubidium clock, 3, 21

Sagnac effect, 38, 189, **190**
Sagnac frequency, 191
Satellite laser ranging (SLR), 7, 10, 27, 96, **181**, 192, 194
Sidereal time, 4, **111**, 205, 236, 237, 288

SLR. *See* Satellite laser ranging (SLR)
Snell's law, 115
SOFA, vi, **235**, 277
Space motion, 7, 92, **133**, 151, 236
Space–time distance, 30
Special relativity (SRT), 129, 150, 239, 273
Spectral power density, 15
Spin quantum number, 18
SRT. *See* Special relativity (SRT)
Stellar constants, **126**, 128, 132, 270
Sundial, 2, 203

TAI. *See* Atomic time (TAI)
TCB. *See* Barycentric coordinate time (TCB)
TCG. *See* Geocentric coordinate time (TCG)
TDB. *See* Barycentric dynamical time (TDB)
Terrestrial time (TT), 4, 36, **39**, 79, 135, 219, 269, 285, 287–289, 291, 292, 294, 295, 299, 301, 304
Time zone, 41
TIO, **229**, 295
Tisserand condition, 193
TT. *See* Terrestrial time (TT)
Tycho, 166, 170, **171**, 174

UCAC, 172
UT0, 67, **205**

UT1, 37, 112, 163, **205**, 206, 230, 231, 285, 287, 289, 310
UTC, **39**, 78, 112, 188, 266, 269, 285–287, 289, 294, 298, 309

Variations séculaire des orbites planétaires (VSOP), 44, 85, 220, 299, 307
Vernal equinox, **6**, 92, 99, 103, 138, 269, 274
Vertical plane, 98, 180
Very long baseline interferometry (VLBI), v, 6, 7, 27, 51, 69, **160**, 172, 174, 192, 194, 211, 214, 217
VLBI stations, 161
VMF, 163
VSOP. *See* Variations séculaire des orbites planétaires (VSOP)

Water clock, 2
White frequency noise, 15
White phase noise, 15
Wiener–Khinchin, 15

Zeeman effect, 19
Zenith distance, 6, **101**, 115, 178, 269, 296, 297

If you have any concerns about our products,
you can contact us on
ProductSafety@springernature.com

In case Publisher is established outside the EU,
the EU authorized representative is:
**Springer Nature Customer Service Center GmbH
Europaplatz 3, 69115 Heidelberg, Germany**

Printed by Libri Plureos GmbH
in Hamburg, Germany